Springer Handbook of Auditory Research

Series Editors: Richard R. Fay and Arthur N. Popper

SPRINGER HANDBOOK OF AUDITORY RESEARCH

Volume 1: The Mammalian Auditory Pathway: Neuroanatomy
Edited by Douglas B. Webster, Arthur N. Popper, and Richard R. Fay

Volume 2: The Mammalian Auditory Pathway: Neurophysiology
Edited by Arthur N. Popper and Richard R. Fay

Volume 3: Human Psychophysics
Edited by William Yost, Arthur N. Popper, and Richard R. Fay

Volume 4: Comparative Hearing: Mammals
Edited by Richard R. Fay and Arthur N. Popper

Volume 5: Hearing by Bats
Edited by Arthur N. Popper and Richard R. Fay

Volume 6: Auditory Computation
Edited by Harold L. Hawkins, Teresa A. McMullen, Arthur N. Popper, and Richard R. Fay

Volume 7: Clinical Aspects of Hearing
Edited by Thomas R. Van De Water, Arthur N. Popper, and Richard R. Fay

Volume 8: The Cochlea
Edited by Peter Dallos, Arthur N. Popper, and Richard R. Fay

Volume 9: Development of the Auditory System
Edited by Edwin W Rubel, Arthur N. Popper, and Richard R. Fay

Volume 10: Comparative Hearing: Insects
Edited by Ronald Hoy, Arthur N. Popper, and Richard R. Fay

Volume 11: Comparative Hearing: Fish and Amphibians
Edited by Richard R. Fay and Arthur N. Popper

Volume 12: Hearing by Whales and Dolphins
Edited by Whitlow W.L. Au, Arthur N. Popper, and Richard R. Fay

Volume 13: Comparative Hearing: Birds and Reptiles
Edited by Robert Dooling, Arthur N. Popper, and Richard R. Fay

Volume 14: Genetics and Auditory Disorders
Edited by Bronya J.B. Keats, Arthur N. Popper, and Richard R. Fay

Volume 15: Integrative Functions in the Mammalian Auditory Pathway
Edited by Donata Oertel, Richard R. Fay, and Arthur N. Popper

Volume 16: Acoustic Communication
Edited by Andrea Simmons, Arthur N. Popper, and Richard R. Fay

Volume 17: Compression: From Cochlea to Cochlear Implants
Edited by Sid P. Bacon, Richard R. Fay, and Arthur N. Popper

Volume 18: Speech Processing in the Auditory System
Edited by Steven Greenberg, William Ainsworth, Arthur N. Popper, and Richard R. Fay

Volume 19: The Vestibular System
Edited by Stephen M. Highstein, Richard R. Fay, and Arthur N. Popper

Volume 20: Cochlear Implants: Auditory Prostheses and Electric Hearing
Edited by Fan-Gang Zeng, Arthur N. Popper, and Richard R. Fay

Volume 21: Electroreception
Edited by Theodore H. Bullock, Carl D. Hopkins, Arthur N. Popper, and Richard R. Fay

Continued after index

Peter M. Narins
Albert S. Feng
Richard R. Fay
Arthur N. Popper
Editors

Hearing and Sound Communication in Amphibians

Springer

Peter M. Narins
Departments of Physiological Science *and* Ecology & Evolutionary Biology
University of California
Los Angeles CA 90095-1606
USA
e-mail: pnarins@ucla.edu

Albert S. Feng
Department of Molecular & Integrative Physiology
University of Illinois
Urbana, IL 61801
USA
afeng1@uiuc.edu

Richard R. Fay
Parmly Hearing Institute and Department of Psychology
Loyola University of Chicago
Chicago, IL 60626
USA
rfay@wpo.it.luc.edu

Arthur N. Popper
Department of Biology
University of Maryland
College Park, MD 20742
USA
apopper@umd.edu

Series Editors:
Richard R. Fay
Parmly Hearing Institute and Department of Psychology
Loyola University of Chicago
Chicago, IL 60626
USA

Arthur N. Popper
Department of Biology
University of Maryland
College Park, MD 20742
USA

Cover illustration: The image includes parts of Figures 3.1, 3.2 and 7.2 appearing in the book.

Library of Congress Control Number: 2006920913

ISBN 10: 0-387-32521-2 Printed on acid-free paper.
ISBN 13: 978-0387-32521-7

9 8 7 6 5 4 3 2 1

springer.com

Dedication

This volume is dedicated to our teacher, colleague, and dear friend, Robert Capranica. Bob is an extraordinary mentor, role model, and scholar, and he continues today to inspire new generations of neuroethologists who follow in his path of doing meticulous and outstanding science. We are tremendously proud to be able to dedicate this book to him and to all he means for our field.

Table of Contents

Contributors

MARK A. BEE
Department of Ecology, Evolution and Behavior, College of Biological Sciences, University of Minnesota, St. Paul, MN 55108, USA

DAVID CANNATELLA
Section of Integrative Biology and Texas, Memorial Museum, Austin, TX 78712, USA

HEIKE ENDEPOLS
Zoologisches Institut der Univsität zu Köln, II. Lehrstuhl, 50923 Köln, Germany

ALBERT S. FENG
Department of Molecular and Integrative Physiology and Beckman Institute, University of Illinois at Urbana-Champaign, 2355 Beckman Institute, Urbana, IL 61801, USA

H. CARL GERHARDT
Division Biological Sciences, University of Missouri, Columbia, MO 65211, USA

DAVID M. GOOLER
Department of Speech and Hearing Science, University of Illinois, Urbana, IL 61820, USA

SETH S. HOROWITZ
Departments of Psychology and Neuroscience, Brown University, Providence, RI 02912, USA

MATTHEW J. MASON
University of Cambridge, Department of Physiology, Development and Neuroscience, Cambridge CB2 3EG, UK

SEBASTIAAN W. F. MEENDERINK
University Hospital Maastricht, Department of Otorhinolaryngology and Head & Neck Surgery, 6202 AZ Maastricht, The Netherlands

PETER M. NARINS
Department of Physiological Science, University of California, Los Angeles, CA 90095-1606, USA

GARY J. ROSE
Department of Biology, University of Utah, Salt Lake City, UT 84112, USA

JOHANNES SCHUL
Department of Biological Sciences, University of Missouri, Columbia, MO 65211, USA

JOSHUA J. SCHWARTZ
Department of Biological Sciences, Pace University, Pleasantville, NY 10570, USA

ANDREA MEGELA SIMMONS
Departments of Psychology and Neuroscience, Brown University, Providence, RI 02912, USA

DWAYNE D. SIMMONS
Departments of Otolaryngology and Anatomy and Neurobiology, Washington University School of Medicine, St. Louis, MO 63110, USA

PANTELIS N. VASSILAKIS
School of Music, De Paul University, Chicago, IL 60614-3296, USA

WOLFGANG WALKOWIAK
Zoologisches Institut der Univ. zu Köln, II. Lehrstuhl (Tierphysiologie), 50931 Köln, Germany

KENTWOOD D. WELLS
Department of Ecology and Evolutionary Biology, University of Connecticut, Storrs, CT 06269-3043, USA

WALTER WILCZYNSKI
Department of Psychology and Center for Behavioral Neuroscience, Georgia State University, Atlanta, GA 30302-3966, USA

Series Preface

Springer Handbook of Auditory Research

The *Springer Handbook of Auditory Research* presents a series of comprehensive and synthetic reviews of the fundamental topics in modern auditory research. The volumes are aimed at all individuals with interests in hearing research including advanced graduate students, postdoctoral researchers, and clinical investigators. The volumes are intended to introduce new investigators to important aspects of hearing science and to help established investigators to better understand the fundamental theories and data in fields of hearing that they may not normally follow closely.

Each volume presents a particular topic comprehensively, and each serves as a synthetic overview and guide to the literature. As such, the chapters present neither exhaustive data reviews nor original research that has not yet appeared in peer-reviewed journals. The volumes focus on topics that have developed a solid data and conceptual foundation rather than on those for which a literature is only beginning to develop. New research areas will be covered on a timely basis in the series as they begin to mature.

Each volume in the series consists of a few substantial chapters on a particular topic. In some cases, the topics will be ones of traditional interest for which there is a substantial body of data and theory, such as auditory neuroanatomy (Vol. 1) and neurophysiology (Vol. 2). Other volumes in the series deal with topics that have begun to mature more recently, such as development, plasticity, and computational models of neural processing. In many cases, the series editors are joined by a co-editor having special expertise in the topic of the volume.

RICHARD R. FAY, Chicago, Illinois
ARTHUR N. POPPER, College Park, Maryland

Volume Preface

In 1999, Volume 11 of the *Springer Handbook of Auditory Research* (*SHAR*) entitled *Comparative Hearing: Fish and Amphibians* presented a direct comparison of the auditory systems of fish and amphibians. Since 1999, there have been significant advances in our understanding of the physiology and behavior of anuran amphibians (the subject of most intense research), and it became apparent that it was time to explore anuran bioacoustics in detail and to present the first comprehensive overview of this topic in many decades.

In Chapter 1, Peter Narins and Albert Feng provide a framework for the volume, discuss the profound influence of Dr. Robert Capranica on the field and on neuroethology, and present a personal view of the future of the field. This is followed by Chapter 2 in which David Cannatella provides a contemporary molecular phylogenetic framework for the amphibia. As Cannatella points out, the study of animal communication has only recently begun to integrate phylogenetic thinking into its practices.

Kentwood Wells and Joshua Schwartz provide a review of the behavioral ecology of anuran vocal communication in Chapter 3. Following a brief description of the mechanisms underlying the production of frog calls, they discuss the energetic costs of call production, the major types of calls produced by males, and the less common female vocal signals. The mechanisms of call production and the neural control of vocalization in frogs are the subjects of Chapter 4 by Wolfgang Walkowiak. He provides detailed descriptions of the anuran larynx and its associated musculature and both expiratory and the less common inspiratory call generation mechanisms.

In Chapter 5, Carl Gerhardt and Mark Bee review the detection, recognition, and localization of acoustic signals by frogs; both the static and dynamic properties of these signals are considered. The ability of frogs to detect, recognize, and localize sound is of paramount importance for their reproductive success, mate choice, and competitive interactions between signalers.

For the first terrestrial vertebrates, hearing airborne sounds requires the creation of specialized anatomical structures that can effectively transfer sound pressure waves from air into the animal. In Chapter 6, Matthew Mason provides a comprehensive overview of the anatomy of these structures and their physiology in amphibians. This is followed in Chapter 7 by Dwayne Simmons, Sebastiaan Meenderink, and Pantelis Vassilakis who summarize the anatomy, physiology, and function of the auditory end organs in the frog inner ear. Emphasis is placed on hair cell morphology and innervation patterns, tuning properties and ion channels, and synaptic ultrastructure.

Walter Wilczynski and Heike Endepols follow in Chapter 8 with a review of the latest anatomical findings, including cytoarchitecture, chemoarchitecture, and connectivity patterns of the major auditory nuclei in the amphibian brain. This is followed in Chapter 9 by Gary Rose and David Gooler in which the authors bring together an enormous body of work on the function of the central auditory system in amphibians.

In Chapter 10, Andrea Megela Simmons and Seth Horowitz delve into the intricacies of a fascinating feature of amphibian development: auditory system plasticity across metamorphosis. In the final chapter (11), Albert Feng and Johannes Schul describe the behavior and physiology of hearing in real-world environments.

Although the chapters in this volume stand alone, additional related material can be found in other *SHAR* volumes. In particular, readers are referred to the aforementioned Volume 11 (*Comparative Hearing: Fish and Amphibians*—Fay and Popper 1999) for additional material and discussions of amphibians. In that volume, Lewis and Narins describe the anatomy and physiology of the ear, McCormick compares the auditory CNS of fish and amphibians, and Feng and Schellart do the same for CNS physiology. Acoustic communication in frogs is discussed in that volume by Zelick, Mann, and Popper. More recently, Christensen-Dalsgaard delved deeply into sound source localization by non-mammalian tetrapods, with an extensive analysis of localization by amphibians in *Sound Source Localization* (Volume 25—Popper and Fay 2005). The evolution of the amphibian ear has recently been considered by Smotherman and Narins in Volume 22 of this series (*Evolution of the Vertebrate Auditory System*—Manley, Popper, and Fay 2004). Finally, a wide range of topics on amphibian communication was discussed in *Acoustic Communication* (Volume 16—Megela Simmons, Popper, and Fay 2003).

PETER M. NARINS, Los Angeles, California
ALBERT S. FENG, Urbana, Illinois
RICHARD R. FAY, Chicago, Illinois
ARTHUR N. POPPER, College Park, Maryland

1
Hearing and Sound Communication in Amphibians: Prologue and Prognostication

Peter M. Narins and Albert S. Feng

1. Prologue

The vertebrate class Amphibia is composed of three orders: the Gymnophiona (caecilians) or legless amphibians (not known to vocalize), the Urodela (salamanders and newts), and the Anura (frogs and toads). With few exceptions, anuran amphibians are the most highly vocal of the amphibia, although the Pacific giant salamander (*Dicamptodon tenebrous*) is among a small group of urodeles known to produce calls. What do these sounds mean? How are they encoded in the nervous system?

In 1999, Volume 11 of the *Springer Handbook of Auditory Research* (*SHAR*) entitled: *Comparative Hearing: Fish and Amphibians* (Fay and Popper 1999) presented a direct comparison between the auditory systems of fish and amphibians. Why then, only seven years later, do we bother to assemble much that is known for the amphibians in a new volume? First, the topics in the present volume are all specific to amphibians and attempt to provide comprehensive coverage of current knowledge of the auditory system and its function in these fascinating animals. Second, there have been significant advances in our understanding of the physiology and behavior of anuran amphibians (the subject of most intense research) since the *SHAR* volume appeared. Third, in December of 2002 at the meeting of the Acoustical Society of America in Cancun, Mexico, one of us (PMN) organized a symposium highlighting the seminal work of Robert Capranica entitled: "Amphibian Bioacoustics Honoring Robert Capranica." In addition to Bob and his wife Pat, many of his former students, friends, and his PhD mentor, Moise Goldstein, were present. The excitement of being able to put together a program that was inspired by, dedicated to, and appreciated by Bob was palpable. That symposium was the genesis of the present monograph, and we believe that the high quality of its content speaks volumes to Bob's character, his dedication to his students, and his insistence on high-quality science by example. These qualities are reflected in the enthusiasm displayed by every contributor to this volume, their fervor for the field, their genuine desire for excellence, and their generosity of spirit.

A little bit about Bob. Bob has devoted his lifetime to research in animal bioacoustics and made extraordinary contributions to this field. He is a distinguished leader in animal bioacoustics. Bob was trained initially in electrical engineering but soon discovered his love for animal bioacoustics, and this interest prompted him to start his journey as a student in this field. For his dissertation research, he carried out what is arguably the most elegant dissertation research work ever completed in animal bioacoustics. This dissertation work, entitled "Evoked Vocal Responses of the Bullfrog," was published in an MIT monograph in 1965 (Capranica 1965) and instantly captured the praise of the entire research community in terms of its systematic and quantitative analysis, and conceptual elegance. Capranica showed that through well thought-out and carefully executed behavioral experiments, the salient feature in an animal's complex vocal signal could be pinned down. In a single stroke, Bob raised the bar for research in bioacoustics forever, and transformed the field; quantitative behavioral analysis became the new standard. What followed was equally remarkable. Bob then teamed up with Larry Frishkopf to elucidate the physiological underpinning of feature detection in bullfrogs. Bob's exemplary work was considered a cornerstone of neuroethology, now a vibrant field in neuroscience. He is in fact recognized as one of the founding fathers of neuroethology.

After joining the faculty rank at Cornell University in 1970, Bob continued to make one breakthrough after another. His insight in sound communication is second to none. As a scholar, he is as brilliant as they come. At Cornell, his laboratory attracted a large cadre of graduate students and postdocs. The authors of this chapter have always felt privileged to be among his first students. The feeling among Bob's students was overwhelmingly and uniformly positive because Bob gave us total freedom to undertake dissertation work of our choices. He encouraged us to think boldly and even unconventionally, and he provided whatever facilities and guidance necessary for successful execution of our dissertation research projects that covered a variety of different topics. A yardstick of success for a research scholar is usually measured by the success of his or her students and postdocs. In this regard, Bob's success is enormous. Students and postdocs trained under him now are distinguished researchers themselves at major institutions throughout the United States and around the world, making major contributions to the fields of animal bioacoustics and neuroethology. Bob is a true giant in the field of animal bioacoustics.

This monograph is subdivided into 11 chapters, each with a theme, but always with an eye toward integration with the other chapters. In Chapter 2, David Cannatella provides a contemporary molecular phylogenetic framework for the amphibia. This exercise is based on comparative methods that use a tree topology with branch lengths estimated by molecular data as a statistical framework for molecular evolution. As Cannatella points out, the study of animal communication has only recently begun to integrate phylogenetic thinking into its practices. Cannatella uses modern theory for defining and naming taxa, namely by patterns of relationship rather than by the possession of certain characters or traits (de Queiroz and Gauthier 1990). The number of living amphibians, about 5780, exceeds that of the mammals (Glaw and Köhler 1998). Nevertheless, amphibian population declines clearly signal environmental degradation (Hanken 1999). Thus, a comprehensive phylogeny of the extant amphibia is an exceedingly useful tool to enable the interpretation of behavioral and functional diversity in terms of evolutionary history. This becomes more critical in the face of habitat destruction and fragmentation, processes known to affect the life histories of many amphibia.

Kentwood Wells and Joshua Schwartz provide a review of the behavioral ecology of anuran vocal communication (Chapter 3). Following a brief description of the mechanisms underlying the production of frog calls, they discuss the energetic costs of call production, the major types of calls produced by males, and the less common female vocal signals. A large portion of the chapter concerns the interactions among males in calling assemblages and the ways in which these interactions affect the ability of males to attract mates; the active male–male interactions remind us that calling is a product of complex auditory–motor integration (a topic covered in the next chapter), rather than a product of pure motor commands in response to seasonal changes in the levels of sex hormones. The chapter ends with a brief consideration of how various features of the anuran auditory system may facilitate communication within a chorus setting, a topic that is discussed in greater detail in Chapter 11. Although they draw substantially from their own work, this chapter provides a thorough review of the field for any student interested in investigating chorus interactions and modeling them as communication networks, the adaptive plasticity in anuran vocalizations, such as socially mediated changes in anuran calling patterns and/or behavior, or the effects of background noise on frog calling.

Frogs typically produce calls by muscular contractions of the body wall, forcing air from the lungs through the larynx into the vocal sac (expiratory call generation). The mechanisms of call production and the neural control of vocalization in frogs are the subjects of Chapter 4 by Wolfgang Walkowiak. He provides detailed descriptions of the anuran larynx and its associated musculature, and both expiratory and the less common inspiratory call generation mechanisms. Some groups, such as the Pipidae of South America and Africa (including *Xenopus*), exhibit intrinsic call generation in which calling underwater is accomplished with no net airflow (Yager 1992; Kelley 2004; Tobias et al. 2004). The chapter additionally provides an extensive account of the neural control and hormonal influences on vocalizations, and sensory–motor integration.

Frog calling is also interesting for being extreme (Gridi-Papp 2003). In several species, calling is the most energetically costly activity performed by the animals (Wells 2001). Some Old World species produce a repertoire of tens of calls (Narins et al. 2000) or more (Feng et al. 2002), yet the neural control of the larynx is only just starting to be understood in these groups. The tissues involved in holding and moving air are surprisingly thin and elastic (McAlister 1961; Jaramillo et al. 1997). And the structures that radiate sound during calling are not restricted to the vocal sac, but may include the tympanic membranes in some species (Purgue 1997; Narins et al. 2001).

Studies of anurans have been especially influential in showing that the reciprocal selection pressures exerted by senders and receivers, the external environment (e.g., their predators and sympatric species), and sensory mechanisms can all influence the evolution of communication systems. In Chapter 5, Carl Gerhardt and Mark Bee review the detection, recognition, and localization of acoustic signals by frogs; both the static and dynamic properties of these signals are considered. The ability of frogs to detect, recognize, and localize sound is of paramount importance for their reproductive success, mate choice, and competitive interactions between signalers. These are discussed in the context of broad-scale evolutionary patterns, including geographical variation, habitat acoustics, reproductive character displacement, and pre-existing sensory biases. For many species, playback experiments using synthetic calls to evoke female phonotaxis and/or male vocal responses have elegantly revealed the sound features in the advertisement calls that are essential for call recognition. Recent evidence has shown that individual males have distinct call signatures and they have the ability to discriminate calls of neighboring males versus distant males, in addition to distinguishing the species-specific advertisement calls. Whether this is universal among frogs remains to be seen, but it is clear that the view of frogs having only a crude perceptual ability is overly simplistic or mistaken altogether. Mechanisms underlying sound localization are considered briefly at the end of the chapter.

For the first terrestrial vertebrates, hearing airborne sounds required the creation of specialized anatomical structures that could facilitate the transfer of sound pressure waves from air into the animal (Lewis and Fay 2004). In Chapter 6, Matthew Mason provides a comprehensive overview of the anatomy of these structures and their physiology in amphibians. In addition, extratympanic sound transmission is defined as "the transmission of airborne sound vibrations to the inner ear of amphibians by routes other than through the tympanic membrane and stapes"; these pathways are also considered. A substantial section of this chapter provides a modern treatment of our understanding of the function of the opercularis system, unique to amphibians, which consists of the operculum within the oval window, and the opercularis muscle connecting it to the shoulder girdle. Experimental evidence leading to several competing hypotheses concerning the function of this system are presented and evaluated. The specialized adaptations of the middle ear and related structures for underwater hearing in larval and adult amphibians are carefully reviewed and provide novel insights into the beautiful complexity of the various solutions to the problem of stimulating the inner ear in a medium with a characteristic impedance closely resembling that of the animal's body tissue.

The amphibian inner ear is unique among vertebrate animals in that it has two sensory organs specialized for the reception of airborne sound: the amphibian papilla (AP) and the basilar papilla (BP). Dwayne Simmons, Sebastiaan Meenderink, and Pantelis Vassilakis summarize the anatomy, physiology, and function of the auditory end organs in the frog inner ear in Chapter 7. Emphasis is placed on hair cell morphology and innervation patterns, tuning properties and ion channels, and synaptic ultrastructure. The literature on auditory nerve responses is reviewed and eighth-nerve fiber characterization techniques including both the traditional methods (frequency-threshold or tuning curves) and more recently developed methods (Wiener kernel analysis) are treated. It is worth noting that compared to the mammalian ear, the frog's AP is relatively simple. Like the mammalian cochlea, the AP exhibits characteristics of an active amplification system (see below) but without prestin-based electromotility. It has been shown to be a useful model system for elucidating the functional significance of individual components of the mammalian inner ear (e.g., hair cells, tectorial membrane, basilar membrane). We expect that it will continue to be a valuable model for exploring other mysteries in the auditory periphery.

One of the remarkable properties of the vertebrate inner ear is its great sensitivity. From early on it was recognized that such sensitivity could not arise solely from passive responses to sound. Rather, some active amplification mechanism would be required to enhance the vibration of inner ear structures in response to low-level acoustic stimuli (Gold 1948). The discovery of low-level sounds corresponding to such vibrations (Kemp 1978) provided the first evidence for the presence of an active amplification mechanism within the inner ear. These sounds are now known as Otoacoustic Emissions (OAEs) and can be measured by placing a sensitive microphone in the ear canal. A thorough review of both Spontaneous Otoacoustic Emissions (SOAEs) and Evoked Otoacoustic Emissions (EOAEs) in anurans is presented, and the implications of these emissions with respect to the presence of active processes and an "inner ear amplifier" in the amphibian are explored.

Walter Wilczynski and Heike Endepols follow in Chapter 8 with a review of the latest anatomical findings, including cytoarchitecture, chemoarchitecture, and connectivity patterns of the major auditory nuclei in the amphibian brain. Studies have shown that auditory connections within the anuran central nervous system are extensive (as they are in birds and mammals). Within the forebrain, relatively direct auditory pathways can be followed to most diencephalic and nearly all telencephalic regions, a characteristic consistent with the importance of acoustic signals in guiding all aspects of anuran social behavior. The largest single center of the auditory system, the midbrain torus semicircularis, serves as a key point in the central auditory pathways, integrating ascending auditory and descending forebrain inputs and serving as a transition from the lower brainstem auditory areas and its forebrain targets, and as an audiomotor interface. The gradual loss of tonotopy and increase in multimodal organization as one proceeds from the brainstem nuclei to the midbrain nuclei, and more rostrally, is of critical importance in understanding the anuran midbrain; this is clearly laid out in this chapter. The authors offer a concise perspective from which to conceptualize the central auditory system of anuran amphibians: it is likened to the difference between "hearing" and "sound communication." That is, "hearing"—the representation, identification, and localization of acoustic stimuli defining the sensory portion of the system—is consistent with the brainstem components of the auditory system. "Sound communication"—the broader context in which the outcome of the auditory sensory analysis is linked to motor, endocrine, motivational, and mnemonic processes connected to social interactions—characterizes the anatomical organization of auditory pathways throughout the forebrain. The midbrain torus semicircularis serves as the central station bringing both functions together.

Chapter 9 by Gary Rose and David Gooler integrates an enormous body of work on the function of the central auditory system in amphibians. As the authors point out, anuran neuroethology has its roots in Capranica's (1965, 1966) evoked calling studies with bullfrogs, supporting the notion of a neural logical AND operation that detects the formant-like simultaneous presence of low- and high-frequency energy peaks in their mating call. Numerous studies suggest that anurans also have neural specializations for analyzing the temporal structure of acoustic communication signals. These include filters for repetition rate and pulse shape (Amplitude Modulation, AM), direction of frequency change (Frequency Modulation, FM) and duration of notes. In addition, it is important in the context of mate selection and aggressive interactions for anurans to localize sound sources. The small interaural distances for most anurans pose formidable challenges both for the animals and for the experimenters interested in understanding the multifaceted mechanisms that underlie sound localization. In this chapter, the authors summarize the current understanding of the neural substrates of spectral and temporal processing, and directionality mechanisms in the central auditory systems of anurans.

In Chapter 10, Andrea Megela Simmons and Seth Horowitz delve into the intricacies of a fascinating feature of amphibian development: auditory system plasticity across metamorphosis. Metamorphosis in anurans that transforms aquatic-dwelling tadpoles into terrestrial frogs is a time of rapid morphological and behavioral changes affecting all sensory, motor, and vegetative systems. Metamorphosis features regression of structures important only in larval forms, transformation of larval structures into adult structures, and development of new structures necessary for the adult. This chapter outlines what is known about auditory system development over metamorphosis in both semi-terrestrial (*Rana*, *Hyla*) and fully aquatic (*Xenopus*) anurans. Much of the discussion focuses on two species that have a well-defined tadpole stage and for which most data have been gathered, *Rana catesbeiana* and *Xenopus laevis*. Changes occurring during early postmetamorphic development are also described, and areas where intensive study is still needed are highlighted.

In the final chapter (11) Albert Feng and Johannes Schul describe the behavior and physiology of hearing in real-world environments. In the first half of the chapter, they outline the physics of sound transmission, and how frog calls are affected by the frog's natural environment. This is followed by a discussion of the challenges frogs face when communicating within a chorus, in terms of signal detection, recognition, and localization, due to signal degradation during sound transmission, complexity of the auditory scene, and masking by background noise. The second half of the chapter concerns the mechanisms of hearing in complex environments focusing on two problems that are relatively well-characterized, spatially mediated masking release and comodulation masking release. Feng and Schul point out many of the unknowns with respect to solving the communication problems in a chorus. One area of opportunity concerns auditory performance (detection, recognition, and localization) in the presence of multiple competing sounds, resembling the frog's natural listening environments. Another pressing issue is to determine whether frogs perform auditory grouping and stream segregation and, if so, to uncover the underlying neural mechanisms.

2. Prognostication

What other areas emerge from these chapters that beg for additional studies? What are the projects left for the students in this field? Fortunately, there are many. Only a few are presented here, so that future investigators can come up with a new list of their own.

1. *The matched filter hypothesis—alive or dead?* This hypothesis, when referring to amphibian call production and detection was first stated by Capranica and Moffat (1983) as follows. "The receiver could try to analyze the distribution of energy (spectrum) at different frequencies (Fourier analysis). To do this requires an organ in the inner ear which maps frequencies systematically. The optimal detection strategy in this case is to employ a 'matched filter' technique." They go on: "The receiver must 'know' the shape of (t)his bimodal spectrum; that is, the frequency template of this signal must somehow be represented in the receiver's auditory system (again, either through learning or else innate in origin)." Now if the receiver has "a frequency response which exactly matches the envelope of the energy spectrum of the sender's call," the receiver then "obtains the highest signal-to-noise ratio in the frequency domain for that particular call."

What does this imply? Unless there are multiple matched filters tuned to a wide range of biologically relevant signals, it implies that the receiver is matched to the incoming call spectrum, at the expense of being matched to any other signal. This would seem to be counterproductive in light of the variability in spectral degradation by frogs' natural environments, especially if frogs must perform individual recognition on top of species recognition; yet this is not impossible. Measurements are needed to determine the spectra of a set of natural sounds to which an individual would be reasonably expected to be exposed in its natural environment. The frequency response of the auditory system could then be determined and the matched filter hypothesis could be directly tested.

2. *Individual recognition by call and/or calling site.* Is individual recognition an ability limited to a few species, or is it common among frogs? Is it a function of chorus density, territorial behavior, or analytical ability of the individual species?

3. *A general theory of call generation in amphibia.* How do the structures in the amphibian larynx function to produce the wide variety of sounds present in

many species' vocalizations? Do the nonlinear features found in the vocalizations contribute to the call diversity and, if so, how are they generated? Do they convey any behavioral significance?

4. *Call radiation by eardrums in males.* Purgue (1997) showed that the eardrums in the North American bullfrog, *Rana catesbeiana*, are capable of radiating considerable sound energy. Is this property unique to this species? What is the distribution of this property among amphibians and how has this feature evolved?

5. *Low-temperature calling in amphibia. Eusophus* species from southern Chile call routinely at 5 to 7 degrees Celsius (Penna et al. 2005). What are the physiological mechanisms that allow such vigorous calling activity to persist in an ectotherm at these low temperatures?

6. *Spectral shaping by environments.* Earlier studies have shown that frogs can develop local dialects (modifications of spectral and/or temporal attributes of vocalizations) in response to selection pressure from competing environmental sounds. Such changes typically evolve over a long period but recent studies show that spectral shaping can occur acutely when frogs adjust their call frequency to that of the resonance frequency of the calling site in order to maximize broadcast power (Lardner and bin Lakim 2002); also birds have been shown to shift their call frequency in response to urban noise (Slabbekoorn and Peet 2003). In a recent field study, Sun and Narins (2005) documented significant call rate changes in chorusing male frogs in response to anthropogenic sounds including airplane flyby and motorcycle engine noise. How common is this among anuran amphibians and how plastic is the frog's call production system?

7. *Function of the hyperextended call repertoire.* The call repertoires of males of *Boophis madagascariensis* (Madagascar; Narins et al. 2000), *Polypedates leucomystax* (Thailand; Christensen-Dalsgaard et al. 2002) and *Amolops tormotus* (Central China; Feng et al. 2002) have been documented and been shown to be relatively extensive compared to other frogs studied. What are the evolutionary advantages conferred by and the biological significance of an extended repertoire in anurans?

8. *Effect of immigration and emigration on chorus structure.* How does the appearance of a new individual (by emigration) or the disappearance of an individual (say, by predation) modify the vocal interactions in a calling assemblage of conspecific males?

9. *Extended high-frequency sensitivity in amphibia.* Recent studies of male arboreal frogs (*Amolops tormotus*) living in central China have revealed that their advertisement calls contain prominent ultrasonic (above 20 kHz) components (Narins et al. 2004). Moreover, recent behavioral and physiological experiments demonstrated that this species (as well as a sympatric species) has the ability to hear and communicate with ultrasound (Feng et al. 2006). The ultrasound sensitivity is due in part to having very thin tympanic membranes that are recessed in the skull at the end of the ear canals (thereby shortening their distances to the inner ear, and resulting in reduced ossicular mass); these adaptations facilitate ultrasound transmission to the inner ear. It is highly likely that these species are

not unique in their ability to sense ultrasound, and so it is incumbent upon the next generation of researchers to discover the distribution of this trait and to determine the factors that contribute to the evolution of ultrasound perception.

10. *Multimodal processing.* Recent field experiments have demonstrated that coordinated visual and auditory cues are involved in guiding aggressive interactions in dendrobatid frogs; these results were obtained with the use of robotic frogs (Narins et al. 2003, 2005). More realistic models that move, fight, adjust call parameters in response to the behavior of nearby conspecifics, and so on have the potential for revealing the underpinnings of a wide variety of natural behaviors in the field and determining whether other behaviors are also guided by multimodal sensory integration.

11. *Slow motility in frog AP hair cells.* Slow motility refers to the changing of the hair cell body length in response to particular stimuli over a period of seconds to minutes. In the mammalian cochlea, slow motility in outer hair cells is actin–myosin-based, and thought to adjust the gain of the auditory sensors (inner hair cells) in the presence of different levels of background noise (Frolenkov et al. 1998). Recent evidence supports the idea that hair cells from the amphibian papilla of the leopard frog (*Rana pipiens pipiens*) also exhibit isometric shortening in response to an increasing concentration of intracellular free calcium (Farahbakhsh and Narins 2005). How does this process affect auditory sensitivity? How is this behavior mapped across frog families or species?

12. *Chemistry of specific auditory projections.* Whereas we have made significant strides in understanding the cytoarchitecture, chemoarchitecture, and connectivity of the frog central auditory system, the neurotransmitters and neuromodulators associated with specific neural connections are poorly understood, and so are their actions on the pre- and postsynaptic terminals.

13. *Biophysical properties of neuronal cell membranes.* Almost nothing is known about the membrane and biophysical properties of central auditory neurons in amphibians. This is a major impediment for modeling work at the cellular and systems levels that is important for the understanding of the neural mechanisms underlying units' response selectivities in the time or frequency domains.

14. *Feature binding.* Single neurons in the frog auditory system show impressive response selectivities to salient temporal, spectral, or directional features. How are these individual features integrated to give coherent perception of a call, among the many calls in a chorus? Do frogs show coherent perception?

Clearly there is more to be done. It is our hope that this volume inspires its readers to answer these questions and others that emerge in the course of our investigations.

References

Capranica RR (1965) The evoked vocal response of the bullfrog: A study of communication by sound. MIT Res Monogr 33, MIT Press, Cambridge, MA.

Capranica RR (1966) Vocal response of the bullfrog to natural and synthetic mating calls. J Acoust Soc Am 40:1131–1139.
Capranica RR, Moffat AJM (1983) Neurobehavioral correlates of sound communication in anurans. In: Ewert JP, Capranica RR, Ingle DJ (eds) Advances in Vertebrate Neuroethology. London, New York: Plenum, pp. 701–730.
Christensen-Dalsgaard J, Ludwig T, Narins PM (2002) Call diversity in an Old World treefrog: Level dependence and latency of acoustic responses. Bioacoustics 13:21–35.
de Queiroz K, Gauthier J (1990) Phylogeny as a central principle in taxonomy: Phylogenetic definitions of taxon names. Syst Zool 39:307–322.
Farahbakhsh N, Narins PM (2005) Slow motility in hair cells of the frog amphibian papilla: Ca^{2+}-dependent shape changes. Hear Res 212:140–159.
Fay RR, Popper AN (eds) (1999) Comparative Hearing: Fishes and Amphibians. New York: Springer-Verlag.
Feng AS, Narins PM, Xu CH (2002) Vocal acrobatics in a Chinese frog, *Amolops tormotus*. Naturwissenschaften 89:352–356.
Feng AS, Narins PM, Xu C-H, Lin W-Y, Yu Z-L, Qiu Q, Xu Z-M, Shen J-X (2006) Ultrasonic communication in frogs. Nature 440:333–336.
Frolenkov GI, Atzori M, Kalinec F, Mammano F, Kachar B (1998) The membrane-based mechanism of cell motility in cochlear outer hair cells. Mol Biol Cell 9:1961–1968.
Glaw F, Köhler J (1998) Amphibian species diversity exceeds that of mammals. Herpet Rev 29:11–12.
Gold T (1948) Hearing. II. The physical basis of the action of the cochlea. Proc Roy Soc Lond B, Biol Sci 135:492–498.
Gridi-Papp M (2003) Mechanism, behavior and evolution of calling in four North American treefrogs. PhD thesis, University of Texas at Austin.
Hanken J (1999) Why are there so many new amphibian species when amphibians are declining? Trends Ecol Evol 14:7–8.
Jaramillo C, Rand AS, Ibáñez R, Dudley R (1997) Elastic structures in the vocalization apparatus of the tungara frog *Physalaemus pustulosus* (Leptodactylidae). J Morphol 233:287–295.
Kelley DB (2004) Vocal communication in frogs. Curr Opin Neurobiol 14:751–757.
Kemp D (1978) Stimulated acoustic emissions from within the human auditory system. J Acoust Soc Am 64:1386–1391.
Lardner B, bin Lakim M (2002) Tree-hole frogs exploit resonance effects. Nature 420:475.
Lewis ER, Fay RR (2004) Environmental variables and the fundamental nature of hearing. In: Manley GA, Popper AN, Fay RR (eds) Evolution of the Vertebrate Auditory System. New York: Springer-Verlag, pp. 27–54.
McAlister WH (1961) The mechanics of sound production in North American *Bufo*. Copeia 1:86–95.
Narins PM, Feng AS, Schnitzler H-U, Denzinger A, Suthers RA, Lin W, Xu C-H (2004) Old World frog and bird vocalizations contain prominent ultrasonic harmonics. J Acoust Soc Am 115:910–913.
Narins PM, Grabul DD, Soma K, Gaucher P, Hödl W (2005) Cross-modality integration in a dart-poison frog. Proc Nat Acad Sci 102:2425–2429.
Narins PM, Hödl W, Grabul DS (2003) Bimodal signal requisite for agonistic behavior in a dart-poison frog, *Epipedobates femoralis*. Proc Nat Acad Sci 100:577–580.
Narins PM, Lewis ER, McClelland BE (2000) Hyperextended call repertoire of the endemic Madagascar treefrog *Boophis madagascariensis* (Rhacophoridae). J Zool Lond 250:283–298.

Narins PM, Lewis ER, Purgue AP, Bishop PJ, Minter LR, Lawson DP (2001) Functional consequences of a novel middle ear adaptation in the West African frog *Petropedetes parkeri* (Ranidae). J Exp Biol 204:1223–1232.

Penna M, Narins PM, Feng AS (2005) Thresholds for evoked vocal responses of *Eusophus emiliopugini* (Amphibia, Leptodactylidae). Herpetologica 61:1–8.

Purgue AP (1997) Tympanic sound radiation in the bullfrog *Rana catesbeiana*. J Comp Physiol 181:438–445.

Slabbekoorn H, Peet M (2003) Birds sing at a higher pitch in urban noise. Nature 424:267.

Sun JWC, Narins PM (2005) Anthropogenic sounds differentially affect amphibian call rate. Biological Conservation 121:419–427.

Tobias ML, Barnard C, O'Hagen R, Horng SH, Rand M, Kelley DB (2004) Vocal communication between male *Xenopus laevis*. Anim Behav 67:353–365.

Wells KD (2001) The energetics of calling in frogs. In: Ryan MJ (ed) Anuran Communication. Washington, DC: Smithsonian Institution Press, pp. 45–60.

Yager DD (1992) A unique sound production mechanism in the pipid anuran *Xenopus borealis*. Zool J Linn Soc 104:351–375.

2
An Integrative Phylogeny of Amphibia

David C. Cannatella

1. Phylogeny Estimation as an Integrative Activity

1.1 Advances in Phylogeny Estimation

The use of molecular sequences in recent years has injected a rich source of information about phylogeny into studies of amphibian evolution. Molecular phylogenies are not ipso facto more accurate than phylogenies based on other data. However, basic principles of data analysis indicate that larger datasets produce more accurate results than smaller ones, and DNA sequence datasets may typically be an order of magnitude larger than those from other data sources. Data from morphology and other sources (nonmolecular, or phenotypic data) are by no means obsolete. The quality and numbers of studies using morphological characters is increasing. Moreover, efforts are being made to synthesize results from these disparate datasets.

Work on amphibian relationships has enjoyed the advances in phylogeny estimation, which is now viewed as a task of statistical estimation, rather than solely logical deduction (Felsenstein 2004). Earlier theory advocated a simple combination of all data, with all characters equally weighted, under a parsimony criterion (minimization of postulated evolutionary changes). Recent theory has yielded probabilistic models of evolution that account for differing processes of molecular evolution. Such models enable the use of maximum likelihood (under which the probability of observing a dataset is maximized, given a particular hypothesis or model of evolution) as a criterion for selecting the best tree. Maximum likelihood also accommodates types of data other than DNA sequences. Critics of probabilistic models argue that the approach is flawed because one cannot know whether a model accurately describes real data. However, even simple models provide a better explanation of data than treating all types of changes equally.

The number of taxa in current datasets has made even approximate searches under the maximum likelihood criterion prohibitively slow; strategies for finding the best tree out of the possible trillions have improved. Markov Chain Monte Carlo methods have been introduced for sampling the universe of possible trees in a Bayesian framework (Larget and Simon 1999). As well, the Bayesian framework permits the interpretation of support for particular clades as probability statements (Lewis 2001).

Lastly, the advent of comparative methods allows valid statistical inference in a phylogenetic context (Harvey and Pagel 1991). Standard methods assume each datum is independent. However, evolution and phylogeny destroy this independence, because species that share a common ancestor exhibit covariance in traits (Pagel 1999). Comparative methods use an estimated tree with branch lengths as a statistical framework for trait evolution (Felsenstein 1985; Garland et al. 1999). The study of animal communication has only recently begun to integrate phylogenetic thinking (e.g., Ryan and Rand 1995).

1.2 Definitions and Names of Taxa

Advances in the theory of defining and naming taxa are also changing systematics. A primary advance is that taxa are defined by patterns of relationship rather than by possession of certain characters or traits (de Queiroz and Gauthier 1992). Although characters are used to estimate the patterns of relationship, it is the patterns of ancestry and descent that are given primacy. It is also generally agreed that taxa should be monophyletic. That is, a group should contain an ancestor and all of its descendants, rather than just some of its descendants. Monophyletic taxa enable the correct interpretation of evolutionary patterns such as homology (synapomorphy), convergence, parallelisms, and reversals.

The distinction between taxon names and parts of the tree is important. That is, names can point to nodes or stems (Fig. 2.1). A node name points to a partic-

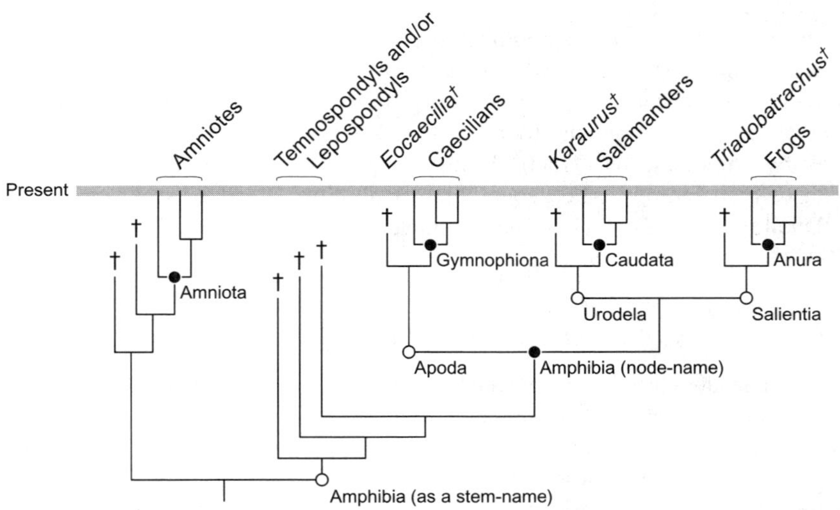

FIGURE 2.1. Node-based and stem-based definitions of Amphibia and the major lineages of amphibians.

ular node, and is useful for designating stable groups, such as those composed of living taxa. A stem-name points to the stem leading up to a node. A stem-name can accommodate changes in fossil taxa without demanding a redefinition of the name.

It is also now widely recognized that the ranks of taxon names (e.g., family, tribe, etc.) contain little information. Two taxa of equal rank need not display equivalent temporal, morphological, or molecular divergence. On the other hand, ranks such as family (ending in –idae) are often used as comparable units for generating lists; a count of the families of salamanders is a convenient, albeit highly imperfect, measure of diversity. Some workers have urged abandoning Linnean ranks (e.g., de Queiroz and Gauthier 1992); whether Anura (frogs) is an order or suborder makes no difference to an understanding of their biology.

2. Modern Amphibians

Modern amphibians include frogs, salamanders, and caecilians, and their Mesozoic (245–65 million years ago, my) and Cenozoic (65 my–present) extinct relatives. Modern amphibians are at times called lissamphibians to distinguish them from the Paleozoic forms, which are better thought of as early tetrapods, rather than amphibians. In the language of phylogenetic taxonomy (de Queiroz and Gauthier 1992), Amphibia is a node-based name that points to the most recent common ancestor of frogs, salamanders, and caecilians, and all the descendants (living and extinct) of that ancestor (Cannatella and Hillis 1993). Thus, modern amphibians comprise the lineage minimally circumscribed by living taxa. This group also includes fossil taxa that are easily recognized as being frogs, salamanders, or caecilians.

The number of living amphibian species, about 5800, exceeds that of the lineage Mammalia (Glaw and Köhler 1998). In the last 20 years, the number of recognized species has increased by 35%. Yet, the decline of amphibian populations tangibly signals environmental degradation (Hanken 1999). The textbook by Duellman and Trueb (1986) remains the most comprehensive treatment of their biology. The geographic distribution of groups was summarized in Duellman (1999). Web resources include Frost (2004) and AmphibiaWeb (2005).

2.1 Features of Modern Amphibians

Amphibians are named for their dual lives: an aquatic larva that metamorphoses into a terrestrial adult. In a loose sense, amphibians bridge the gap between fishes, which are fully aquatic, and amniotes, which have fled the aquatic environment and abandoned metamorphosis. However, amphibians are not in any sense trapped in an evolutionary cul-de-sac. Each type of living amphibian—frog, salamander, and caecilian—is highly distinctive. Frogs are squat, four-legged creatures with large mouths and eyes, and elongate hind limbs used for jumping. There is no tail (hence, Anura, lacking a tail), and the caudal vertebrae are reduced into a bony strut. About 90% of living amphibian species are frogs; these rely mostly on visual and acoustic cues.

Salamanders are more typical-looking tetrapods, all with a tail (the meaning of Caudata) and most have four legs. Some completely aquatic or fossorial species are elongate with reduced limbs and girdles. In general, salamanders use olfactory cues.

Living caecilians are all limbless and elongate. Grooved rings encircle the body, evoking the image of an earthworm. Most caecilians are fossorial, but some have evolved into aquatic habitats. The Latin root *caecus*, meaning blind, is a misnomer; all have eyes, but they are reduced. Just below the eye is a unique protrusible tentacle used for olfaction. The tail is essentially absent in most species.

Several features set modern amphibians apart from other vertebrates, and support monophyly of amphibians relative to amniotes and fishes. For example, the teeth are bicuspid and have a zone of reduced mineralization between the crown and the base (pedicel). These pedicellate teeth suggest relationships to a few temnospondyl labyrinthodonts (Bolt 1977). Modern amphibians also share the absence or reduction of several bones on the skull roof and the palate (Reiss 1996), generally ascribed to paedomorphosis (Alberch et al. 1979). Paedomorphosis is a pattern produced by a change in the developmental process; specifically, a species becomes sexually mature (adult) at an earlier stage of development than its immediate ancestor. As a result, the adult of amphibians resembles the juvenile (or larval) stage of Paleozoic relatives. A corollary is small size (Hanken 1985)—living amphibians are diminuitive relative to the Paleozoic forms (Bolt 1977). Beyond this, some are truly miniaturized (Trueb and Alberch 1985) with profound effects to the nervous system and sensory organs.

A force-pump mechanism is used for breathing (Gans et al. 1969; Brainerd et al. 1993). The buccal cavity forces air into the lungs by positive pressure. In contrast, amniotes fill their lungs using aspiration; the rib cage and/or diaphragm creates negative pressure in the thorax. Amphibian ribs are short and do not form a rib cage as in amniotes, so aspiration is not possible. In addition to the stapes-basilar papilla sensory system of tetrapods, living amphibians have a second acoustic pathway, the opercular–amphibian papilla system. This system is more sensitive to lower frequency vibrations than is the stapes-basilar papilla pathway. The operculum (a bone of the posterior braincase) is connected to the shoulder girdle by the opercularis muscle, derived from the levator scapulae complex. This muscle may transmit vibrations from the ground through the forelimb and shoulder girdle to the inner ear. Alternatively, it may serve as a protective system against loud sounds or, with the operculum, to reduce the effects of high intraoral pressures present during positive pressure breathing and vocalization (see Chapter 6, this volume).

2.2 *Amphibians and Early Tetrapods*

The early tetrapods traditionally included two groups, the Labyrinthodontia and Lepospondyli. Heatwole and Carroll (2000) provided a summary. Labyrinthodonts ranged from the Upper Devonian (375 my) through the Permian (290 my), declined in the Triassic, and persisted to the Cretaceous. Labyrinthodonts are a paraphyletic group that also gave rise to living amniotes. Lepospondyls ranged from the Lower Carboniferous (340 my) to the base of the Upper Permian (250 my); they are a heterogenous group of uncertain monophyly (Carroll et al. 1999; Anderson 2001). The exact relationship of modern amphibians to these two is strongly debated.

The currently favored Temnospondyl Hypothesis (Fig. 2.2A), suggests that the group of frogs, salamanders, and caecilians is monophyletic and is nested within dissorophoid temnospondyls (Bolt 1977; Milner 1988; Bolt 1991; Trueb and Cloutier 1991a; Milner 1993; Ruta et al. 2003). Dissorophoids include some small, paedomorphic forms, such as *Doleserpeton*, that share many derived features with living amphibians.

The Lepospondyl Hypothesis (Fig. 2.2B) holds that modern amphibians are a clade, but nested within the lepospondyls (Anderson 2001), particularly within the Microsauria (Laurin and Reisz 1997; but see Coates et al. [2000] and Ruta et al. [2003]). Because temnospondyls are distantly related to amphibians under this hypothesis, the shared similarities with dissorophoid temnospondyls are interpreted as convergences. Under a stem-name definition of Amphibia, lepospondyls would be included in Amphibia under the Lepospondyl Hypothesis. But under the Temnospondyl Hypothesis, temnospondyls but not lepospondyls are part of Amphibia. A third arrangement, the Polyphyly Hypothesis (Fig. 2.2C), claims that caecilians are derived from microsaurs (Carroll and Currie 1975; Carroll 2000a,b and salamanders and frogs from temnospondyls. The Polyphyly Hypothesis gained some strength from the discovery of *Eocaecilia* fossils (see below).

DNA sequences and features of soft anatomy ally frogs, salamanders, and caecilians as a clade relative to living amniotes and fishes. Because fossils do not easily yield information about nucleotides or soft anatomy, these characters have provided no direct evidence for the monophyly of Amphibia vis-à-vis Paleozoic tetrapods (Trueb and Cloutier 1991a; b).

2.3 *Interrelationships of Modern Amphibians*

Two hypotheses of modern amphibian relationships have been advanced. One tree, based on nonmolecular data, allies frogs and salamanders (= Batrachia), with caecilians as the odd one out (Fig. 2.1). In the second hypothesis, early analyses of DNA data slightly favored salamanders and caecilians (= Procera) as closest relatives (Hedges and Maxson 1993; Feller and Hedges 1998). However, Zardoya and Meyer (2001) analyzed complete mitochondrial sequences of a frog, salamander, and caecilian, and found the frog and salamander to be sister-groups.

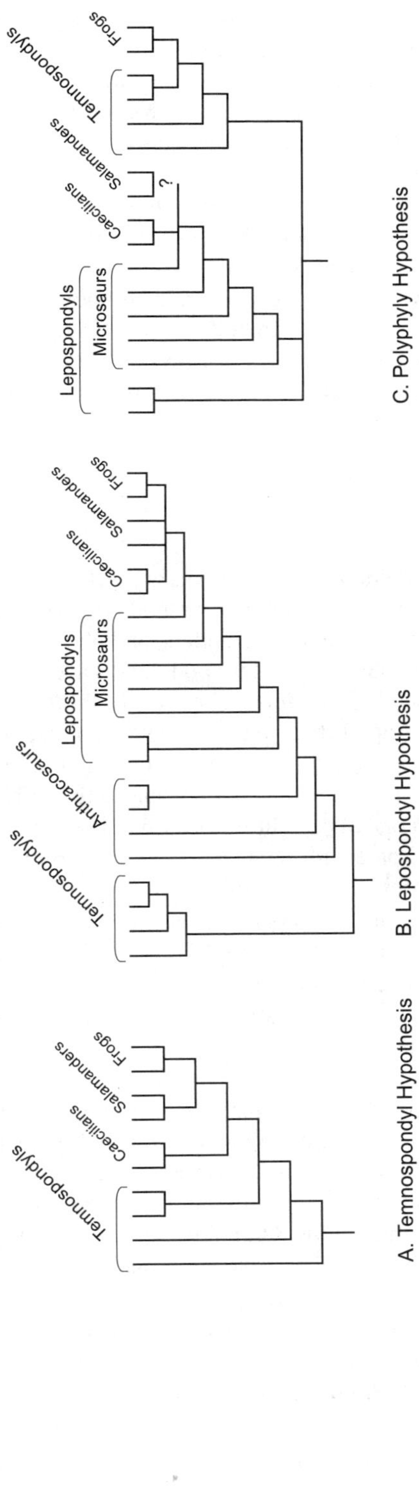

FIGURE 2.2. Alternative hypotheses of relationships among modern amphibians (caecilians, frogs, and salamanders) and Paleozoic groups (temnospondyls, microsaurs, and lepospondyls).

Zhang et al. (2003), in contrast, found salamanders to be the sister-group of the one caecilian. Although taxon sampling in both studies was limited, the results suggest significant uses for large datasets such as mitochondrial genomes. San Mauro et al. (2005) also found a salamander + frog relationship, and placed the split between caecilians and (frogs + salamanders) at 367 my, and that between frogs and salamanders at 357 my.

A fourth group of amphibians is the Albanerpetontidae, known only as fossils from the Jurassic to the Miocene (Milner 2000). The name Allocaudata has been used infrequently for these animals, which closely resemble salamanders in skull shape and in the primitive features of a generalized body shape, four limbs and a tail. Albanerpetontids do not have pedicellate teeth. They have been considered to be nested within salamanders, or the sister-group of Batrachia (McGowan and Evans 1995); an extensive analysis (Gardner 2001) places them in the latter position.

2.4 The Name Amphibia

Ernst Haeckel divided Amphibia into Lissamphibia (salamanders and frogs), and Phractamphibia (caecilians and fossil labyrinthodonts) (Haeckel 1866). Liss- refers to the naked skin of frogs and salamanders, and phract- refers to the helmet of dermal skull bones found in early tetrapods and caecilians. Gadow (1901) transferred the caecilians to Lissamphibia. For most of the 20th century, the name Amphibia was used for tetrapods that were not reptiles, birds, or mammals. Thus, the oldest known tetrapods (Devonian labyrinthodonts) as well as the Lepospondyli, were included in Amphibia. In part due to the influence of the paleontologist Alfred S. Romer, this rendition of Amphibia appeared in almost all comparative anatomy and paleontology texts. Modern amphibians were believed to be derived from different lineages; frogs from Labyrinthodontia, and salamanders and caecilians from Lepospondyli. Parsons and Williams (1962; 1963) synthesized evidence supporting the monophyly of modern amphibians and resurrected Gadow's term Lissamphibia. However, this term is used mainly among specialists; most biologists and most textbooks refer to frogs, salamanders, and caecilians simply as amphibians.

The use of Amphibia in the Romerian sense has been largely abandoned and the name has been redefined as a monophyletic group, but in two contrasting ways. The name Amphibia is applied to the node in the tree that is the last (most recent) ancestor common to living frogs, salamanders, and caecilians (de Queiroz and Gauthier 1992). Amphibia includes the modern forms and their close fossil relatives, including albanerpetontids. On the other hand, Amphibia is defined as the stem containing frogs, caecilians, and salamanders, as well as all other (extinct) taxa more closely related to these living taxa than to amniotes. In other words, the stem-name Amphibia subsumes all taxa branching from the stem leading to modern amphibians; this includes either the temnospondyls, the lepospondyls, or perhaps both (Laurin and Reisz 1997; Anderson 2001; Ruta et al.

2003). Depending on which hypothesis one accepts, the set of taxa included in Amphibia changes dramatically.

The stem-based definition of Amphibia has an undesirable effect. Generalizations about the biology of modern amphibians may be wrongly extended to extinct temnospondyls and/or lepospondyls, which bear little biological resemblance to the living forms. Under a stem-based definition, the statement "amphibians have mucous glands," commonly found in biology texts, would mean that lepospondyls had mucous glands, an extrapolation for which there is no evidence. In contrast, under the node-name definition of Amphibia, one can infer that extinct frogs, salamanders, and caecilians have mucous glands without inappropriately extending the inference to extinct temnospondyls and lepospondyls.

In the following treatment, the major lineages are discussed in turn, starting arbitrarily with one of the terminal clades, and working toward the base of the tree.

3. Caecilians

Caecilians include nearly 165 living species, in five or six families, concentrated in the tropical regions of America, Africa (excluding Madagascar), the Seychelles Islands, and much of southeast Asia. The name Gymnophiona, which means naked (i.e., lacking scales) snake, refers to the node of the tree that is the most recent ancestor of all living species. Being fossorial, caecilians are rarely seen in the wild even by dedicated herpetologists. Occasional individuals may surface after rains. Most species are 0.3 to 0.5 meters long, although one reaches 1.5 meters. All caecilians are elongate, with 86 to 205 vertebrae. Caecilians are almost unique among amphibians in having a phallodeum, or male intromittent organ, for internal fertilization.

Relative to other amphibians, caecilian skulls are highly ossified, with many fused bones. The bullet-shaped cranium is used for digging and compacting the soil. Most caecilians are fossorial. However, one group is aquatic with a laterally compressed posterior body. Like that of an earthworm, the body is highly annulated; the primitive groups are the most heavily ringed. Living caecilians have small eye sockets with reduced eyes; in some the eyes are concealed under the dermal skull bones. Most caecilians lay eggs that hatch into free-living larvae (oviparity). Live birth (viviparity) has evolved independently in different families.

Gymnophiona is the least well understood of all major vertebrate lineages. Compared to salamanders and frogs, the evolutionary relationships among caecilian families have not been as contentious. However, the sampling of genera is poor, simply because specimens are hard to come by. Taylor's (1968) landmark volume stimulated work on this poorly known group. Lescure et al. (1986) radically revised caecilian taxonomy. But because those conclusions were based on sparse data, Nussbaum and Wilkinson (1989) urged the retention of the orthodox taxonomic groups.

3.1 Caecilian Phylogeny

DNA sequences for 12S and 16S rRNA genes were analyzed for 13 species in ten genera by Hedges et al. (1993), and by Wilkinson et al. (2002b) for a few Indian species. A study of complete mitochondrial genomes for a representative group of species (San Mauro et al. 2004b) yielded relationships comparable to those found using smaller datasets. New morphological datasets have been also important (e.g., Wilkinson 1997).

The relationships of caecilians are summarized in Figure 2.3, which reflects both molecular and morphological datasets. Scolecomorphidae is an African lineage with bizarre features; in some a layer of bone covers the eye, and in at least one the eye is protrusible because of its attachment to the tentacle. The Typhlonectidae, a South American group, is modified for an aquatic lifestyle, with features such as a laterally compressed posterior body (Wilkinson 1989). Typhlonectid phylogeny has been elucidated using morphological characters (Wilkinson and Nussbaum 1999).

Both Scolecomorphidae and Typhlonectidae are derived from within the larger group Caeciliidae. This geographically and biologically diverse group comprises most of the phylogenetic uncertainty. Caeciliids are pantropical (Mexico, Central, and South America; Africa and the Seychelles, India, and Southeast Asia), and include a great variety of taxa—including the smallest and largest species—and the gamut of reproductive modes, such as egg-laying with free-living larvae, egg-laying with direct development, viviparous forms, and maternal care.

The semi-fossorial species of Ichthyophiidae inhabit India, Sri Lanka, and Southeast Asia. Its closest relative is Uraeotyphlidae, from southern India (Wilkinson and Nussbaum 1996). Based on morphological and molecular datasets, Rhinatrematidae is considered as the sister-taxon to all other living taxa (Nussbaum 1977; Hedges et al. 1993), a clade named Stegokrotaphia (Cannatella

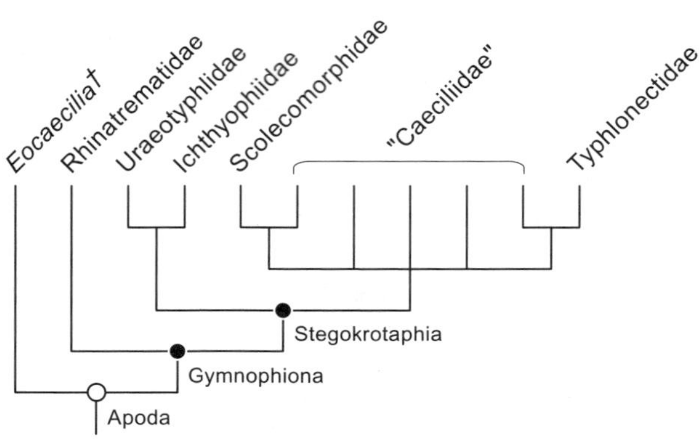

FIGURE 2.3. A generally accepted phylogenetic hypothesis of relationships among caecilians. The quotes around "Caeciliidae" indicates that the group is paraphyletic with respect to Scolecomorphidae and Typhlonectidae.

and Hillis 1993). Rhinatrematidae are found in northern South America. Reflecting their phylogenetic position, these species retain a very short tail as a primitive feature; other caecilians lack a tail. Rhinatrematids are also the most heavily annulated species.

Fossil caecilians are few. Vertebrae are known from the Upper Cretaceous, Tertiary, and Quaternary (Wake et al. 1999), but do not add much information to the higher-level phylogeny of caecilians. Although living caecilians lack limbs and are nearly or completely tailless, a Jurassic fossil with legs and a tail has been identified as a caecilian: *Eocaecilia micropodia* has an elongate body and small but robust limbs (Jenkins and Walsh 1993; Carroll 2000a). The margin of the orbit bears a groove, interpreted as a space for the tentacle; thus, *Eocaecilia* supposedly shares a derived character with extant caecilians. The possession of primitive characters (in addition to limbs and tail) places *Eocaecilia* as the sister-group of living caecilians. Apoda is the stem-name for the clade of *Eocaecilia* + Gymnophiona (Fig. 2.1).

4. Salamanders

Caudata is the name applied to the node that is the most recent common ancestor of living salamanders. The approximately 500 species of living salamanders are placed in ten families (Fig. 2.4). The smallest (*Thorius*; Plethodontidae) may

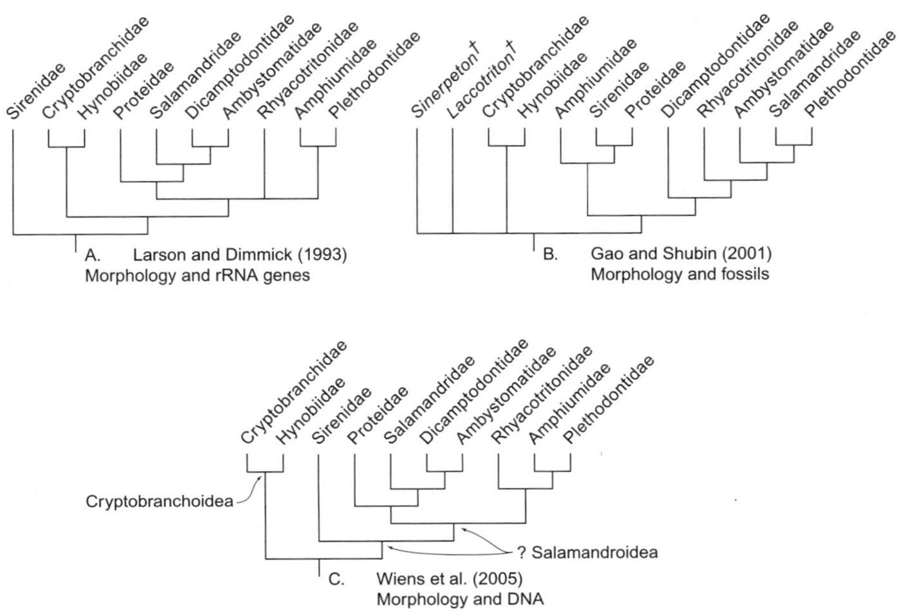

FIGURE 2.4. Alternative relationships among the families of salamanders.

only reach 30 mm total length. The largest salamanders (*Andrias*; Cryptobranchidae) reach 1.5 meters.

Several salamanders have undergone elongation with associated limb reduction or loss. Some of these are larger, fully aquatic, neotenic forms (retaining larval features as adults), such as Sirenidae and Amphiumidae. Fully aquatic salamanders, whether neotenic or not, typically retain gill slits, and some have red, plumose external gills. Limb reduction and elongation are also correlated in some terrestrial salamanders that occupy a semifossorial niche. Internal fertilization by way of a spermatophore, typically a mushroom-shaped mass of spermatozoa and mucous secretions characterizes a large clade of most of the major groups of salamanders. Fertilized eggs develop directly—into a small salamander—or indirectly, with a larva that metamorphoses.

4.1 Salamander Phylogeny

The content of the Linnean families of salamanders is without much controversy. Ten families of living salamanders are recognized; each is monophyletic. Some phylogenies are well sampled at the species level (i.e., Titus and Larson 1995), in contrast to frogs and caecilians. In several families nearly all species have been examined using DNA sequences.

Until recently, the phylogeny of family-level groups of salamanders lacked synthesis, and different datasets (sequences, morphology, and fossils) and combinations of these produce very different trees (e.g., Duellman and Trueb 1986; Good and Wake 1992; Hay et al. 1995). Generally, the Sirenidae and Cryptobranchoidea have been thought to be among the most primitive salamanders. Analysis of nuclear-encoded rRNA genes (Larson 1991) placed Plethodontidae and Amphiumidae at the base of the salamander tree, a dramatic departure from previous hypotheses. In a combined analysis with morphological and molecular data (Larson and Dimmick 1993), Larson's (1991) tree was effectively rerooted so that Sirenidae and Cryptobranchoidea were basal (Fig. 2.4A). Analyses of 12S and 16S mtDNA (Hedges and Maxson 1993; Hay et al. 1995) also placed Sirenidae at the base of the tree.

Wiens et al. (2005) integrated characters from osteology with DNA sequences and published data from neurobiology and the reproductive system. The parsimony and Bayesian analyses of Wiens et al. (2005) were in general agreement about relationships among the families, and the differences not in strong conflict. The Bayesian tree (Fig. 2.4C) is used as the framework for the following discussion.

Plethodontidae is the largest family, with 27 genera and about 360 species of 500 total species of salamanders. These lungless salamanders use primarily cutaneous respiration. They are the most diverse in ecomorphology and life history, with arboreal, aquatic, terrestrial, rock-dwelling, and burrowing forms. Four major clades of Plethodontidae were recognized following Wake (1966): Desmognathinae, Plethodontini, Bolitoglossini, and Hemidactyliini (the last three in the subfamily Plethodontinae). Although salamanders are primarily of the north temperate regions, the clade Bolitoglossini, with more than 255 species, has diversified in the Neotropics.

Most work has concentrated on relationships within each group, however, an analysis of mitochondrial genomes (Mueller et al. 2004) has found unprecedented novel relationships, rejecting the monophyly of most of these groups. Similarly, a Bayesian analysis of parts of two mitochondrial genes, one nuclear gene, and morphological characters by Chippindale et al. (2004: Fig. 3) found the clade Desmognathinae nested within the traditional Plethodontini. Mueller et al. (2004) found comparable results, except that the inclusion of *Hydromantes* (a bolitoglossine) was also included within the Plethodontini. Despite the differences, the results of the two analyses are impressively in agreement. Mueller et al. (2004) included complete mitochondrial genomes, and Chippindale et al. (2004) used mitochondrial genes, one nuclear gene, and morphology. In neither analysis is the monophyly of Plethodontidae in doubt.

Although these new results jar the taxonomy of well-accepted groups, they more profoundly affect the interpretation of the organisms' biology. These phylogenies indicate that in at least one case larval (indirect) development has reevolved from direct development, in which the larval stage is bypassed. It has been often assumed that if a lineage loses the larval stage in favor of developing directly to a juvenile, the larval stage cannot be reevolved. Yet these phylogenies call into question that assumption about the way development evolves (Chippindale et al. 2004; Mueller et al. 2004). A reversal of life-history mode in this way is unknown in other vertebrates.

The discussion of the phylogeny of the remaining groups derives from Wiens et al. (2005). The sister group of the Plethodontidae, Amphiumidae, is not an intuitive choice. Amphiumidae includes a few species in the southeast United States. These elongate neotenic species lack external gills, although they have gill slits. The spindly limbs retain only remnants of digits. Amphiumidae superficially resembles two other groups of elongate, neotenic salamanders that have independently acquired this habitus, the Proteidae and Sirenidae, discussed below.

The sister-group of the Plethodontidae + Amphiumidae is *Rhyacotriton* (Rhyacotritonidae), which includes a few very similar species in the Pacific Northwest. *Rhyacotriton* was at times considered as the sister-group of Dicamptodontidae, but recent analyses (Good and Wake 1992; Larson and Dimmick 1993) placed the two as a paraphyletic group or more distantly related (Wiens et al. 2005).

A large clade includes Ambystomatidae, Dicamptodontidae, Salamandridae, and (based on the Bayesian analysis) probably Proteidae. Salamandridae are found in Eastern and western North America, Europe and adjacent western Asia, northwest Africa, and eastern Asia. Many have a bright warning coloration and have poison skin glands to deter predators; some of these aposematic forms are newts, species with drier skin during a terrestrial existence as part of their life. At least two species are viviparous, a rare occurrence. Salamandridae are also diverse in morphology and life history, although not as speciose as Plethodontidae. Salamandrid phylogeny has been assessed using morphology (Özeti and Wake 1969; Wake and Özeti 1969) and DNA sequences (Titus and Larson 1995).

Ambystomatidae includes several species of *Ambystoma* in North America and Mexico. Some are facultatively neotenic and retain the ability to metamorphose; others are constrained to the larval morphology, and are perennially aquatic. Most species have been examined using mtDNA sequences and allozymes (Shaffer 1984; Shaffer et al. 1991). Its sister-group is *Dicamptodon*, the Pacific Giant salamander (Dicamptodontidae). Like Rhyacotritonidae, *Dicamptodon* is found in the Pacific Northwest and adjacent Canada, and includes one genus with a few species. Proteidae includes species both in the eastern United States and Canada (*Necturus*) and the Adriatic region of Europe (*Proteus*). The large paedomorph *Necturus*, the mudpuppy, has prominent external gills, as does *Proteus*, a very elongate and aquatic cave-dweller.

The name Salamandroidea has been used for the clade including at least Plethodontidae, Rhyacotritonidae, Dicamptodontidae, Ambystomatidae, Proteidae, Salamandridae, and Amphiumidae, all the taxa discussed so far. The inclusion of Sirenidae is controversial; compare Figures 2.4A to C. The name Salamandroidea has not been phylogenetically defined, so inclusion of Sirenidae is an open question. Evans et al. (2005) described the Jurassic fossil *Iridotriton hechti* and hypothesized it to be the sister of *Valdotriton* (Evans and Milner 1996) + Salamandroidea.

Some analyses have placed sirens as the sister-group of all other salamanders (e.g., Larson and Dimmick 1993). Species of Sirenidae are found in the eastern United States and adjacent Mexico. These are nonmetamorphosing forms with external gills. Strangely, the front limbs are present and robustly developed, but the pelvic girdle and hind limbs are absent.

The clade Cryptobranchoidea is acknowledged to be among the most plesiomorphic of salamanders; it includes Cryptobranchidae and Hynobiidae. Cryptobranchoidea is the sister-group of Salamandroidea, regardless of whether sirens are included in the latter. Cryptobranchidae includes two genera with a disjunct temperate distribution in eastern Asia (China and Japan) and the eastern United States, similar to alligators and sturgeon. Cryptobranchids are very large (*Andrias* up to 1.5 meters), retain external gill slits without the gill filaments, and are highly aquatic. Cryptobranchid fossils (Gao and Shubin 2003) from the Jurassic represent the oldest crown-group salamanders (Caudata).

Hynobiidae are distributed from continental Asia to Japan. These are the most plesiomorphic of salamanders, and all live in temperate to subarctic regions. These are mostly small species; like cryptobranchids, fertilization is external, in contrast to other salamanders, which have internal fertilization.

The oldest fossil salamander is an articulated skeleton, *Karaurus sharovi*, from the Jurassic. Urodela is the stem-based name for the clade that includes *Karaurus* + Caudata, so *Karaurus* is part of Urodela but not part of Caudata. Although this fossil established the presence of salamanders in the Jurassic, fossil salamanders have not been useful in elucidating the phylogeny of extant taxa. However, Gao and Shubin's (2001) analysis of Jurassic urodeles (Fig. 2.4B) placed these fossils at the base of the clade of extant salamanders, perhaps due to adverse effects of paedomorphic characters in the dataset (Evans et al. 2005; Wiens et al. 2005).

5. Frogs

Anura (frogs and toads) is a large clade comprising more than 5000 species (AmphibiaWeb 2005). The largest frogs are *Conraua goliath* of Cameroon, which may weigh up to 3.3 kg; the smallest are *Psyllophryne didactyla* of Brazil and *Eleutherodactylus iberia* of Cuba, with adults about 8 to 9 mm in length. Unlike salamanders and caecilians, frogs have reduced or lost only some digits; the limbs remain well developed. Moreover, frogs tend to reduce the number of vertebrae, rather than increase it as in caecilians and salamanders. Associated with this reduction is a highly modified hip girdle, fused tail vertebrae, and an elaborated ankle joint that function during jumping, clearly an evolutionarily salient novelty.

Although some frogs have escaped an aquatic existence, as a whole the lineage has embraced it. In contrast to caecilian and salamander larvae, frog tadpoles are highly morphologically specialized to exploit an often unpredictable larval environment. The tadpole is mostly a locomotor mechanism in the tail and a feeding apparatus in the head; the latter includes a highly efficient pump and filters that sequester organic minisculi from the water. Tadpoles basically eat until their quick and awkward metamorphosis to a froglet.

Frogs have a dazzling array of features associated with reproduction; however, there are no neotenic or paedomorphic tadpoles. The diverse male vocal signals are used for mate advertisement and territorial displays. Parental care is highly developed in many lineages, including brooding of developing larvae on a bare back, in pouches on the back of females, in the vocal sacs of males, and in the stomach of females. Females in some unrelated lineages raise their tadpoles in the aquatic mesocosm of a bromeliad axil and supply their own unfertilized eggs as food. Whereas amniotes escaped from the watery environment once by evolving a unique shelled egg, frogs have escaped at least 20 times by evolving direct development of terrestrial eggs, in which the free-swimming tadpole stage is bypassed.

5.1 Overview of Frog Phylogeny

The lineage leading to frogs diverged from other tetrapods at least 250 million years ago (Rage and Rocek 1989). Many of the putatively most basal groups are currently represented by only one or a few species. An historical understanding of frog phylogeny rests primarily on morphological data. In general, morphological characters resolved the plesiomorphic basal branches known as archaeobatrachians (Trueb 1973; Cannatella 1985; Duellman and Trueb 1986; Haas 1997), approximately 200 species representing several ancient, generally species-poor lineages that branched off early in the evolutionary history of anurans. They contain some of the most interesting and enigmatic frog species. Using a dataset

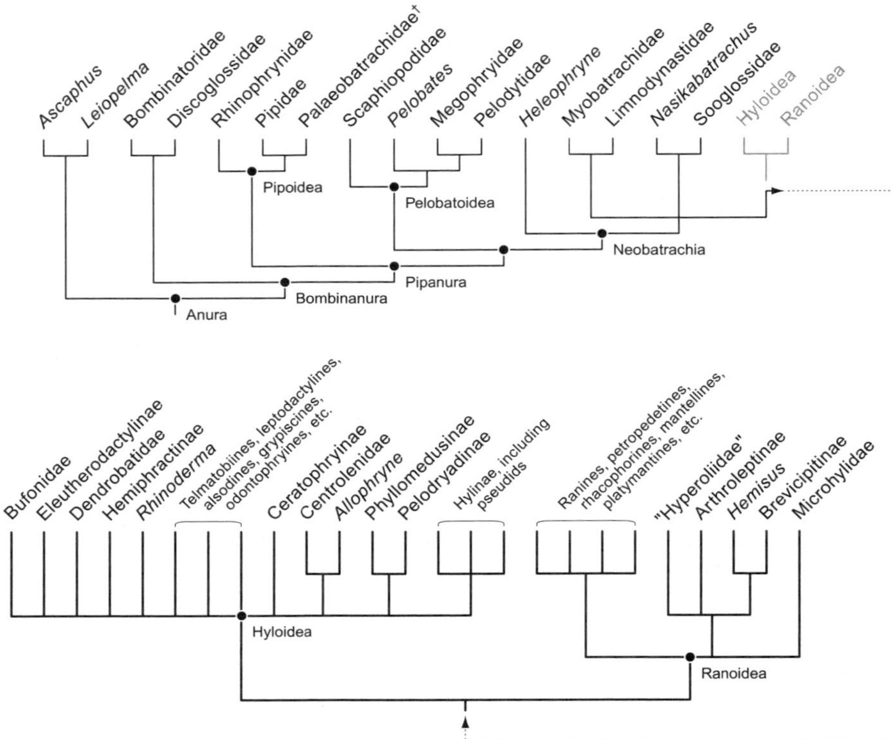

FIGURE 2.5. Summary diagram of the relationships of among the major lineages of frogs.

from morphology, Cannatella (1985) found that archaeobatrachians were a paraphyletic assemblage of several lineages. In contrast, Hay et al. (1995) concluded that a monophyletic Archaeobatrachia was the sister clade of Neobatrachia. Using nuclear genes, Hoegg et al. (2004) found strong support for paraphyly of archaeobatrachians. Analyses or summaries of fossils (Sanchíz 1998; Gao and Wang 2001) also found archaeobatrachians to be paraphyletic.

Living frogs are arranged in 20 to 25 families (Fig. 2.5). However, a definitive listing of the families would be arbitrary and have little general agreement, not because of strong controversy, but because this is a dynamic area of investigation (Duellman 2003). For consistency with textbooks, generally recognized families are used here. This does not indicate this author's support of this classification; the reader should understand that ranks are generally meaningless. Some families include hundreds of species, and several include only one or two. The names of higher taxa are used following the guidelines of Ford and Cannatella (1993), which include formal recognition of only monophyletic groups and identification of clearly nonmonophyletic groups as such.

The tree presented here (Fig. 2.5) is not a definitive phylogeny, and should not be cited as such. Rather, it is a fusion of various published and soon-to-be published subtrees. It is conservative in that some nodes that are resolved with low support in published works (e.g., Darst and Cannatella 2004) are presented as unresolved, in anticipation of research in progress.

5.2 Ranoidea

Ranoidea is primarily Old World, in contrast to Hyloidea, which is mostly New World. For years Ranoidea included Microhylidae, "Ranidae", Hyperoliidae, and Rhacophoridae, the latter two being treefrog morphs independently derived from within Ranidae. Ford and Cannatella (1993) defined Ranoidea as a node name for the clade containing the aforementioned taxa, plus Dendrobatidae, the New World poison frogs, based on Ford (1989). Subsequent analyses (Hay et al. 1995; Ruvinsky and Maxson 1996; Darst and Cannatella 2004) found dendrobatids to be more closely related to hyloids (bufonoids). The implications for the definition of Ranoidea were discussed by Darst and Cannatella (2004). Here, Ranoidea excludes Dendrobatidae.

Based on several studies (Emerson et al. 2000; Darst and Cannatella 2004; Hoegg et al. 2004), the emerging ranoid phylogeny contains three clades: (1) Microhylidae (excluding Brevicipitinae; see below); (2) a clade that includes Arthroleptinae, Astylosterninae, "Hyperoliidae" (including *Leptopelis*), *Hemisus*, and Brevicipitinae; (3) a clade that includes the Malagasy and asian treefrogs formerly considered Rhacophoridae, Mantellinae, and a series of less-well defined groups: ranines, cacosternines, micrixalines, phrynobatrachines, ptychadenines, petropedetines, platymantines, nyctibatrachines, dicroglossines, and others. The appropriate names for clades 2 and 3 are not well established; clade 2 could be Brevicipitidae or Brevicipitoidea, depending on whether the family or superfamily rank is desired. Similarly, the name for clade 3 could be Ranidae or Ranoidea. If the latter is used, then another name must be used for the larger lineage that includes clades 1, 2, and 3; the historically used Firmisternia would be appropriate.

Microhylidae includes a plethora of forms found on almost all continents. The New World species are fossorial or leaf-litter dwellers of generally similar shape and habitus. However, Old World taxa in Madagascar and New Guinea are quite diverse and include arboreal, fossorial, terrestrial, cunicular, and saxicolous species. Accruing evidence suggests clades in the New World (centered in South America), Madagascar, Papua-New Guinea radiation, and Southeast Asia. Microhylid osteology has been surveyed comprehensively by Wu (1994), although his phylogenetic analysis yielded some unorthodox results. However, the relationships among the microhylid subclades are actively under investigation by several groups using DNA sequences.

The second major ranoid clade includes a diverse group of taxa restricted to Africa: Arthroleptidae (including astylosternines), "Hyperoliidae" (including *Leptopelis*), *Hemisus*, and Brevicipitinae. Darst and Cannatella (2004) and van der Meijden et al. (2004) found that brevicipitines were actually not related to microhylids but were more closely related to other ranoids, specifically *Hemisus* (Darst and Cannatella 2004). An overlooked morphological analysis (Blommers-Schlösser 1993) found similar results. *Hemisus* has often been placed in its own family (Hemisotidae) in ignorance of its relationships. These are small frogs with intensely ossified skull bones, and are the only frogs that burrow headfirst.

The Brevicipitinae, formerly considered microhylids, are an exclusively African group of smallish frogs. The skull is severely hypo-ossified, and *Breviceps* derives its name from the very short, bulldoglike head. A phylogeny of all species was realized by Loader et al. (2004). "Hyperoliidae" are a primarily African group of mostly small treefrogs, including many brightly colored and boldly patterned species of *Hyperolius*. Some genera occur on Madagascar and the Seychelles. Phylogenetic analyses of DNA data and morphological characters exist for Hyperoliidae including *Leptopelis* (Liem 1970; Drewes 1984; Channing 1989; Richards and Moore 1996; Wilkinson et al. 2002a). The name is used in quotes here to indicate its nonmonophyly, because some evidence indicates that *Leptopelis* is not part of this clade. Arthroleptidae has in recent years included the subfamilies Arthroleptinae and Astylosterninae. The trend is to treat these as distinct families. The arthroleptines are mostly smaller species, inhabiting sub-Saharan Africa. Most have direct development. Astylosterninae are mostly restricted to west Africa; these frogs are associated with flowing water. One species, *Trichobatrachus robustus*, is called the hairy frog because its skin has hairlike vascularized appendages that function in cutaneous respiration.

The third major clade of Ranoidea might be called Ranidae, but its content is very different from the "Ranidae" of Ford and Cannatella (1993) who used quotes to indicate nonmonophyly. Various arrangements of subfamilies in this clade have been proposed (Dubois 1981; Blommers-Schlösser 1993), but no consensus has been reached. Some loosely defined subfamilies of "Ranidae" have been elevated to family status; these actions were typically arbitrary and not due to any new discovery of phylogenetic affinity; for example, the recognition of Arthroleptidae (Dubois 1984). These changes have unfortunately been incorporated in textbooks and checklists.

This clade includes the Malagasy and Asian treefrogs in Rhacophoridae and Mantellidae (Mantellinae). Phylogenetic analyses of DNA data and morphological characters exist for the traditional Rhacophoridae (Liem 1970; Channing 1989; Richards and Moore 1998), performed under the assumption that it was monophyletic. Evidence now indicates that Malagasy "rhacophorids" are not the closest relatives of the Asian rhacophorids (Wilkinson et al. 2002a; Vences and Glaw 2003). The clade also includes a series of less-well defined groups, some of which have been perhaps arbitrarily recognized as families: African groups such as cacosternines, petropedetines, phrynobatrachines, ptychadenines, tomopternines; South and Southeast Asian groups such as platymantines, nyctibatrachines, dicroglossines, and micrixalines; and the Eurasian and New World ranines. This clade includes the largest frog (*Conraua goliath*), as well as arboreal species, highly aquatic forms, carnivorous types, and fossorial and deserticolous species. Some have radiated on archipelagoes such as the Philippines, Indonesia, and New Guinea. The only New World representatives form a clade of *Rana* in North, Central, and South America (Hillis and Wilcox 2005).

5.3 Hyloidea

Hyloidea generally refers to neobatrachians with an arciferal pectoral girdle (the epicoracoid cartilages not fused, the plesiomorphic condition), in contrast to those with a firmisternal girdle (fused cartilages), the ranoids. Darst and Cannatella (2004) formally defined the name Hyloidea. Hay et al. (1995) first reported character data supporting the monophyly of Hyloidea (as Bufonoidea), but including Myobatrachidae and Heleophrynidae, a more inclusive group than is used here. Hyloidea is primarily a New World clade and Ranoidea an Old World group, although the hyloids have significant radiations in the Australopapuan region as do the Ranidae and Microhylidae in the New World.

"Leptodactylidae" is a hodge-podge of hyloids that lack distinctive apomorphies. In general, the derived characters of other hyloid families separate them from the traditional family "Leptodactylidae" (Lynch 1971). Leptodactylids are a primarily South American group. Some clades of eleutherodactylines and some species of *Leptodactylus* have radiated into Central America, Mexico, and the West Indies.

Phylogenetic relationships of leptodactylid genera were analyzed using morphology (Heyer 1975). Generally accepted groups of leptodactylids are Leptodactylinae, Hylodinae (Grypiscinae), Ceratophryinae, and Telmatobiinae. Using ribosomal mitochondrial genes, Basso and Cannatella (2001) found Leptodactylidae to be polyphyletic; embedded within leptodactylids are most of the groups discussed below: all taxa of hylids, bufonids, dendrobatids, centrolenids. Of the traditional groups listed above, only Ceratophryinae is clearly monophyletic. Leptodactylinae and Hylodinae are questionably monophyletic, and "Telmatobiinae" is polyphyletic. A major reclassification recognized most of these groups as distinct families (Basso et al. 2006).

Pseudidae and Brachycephalidae were treated as families and were defined as node-names by Ford and Cannatella (1993). Brachycephalidae include two genera of very small frogs that are leaf-litter dwellers of southeastern Brazil; these had been considered classified as a family because they lack a sternum, and have a distinctive pattern of digital reduction. Darst and Cannatella (2004) found Brachycephalidae to be within eleutherodactylines ("Leptodactylidae"). Thus, recognition as at the family rank is no longer justified.

Centrolenidae are the glass-frogs of Mexico, Central America, and South America. These mostly small, fragile frogs live typically high in vegetation overhanging streams where they deposit their egg clutches. *Allophryne ruthveni* is an enigmatic hyloid (Fabrezi and Langone 2000) that has been placed in a monotypic family Allophrynidae; it is likely the sister-group of Centrolenidae (Austin et al. 2002). The two species of *Rhinoderma* have been placed in Rhinodermatidae, a family name that is redundant with the genus name. Were it not for the apomorphic life history of the two species, in which the male broods the developing larvae in his vocal sac, *Rhinoderma* would be included in "Leptodactylidae."

Hylidae are treefrogs that have classically been defined as having intercalary elements—extra cartilages between the ultimate and penultimate phalanges of each digit—and claw-shaped terminal phalanges that support a round digital disc. Hylids include Hemiphractine, Phyllomedusinae, Pelodryadinae, and Hylinae. The content of each is not controversial. Hemiphractines are all Neotropical and mostly South American; they are distinctive in that females brood embryos on the back, either exposed or in a pouch. Phyllomedusinae are the Neotropical leaf frogs; these are slow-moving species that walk along branches rather than jump. They deposit eggs on leaves above standing water or over streams. Pelodryadines are typical-looking treefrogs that are restricted to the Australopapuan region. Each of these three groups is monophyletic. The fourth group, Hylinae, shows weak evidence for monophyly. This group has primary distribution in South and Central America, with some lineages in North America, and a few species in Eurasia.

Morphology-based phylogenies of Hylinae and Hemiphractinae exist (da Silva 1998; Mendelson et al. 2000). Darst and Cannatella (2004) did not find hylids to be monophyletic—two species of Hemiphractinae were highly divergent in sequence distance, and did not cluster with other hylids. However, given the small sample of species their results should be considered tentative. They also found Phyllomedusinae to be the sister-group of Pelodryadinae.

Pseudidae are highly aquatic frogs inhabiting mostly open swampy areas in South America. These have bony intercalary elements, in contrast to the cartilaginous ones found in most hylids. Darst and Cannatella (2004) and da Silva (1998) found Pseudidae to be nested within hylines. Thus, Hylinae is paraphyletic with respect to Pseudidae. Duellman (2001), following da Silva's work, reduced Pseudidae to a subfamily, but this action does not resolve the paraphyly problem. Thus, continued use of Pseudidae as a name of family or subfamily rank is not justified.

The true toads belong to Bufonidae. These are native to all continents except Australia. Bufonids can be thought of as having two morphs: rather generalized toads with parotoid glands, as in *Bufo*; and more specialized forms that lack parotoid glands. The latter group includes the most primitive (basal) bufonids; (Graybeal and Cannatella 1995; Graybeal 1997; Gluesenkamp 2001; Pauly et al. 2004). Dendrobatidae are found in tropical areas of Central and South America. These are small- to medium-sized frogs; some are cryptic and nontoxic, and others are aposematic—brightly colored and toxic; these latter are the poison frogs (also called dart-poison frogs, or poison arrow frogs). Recent work (Santos et al. 2003; Vences et al. 2003) has shown that the aposematic species evolved several times, and that the alkaloids used for chemical defense are derived from their specialized diet on certain groups of insects, usually ants (Daly et al. 1994, 2000; Darst et al. 2005).

5.4 *Neobatrachia*

Neobatrachia consists of the "advanced" frogs and includes 95% of living species. Except for the late Tertiary and Quaternary, they are not well represented in the fossil record. Morphological and molecular analyses have supported monophyly of Neobatrachia (Cannatella 1985; Ruvinsky and Maxson 1996). As discussed in the preceding paragraphs, the two largest groups of Neobatrachia been are Hyloidea and Ranoidea. However, certain other neobatrachian groups are basal to both hyloids and ranoids, but cannot be placed with certainty. Until recently Limnodynastinae and Myobatrachinae were usually treated as subfamilies of Myobatrachidae (Heyer and Liem 1976); these groups are restricted to Australia and New Guinea. Ford and Cannatella (1993) found no synapomorphies for "Myobatrachidae." However, Lee and Jamieson (1992) reported spermatozoon ultrastructural characters supporting myobatrachid monophyly. Ruvinsky and Maxson (1996) placed Myobatrachinae, Limnodynastinae, and *Heleophryne* in a clade at the base of Hyloidea. Some textbooks (Zug et al. 2001) have recognized each group as a family.

The extinct gastric-brooding frog *Rheobatrachus* (two species) has been considered closely related to myobatrachines. Females swallow the tadpoles, which continue development in the stomach and emerge as froglets. Digestion is probably inhibited by a prostaglandin. Recognition of *Rheobatrachus* as a distinct family (e.g., Frost 2004) carries no added information.

Sooglossidae is a family of small species of enigmatic relationships, found on the Seychelles. It has been placed as the sister-group of Hyloidea (Ruvinsky and Maxson 1996), of Ranoidea (Emerson et al. 2000), basal to both (Hay et al. 1995), or as the sister of Myobatrachidae (Duellman and Trueb 1986) or Myobatrachinae (Ford and Cannatella 1993). Nasikabatrachidae, from India, is the most recently described frog family, with one monotypic genus, *Nasikabatrachus* (Biju and Bossuyt 2003). Their analysis placed *Nasikabatrachus* + Sooglossidae as the sister-taxon of all other neobatrachians, a position not supported by other work in progress. *Heleophryne* is the sole genus of Heleophrynidae, restricted to the Cape region of South Africa. Although treefroglike in appearance, these frogs live along boulder-strewn streams. *Heleophryne* is probably the sister-taxon of all other neobatrachians (Darst and Cannatella 2004).

5.5 *Pipanura*

Pipanura is the node name for the clade that includes Neobatrachia, plus Pipoidea and Pelobatoidea; that is, all frogs minus discoglossoids. Historically, Pipoidea and Pelobatoidea have been seen as intermediate between discoglossoids and Neobatrachia, and are represented by numerous Cretaceous and Tertiary fossils (Sanchíz 1998; Rocek 2000). The clades of living pelobatoids are Pelobatidae, Megophryidae, Pelodytidae, and Scaphiopodidae. The content of Pelobatoidea is not controversial. The evidence indicates that pelobatoids are the sister of Neobatrachia.

Until recently, Pelobatidae included a sister-group relation between the European (*Pelobates*) and American Spadefoots (*Scaphiopus* + *Spea*), which were united by synapomorphies related to fossoriality (Cannatella 1985; Maglia 1998). However, García-París et al. (2003) used DNA data to demonstrate statistical support for the nonmonophyly of spadefoot toads, and endorsed the recognition of *Scaphiopus* and *Spea* as a fourth family, Scaphiopodidae. They inferred the fossorial habitus of the two groups to be convergent. Thus, Pelobatidae includes only *Pelobates* among living taxa. Megophryidae is a node name for a diverse clade of primarily Asian taxa. Relationships within this group have been analyzed in a preliminary fashion (Lathrop 1997). Pelodytidae includes two living species, from southwest Europe and the Caucasus mountains, as well as several taxa of fossil pelodytids (Henrici 1994). These are the parsley frogs, so-named because of their unusual smell.

Pipoidea is the node-name for the clade including Pipidae and Rhinophrynidae, as well as the fossil family Palaeobatrachidae (Spinar 1972). Relationships among pipoids have been examined by studies using fossils, morphology of adult forms, and DNA sequences (Cannatella and Trueb 1988; de Sá and Hillis 1990; Cannatella and de Sá 1993; Báez and Trueb 1997; Evans et al. 2004). The tadpole of pipoids is highly derived (Starrett 1973; Sokol 1975; Haas 2003). At one time it was thought that the larval morphology of pipoid frogs argued for a position as the most primitive (early branching, in this context), but highly specialized, group (Starrett 1973); this theme has resurfaced (Púgener et al. 2003). However, other interpretations (Sokol 1975; Haas 1997; Cannatella 1999; Haas 2003) indicated that although pipoids are highly specialized, discoglossoids are the earliest-branching frog lineages (see below). This agrees with analyses of adult morphology.

Pipidae are among the most aberrant living frogs. Their unsurpassed aquatic adaptations include lateral line organs, modification of the larynx into a clicking noisemaker, and complete loss of the tongue, with a novel reliance on suction feeding (Cannatella and Trueb 1988; Trueb 1996). Living species are found in tropical South America and adjacent Panama, and Africa. Several fossil forms are known (Báez 1996).

The single species of the highly fossorial frog *Rhinophrynus* is the closest relative of Pipidae among living forms. This microcephalic frog with thick skin has been appropriately referred to as a "bag of bones." *Rhinophrynus* lives in southern Texas, Mexico, and Central America. It is usually seen only when it emerges to breed after very heavy rains. As with Pelodytidae, Rhinophrynidae was defined as a stem-name by Ford and Cannatella (1993) to accommodate a variety of fossil species.

5.6 The "Basal" Frogs: Discoglossoids

A group of plesiomorphic lineages, the discoglossoids, includes *Ascaphus*, *Leiopelma*, Bombinatoridae, and Discoglossidae (Ford and Cannatella 1993); this group is paraphyletic to the clade of all other frogs, Pipanura. As an informal term, discoglossoid is useful to denote Anura that are not part of Pipanura. One primitive feature of discoglossoids is the rather rounded, disclike tongue; hence the name.

Alytes and *Discoglossus* are included in the Discoglossidae, although the two genera are fairly divergent and evidence of monophyly is not overwhelming.

Bombinatoridae includes *Bombina* and *Barbourula* (Ford and Cannatella 1993). Previously, Discoglossidae was used to include Bombinatoridae and Discoglossidae (e.g., Duellman and Trueb 1986). Some evidence indicates that Discoglossidae is more closely related to other frogs than to Bombinatoridae, *Ascaphus*, or *Leiopelma* (Ford and Cannatella 1993). San Mauro et al. (2004a) found that *Bombina*, *Discoglossus*, and *Alytes* formed a clade, and that *Xenopus* was more closely related to this clade than to neobatrachians, a result strongly at odds with other datasets.

Bombinanura is the node-name for the clade that includes all living frogs except *Ascaphus* and *Leiopelma*. *Leiopelma* is the only genus of Leiopelmatidae and includes four species restricted to cool forests on islands off the coast of New Zealand. However, Pleistocene fossils of *Leiopelma* are extensive on the North and South Islands, indicating recent extinctions. Only a few eggs are laid and the embryos hatch as nonfeeding larvae; males attend the larvae during development.

Ascaphus (Ascaphidae) comprises two species in the Pacific Northwest of the United States; these are stream dwellers with tadpoles highly specialized for lotic habitats. Known as the tailed frog, the caudal appendage is not a tail but rather an everted portion of the cloaca used for internal fertilization. Leiopelmatidae has been used to include both *Ascaphus* and *Leiopelma*. Although these two taxa share many primitive characteristics, morphological evidence for a sister-taxon relationship between them has not been found (Cannatella 1985; Green and Cannatella 1993), and some characters ally *Leiopelma* more closely to other frogs. If this relationship were borne out, Leiopelmatidae *sensu lato* would be paraphyletic. Molecular data usually ally the *Ascaphus* and *Leiopelma* in a clade, although this could be an artifact of the long separation of the two lineages from other frogs. Whether they prove to be sister-taxa or not, *Ascaphus* and *Leiopelma* are plesiomorphic relicts of a once more widely distributed Mesozoic frog fauna (Roelants and Bossuyt 2005).

The fossil record of frogs was thoroughly reviewed by Sanchíz (1998). Among the oldest forms considered as genuine frogs are *Notobatrachus* and *Vieraella* (Middle Jurassic; Báez and Basso 1996). *Prosalirus vitis* (Lower Jurassic; Shubin and Jenkins 1995; Jenkins and Shubin 1998) is fragmentary, but had skeletal features indicative of saltatory locomotion. Gao and Wang (2001) reached different conclusions about the phylogeny of early frogs than did Ford and Cannatella (1993).

Salientia is the stem-based name for the taxon including Anura and taxa (all fossils) more closely related to Anura than to other living amphibians. The use of Salientia for *Triadobatrachus* plus all other frogs is widespread and not controversial. The sister-group of all frogs is *Triadobatrachus massinoti*, known from a single fossil from the Lower Triassic. Often referred to as a proanuran, it retains many plesiomorphic features, such as 14 presacral vertebrae (living frogs have 9 or fewer) and lack of fusion of the radius and ulna, and of the tibia and fibula (living frogs have fused elements, radioulna and tibiofibula) (Rage and Rocek 1989; Rocek and Rage 2000).

6. Prospects for an Integration of Phylogeny and Acoustic Communication

6.1 The Origin of the Tympanum

The function of the frog tympanum has received much attention, but not its evolutionary origin. Some of the "basal frogs" (*Ascaphus*, *Leiopelma*, *Bombina*) lack the tympanum and stapes. Mapping of these states onto the phylogeny suggests that the absence of the tympanum and associated stapes might be a feature of the earliest anurans. But the presence of the columella is difficult to interpret in the relevant fossils, and the optimization of the states under parsimony is equivocal. This begs the question: Is this absence primary or secondary? Can this inference be strengthened by examination of the function of acoustic hair cells (as one of many possibilities)? Certainly the structure and arrangement of stereocilia and kinocilium (Lewis and Narins 1999) offer some hints.

6.2 The Vocal Sac

Vocal sacs (with vocal slits and the typical invagination of the lining of the buccal cavity into the space dorsal to the interhyoideus muscle) are absent not only in *Ascaphus* and *Leiopelma*, but also in *Bombina*, *Barbourula*, *Alytes*, and *Discoglossus*, all of the living "discoglossoids." Thus, it seems that the vocal sac appeared in evolution after the origin of the tympanum. Did the appearance of the tympanum facilitate sound transduction such that evolution of a novel communication system based on the vocal apparatus was favored? What predictions can be made about the differences in the stucture of the central auditory pathways that would test this hypothesis? Given the numerous possible functions of the vocal sac (Rand and Dudley 1993), is it possible that one function evolved before others? Or that the same function (e.g., visual signaling) has evolved independently several times?

6.3 The Tympanum of *Rana catesbeiana*

An apparently novel function of the tympanum of *Rana catesbeiana* is its use in radiating sound (Purgue 1997). Perhaps not coincidentally, the diameter of the male tympanum is significantly larger than that of females, in *Rana catesbeiana* as well as its close relatives (*oskaloosae*, *clamitans*, *heckscheri*, *catesbeiana*, *grylio*; Hillis and Wilcox 2005). Is this novel function found in these other species?

6.4 The Inner Ear of Salamanders

The suite of inner ear characters studied by Lombard (1971) has been used repeatedly for salamander phylogeny. But what can a phylogeny based on several datasets tell us about evolution of these inner ear structures? Are characters of the inner ear less labile than those associated with the tympanum?

6.5 Phylogeny and Advertisement Calls

The anuran advertisement call is of paramount importance in speciation (Blair 1958). But ironically, this character suite so directly involved with generating diversity is poorly studied for its phylogenetic patterns. Perhaps this stems from a mistaken notion that behavior is too plastic to be useful in phylogeny estimation at higher levels. Call parameters might readily respond to selective pressure to avoid hybridization. Thus variation in calls is thought to be associated with species formation, but not really involved in diversification at higher levels.

The few phylogenetic studies of interspecific call evolution (Ryan 1988; Cocroft 1994; Cocroft and Ryan 1995; Ryan and Rand 1995; Cannatella et al. 1998) suggest a rich field of research directions. What is the "character" in a frog call? What are the comparable (homologous) parts (Greene 1994)? Is a note always equivalent to a note? Can an amplitude-modulated call such as a trill evolve into a frequency-modulated sound such as a whistle? Does a call evolve as an array of independent variables or are there genetic correlations that produce phenotypic correlations?

As mentioned in the first paragraphs of this chapter, new strides in phylogenetic theory and comparative methods can enable the interpretation of behavioral and functional diversity in terms of evolutionary history. Comparative methods can provide a framework for answering these questions. Many of the analytical methods are well-worked out. Some initial steps have been taken (e.g., Ryan 1986), but much remains to be accomplished. Perhaps what is needed most is the interaction among workers to generate the questions that will lead to the desired synthesis.

Acknowledgments. I thank Cat Darst, David Bickford, and Meredith Mahoney for comments on the manuscript. This work was funded by National Science Foundation grant 9981631.

References

Alberch P, Gould SJ, Oster G, Wake DB (1979) Size and shape in ontogeny and phylogeny. Paleobiology 5:296–317.

AmphibiaWeb (2005) Information on amphibian biology and conservation. Available: http://amphibiaweb.org. (Accessed: 2005).

Anderson JS (2001) The phylogenetic trunk: Maximum inclusion of taxa with missing data in an analysis of the Lepospondyli (Vertebrata, Tetrapoda). Syst Biol 50:170–193.

Austin JD, Lougheed SC, Tanner K, Chek AA, Bogart JP, Boag PT (2002) A molecular perspective on the evolutionary affinities of an enigmatic neotropical frog, *Allophryne ruthveni*. Zool J Linn Soc Lond 134:335–346.

Báez AM (1996) The fossil record of the Pipidae. In: Tinsley RC, Kobel HR (eds) The Biology of *Xenopus*. Oxford: Clarendon, pp. 329–347.

Báez AM, Basso NG (1996) The earliest known frogs of the Jurassic of South America: Review and cladistic appraisal of their relationships. Münchner Geowiss Abh (a) 30:131–158.

Báez AM, Trueb L (1997) Redescription of the Paleogene *Shelania pascuali* from Patagonia and its bearing on the relationships of fossil and Recent pipoid frogs. Sci Pap Mus Nat Hist Univ Kans 4:1–41.

Basso N, Cannatella DC (2001) The phylogeny of leptodactylid frogs based on 12S and 16S mtDNA. Abstract, SSAR-HL meetings, Indianapolis, Indiana, July 2001.

Basso N, Hillis DM, Cannatella DC (2006) Phylogeny of the frog family "Leptodactylidae" based on 12S and 16S mtDNA (Anura: Neobatrachia). Mol Phylogenet Evol. In Press.

Biju SD, Bossuyt F (2003) New frog family from India reveals an ancient biogeographical link with the Seychelles. Nature 425:711–714.

Blair WF (1958) Mating call in the speciation of anuran amphibians. Am Nat 92:27–51.

Blommers-Schlösser RMA (1993) Systematic relationships of the Mantellinae Laurent 1946 (Anura Ranoidea). Ethol Ecol Evol 5:199–218.

Bolt JR (1977) Dissorophoid relationships and ontogeny, and the origin of the Lissamphibia. J Paleontol 51:235–249.

Bolt JR (1991) Lissamphibian origins. In: Schultze H-P, Trueb L (eds) Origins of the Higher Groups of Tetrapods. Ithaca, NY: Cornell University Press, pp. 194–222.

Brainerd EL, Ditelberg JS, Bramble DM (1993) Lung ventilation in salamanders and the evolution of vertebrate air-breathing mechanisms. Biol J Linn Soc 49:163–183.

Cannatella DC (1985) A phylogeny of primitive frogs (Archaeobatrachians). (Doctoral dissertation) Lawrence: University of Kansas.

Cannatella DC (1999) Architecture: Cranial and axial musculoskeleton. In: McDiarmid RW, Altig R (eds) Tadpoles The Biology of Anuran Larvae. Chicago: The University of Chicago Press, pp. 52–91.

Cannatella DC, de Sá RO (1993) *Xenopus laevis* as a model organism. Syst Biol 42:476–507.

Cannatella DC, Hillis DM (1993) Amphibian phylogeny: Phylogenetic analysis of morphology and molecules. Herp Monogr 7:1–7.

Cannatella DC, Hillis DM, Chippindale PT, Weigt L, Rand AS, Ryan MJ (1998) Phylogeny of frogs of the *Physalaemus pustulosus* species group, with an examination of data incongruence. Syst Biol 47:311–335.

Cannatella DC, Trueb L (1988) Evolution of pipoid frogs: Intergeneric relationships of the aquatic frog family Pipidae (Anura). Zool J Linn Soc Lond 94:1–38.

Carroll RL (2000a) *Eocaecilia* and the origin of caecilians. In: Heatwole H, Carroll RL (eds) Amphibian Biology Volume 4. Chipping Norton, UK: Surrey Beatty, pp. 1402–1411.

Carroll RL (2000b) The lissamphibian enigma. In: Heatwole H, Carroll RL (eds) Amphibian Biology Volume 4. Chipping Norton: Surrey Beatty, pp. 1270–1273.

Carroll RL, Currie PJ (1975) Microsaurs as possible apodan ancestors. Zool J Linn Soc Lond 57:229–247.

Carroll RL, Kuntz A, Albright K (1999) Vertebral development and amphibian evolution. Evol Dev 1:36–48.

Channing A (1989) A re-evaluation of the phylogeny of Old World treefrogs. S Afr J Sci 24:116–131.

Chippindale PT, Bonett RM, Baldwin AS, Wiens JJ (2004) Phylogenetic evidence for a major reversal of life-history evolution in plethodontid salamanders. Evolution 58:2809–2822.

Coates MI, Ruta M, Milner AR (2000) Early tetrapod evolution. Trends Ecol Evol 15:327–328.

Cocroft RB (1994) A cladistic analysis of chorus frog phylogeny (Hylidae: *Pseudacris*). Herpetologica 50:420–437.
Cocroft RB, Ryan MJ (1995) Patterns of advertisement call evolution in toads and chorus frogs. Anim Behav 49:283–303.
da Silva HR (1998) Phylogenetic relationships of the family Hylidae with emphasis on the relationships within the subfamily Hylinae (Amphibia: Anura). (Doctoral dissertation) Lawrence: University of Kansas.
Daly JW, Garraffo HM, Jain P, Spande TF, Snelling RR, Jaramillo C, Rand AS (2000) Arthropod-frog connection: Decahydroquinoline and pyrrolizidine alkaloids common to microsympatric myrmicine ants and dendrobatid frogs. J Chem Ecol 26:73–85.
Daly JW, Garraffo HM, Jaramillo C, Rand AS (1994) Dietary source for skin alkaloids of poison frogs (Dendrobatidae)? J Chem Ecol 20:943–955.
Darst CR, Cannatella DC (2004) Novel relationships among hyloid frogs inferred from 12S and 16S mitochondrial DNA sequences. Mol Phylogenet Evol 31:462–475.
Darst CR, Menendez-Guerrero PA, Coloma LA, Cannatella DC (2005) Evolution of dietary specialization and chemical defense in poison frogs (Dendrobatidae): A comparative analysis. Am Nat 165:56–69.
de Queiroz K, Gauthier J (1992) Phylogenetic taxonomy. Ann Rev Ecol Syst 23:449–480.
de Sá RO, Hillis DM (1990) Phylogenetic relationships of the pipid frogs *Xenopus* and *Silurana*: An integration of ribosomal DNA and morphology. Mol Biol Evol 7:365–376.
Drewes RC (1984) A phylogenetic analysis of the Hyperoliidae (Anura): treefrogs of Africa, Madagascar, and the Seychelles Islands. Occ Pap Calif Acad Sci 139:1–70.
Dubois A (1981) Liste des genres et sous-genres nominaux de Ranoidea (Amphibiens Anoures) du monde, avec identification de leurs espèces-types: conséquences nomenclaturales. Monit Zool Ital 15:225–284.
Dubois A (1984) La nomenclature supragénérique des amphibiens anoures. Mémoires Museum Nat d'Histoire Naturelle, SerA Z 131:1–64.
Duellman WE (ed) (1999) Patterns of Distribution of Amphibians. A Global Perspective. Baltimore: The Johns Hopkins University Press.
Duellman WE (2001) Hylid Frogs of Middle America, Contributions to Herpetology, Volume 18. Society for the Study of Amphibians and Reptiles.
Duellman WE (2003) An overview of anuran phylogeny, classification, and reproductive modes. In: Jamieson BGM (ed) Reproductive Biology and Phylogeny of Anura. Enfield, NH: Science, pp. 1–18.
Duellman WE, Trueb L (1986) Biology of Amphibians. New York: McGraw-Hill.
Emerson SB, Richards C, Drewes RC, Kjer KM (2000) On the relationships among ranoid frogs: A review of the evidence. Herpetologica 56:209–230.
Evans BJ, Kelley DB, Tinsley R, Melnick DJ, Cannatella DC (2004) A mitochondrial DNA phylogeny of clawed frogs: Phylogeography on sub-Saharan Africa and implications for polyploid evolution. Mol Phylogenet Evol 33:197–213.
Evans SE, Lally C, Chure DC, Elder A, Maisano JA (2005) A Late Jurassic salamander (Amphibia: Caudata) from the Morrison Formation of North America. Zool J Linn Soc Lond 143:599–616.
Evans SE, Milner AR (1996) A metamorphosed salamander from the Early Cretaceous of Las Hoyas, Spain. Phil Trans Royal Soc B 351:627–646.
Fabrezi M, Langone JA (2000) Los caracteres morfológicos del controvertido Neobatrachia arborícola *Allophryne ruthveni* Gaige, 1926. Cuadernos de Herpetología 14:47–59.

Feller AE, Hedges SB (1998) Molecular evidence for the early history of living amphibians. Mol Phylogenet Evol 9:509–516.

Felsenstein J (1985) Phylogenies and the comparative method. Am Nat 125:1–15.

Felsenstein J (2004) Inferring phylogenies. Sunderland, MA: Sinauer Associates.

Ford LS (1989) The phylogenetic position of poison-dart frogs (Dendrobatidae): reassessment of the neobatrachian phylogeny with commentary on complex character systems. (Doctoral dissertation) Lawrence: University of Kansas.

Ford LS, Cannatella DC (1993) The major clades of frogs. Herp Monogr 7:94–117.

Frost DR (ed) (1985) Amphibian species of the world: A taxonomic and geographic reference. Lawrence, KS: Allen Press and the Association of Systematics Collections.

Frost DR (2004) Amphibian species of the world: an online reference. Version 3.0 (22 August, 2004). Electronic Database accessible at http://research.amnh.org/herpetology/amphibia/index.html. American Museum of Natural History, New York.

Gadow H (1901) Amphibia. In: Amphibia and Reptiles. London: Macmillan, pp. 3–10.

Gans C, Dejongh HJ, Farber J (1969) Bullfrog (*Rana catesbeiana*) ventilation: How does the frog breathe? Science 163:1223–1225.

Gao K-Q, Shubin NH (2001) Late Jurassic salamanders from northern China. Nature 410:574–577.

Gao K-Q, Shubin NH (2003) Earliest known crown-group salamanders. Nature 422:424–428.

Gao K-Q, Wang Y (2001) Mesozoic anurans from Liaoning province, China, and phylogenetic relationships of archaeobatrachian anuran clades. J Vert Paleontol 21: 460–476.

García-París M, Buchholz DR, Parra-Olea G (2003) Phylogenetic relationships of Pelobatoidea re-examined using mtDNA. Mol Phylogenet Evol 28:12–23.

Gardner JD (2001) Monophyly and affinities of albanerpetontid amphibians (Temnospondyli; Lissamphibia). Zool J Linn Soc Lond 131:309–352.

Garland Jr. T, Midford PE, Ives AR (1999) An introduction to phylogenetically based statistical methods, with a new method for confidence intervals on ancestral values. Amer Zool 39:374–388.

Glaw F, Köhler J (1998) Amphibian species diversity exceeds that of mammals. Herpetol Rev 29:11–12.

Gluesenkamp AG (2001) Developmental mode and adult morphology in bufonid frogs: A comparative analysis of correlated traits. (Doctoral dissertation). Austin, Texas: University of Texas.

Good DA, Wake D (1992) Geographic variation and speciation in the Torrent Salamanders of the genus *Rhyacotriton* (Caudata: Rhyacotritonidae). Publ Zool Univ Calif 126:1–91.

Graybeal A (1997) Phylogenetic relationships of bufonid frogs and tests of alternate macroevolutionary hypotheses characterizing their radiation. Zool J Linn Soc Lond 119:297–338.

Graybeal A, Cannatella DC (1995) A new taxon of Bufonidae from Peru, with descriptions of two new species and a review of the phylogenetic status of supraspecific bufonid taxa. Herpetologica 51:105–131.

Green DM, Cannatella DC (1993) Phylogenetic significance of the amphicoelous frogs, Ascaphidae and Leiopelmatidae. Ecol Ethol Evol 5:233–245.

Greene HW (1994) Homology and behavioral repertoires. In: Hall BK (ed) Homology The Hierarchical Basis of Comparative Biology. San Diego: Academic, pp. 370–391.

Haas A (1997) The larval hyobranchial apparatus of discoglossoid frogs: its structure and bearing on the systematics of the Anura (Amphibia: Anura). J Zool Syst Evol Res 35:179–197.

Haas A (2003) Phylogeny of frogs as inferred from primarily larval characters (Amphibia: Anura). Cladistics 19:23–89.

Haeckel E (1866) Generelle Morphologie der Organismen. Berlin.

Hanken J (1985) Morphological novelty in the limb skeleton accompanies miniaturization in salamanders. Science 229:871–874.

Hanken J (1999) Why are there so many new amphibian species when amphibians are declining? Trends Ecol Evol 14:7–8.

Harvey PH, Pagel MD (1991) The Comparative Method in Evolutionary Biology. Oxford Series in Ecology and Evolution. Oxford: Oxford University Press.

Hay JM, Ruvinsky I, Hedges SB, Maxson LR (1995) Phylogenetic relationships of amphibian families inferred from DNA sequences of mitochondrial 12S and 16S ribosomal RNA genes. Mol Biol Evol 12:928–937.

Heatwole H, Carroll RL (eds) (2000) Paleontology. The evolutionary history of amphibians. Chipping Norton, UK: Surrey Beatty.

Hedges SB, Maxson LR (1993) A molecular perspective on lissamphibian phylogeny. Herp Monogr 7:27–42.

Hedges SB, Nussbaum RA, Maxson LR (1993) Caecilian phylogeny and biogeography inferred from mitochondrial DNA sequences of the 12S rRNA and 16S rRNA genes (Amphibia: Gymnophiona). Herp Monogr 7:64–76.

Henrici A (1994) *Tephrodytes brassicarvalis*, new genus and species (Anura: Pelodytidae), from the Arikareean Cabbage Patch beds of Montana, USA, and pelodytid-pelobatid relationships. Ann Carnegie Museum 63:155–183.

Heyer WR (1975) A preliminary analysis of the intergeneric relationships of the frog family Leptodactylidae. Smiths Contrib Zool 199:1–55.

Heyer WR, Liem DS (1976) Analysis of the intergeneric relationships of the Australian frog family Myobatrachidae. Smiths Contrib Zool 233:1–29.

Hillis DM, Wilcox TP (2005) Phylogeny of the New World true frogs. Mol Phylogenet Evol 34:299–314.

Hoegg S, Vences M, Brinkmann H, Meyer A (2004) Phylogeny and comparative substitution rates of frogs inferred from sequences of three nuclear genes. Mol Biol Evol 21:1188–1200.

Jenkins FA, Shubin NH (1998) *Prosalirus bitis* and the anuran caudopelvic mechanism. J Vert Paleontol 18:495–510.

Jenkins FAJ, Walsh DM (1993) An Early Jurassic caecilian with limbs. Nature 365:246–250.

Larget B, Simon DL (1999) Markov chain Monte Carlo algorithms for the Bayesian analysis of phylogenetic trees. Mol Biol Evol 16:750–759.

Larson A (1991) A molecular perspective on the evolutionary relationships of the salamander families. Evol Biol 25:211–277.

Larson A, Dimmick WW (1993) Phylogenetic relationships of the salamander families: An analysis of congruence among morphological and molecular characters. Herp Monogr 7:77–93.

Lathrop A (1997) Taxonomic review of the megophryid frogs (Anura: Pelobatoidea). Asiatic Herpt Res 7:68–79.

Laurin M, Reisz RR (1997) A new perspective on tetrapod phylogeny. In: Sumida SS, Martin KLM (eds) Amniote Origins. Completing the Transition to Land. San Diego: Academic, pp. 9–59.

Lee MSY, Jamieson BGM (1992) The ultrastructure of the spermatozoa of three species of myobatrachid frogs (Anura, Amphibia) with phylogenetic considerations. Acta Zool Stockholm 73:213–222.

Lescure J, Renous S, Gasc J-P (1986) Proposition d'une nouvelle classification des amphibiens gymnophiones. Mém Soc Zool Fr 1986 (43):145–177.

Lewis ER, Narins PM (1999) The acoustic periphery of amphibians: anatomy and physiology. In: Fay RR, Popper AN (eds) Comparative Hearing: Fish and Amphibians. New York: Springer, pp. 101–154.

Lewis PO (2001) Phylogenetic systematics turns over a new leaf. Trends Ecol Evol 16:30–37.

Liem SS (1970) The morphology, systematics, and evolution of the Old World Treefrogs (Rhacophoridae and Hyperoliidae). Fieldiana Zool 57:1–145.

Loader SP, Gower DJ, Howell KM, Doggart N, Rödel M-O, Clarke BT, de Sá RO, Cohen BL, Wilkinson M (2004) Phylogenetic relationships of African microhylid frogs inferred from DNA sequences of mitochondrial 12S and 16S rRNA genes. Organisms Divers Evol 4:227–235.

Lombard RE (1971) A comparative morphological analysis of the salamander inner ear. (Doctoral dissertation) Chicago: University of Chicago.

Lynch JD (1971) Evolutionary relationships, osteology, and zoogeography of leptodactyloid frogs. Misc Publ Mus Nat Hist Univ Kansas 53:531–238.

Maglia AM (1998) Phylogenetic relationships of the extant pelobatoid frogs (Anura: Pelobatoidea): Evidence from adult morphology. Scient Pap Nat Hist Mus Univ Kansas 10:1–19.

McGowan G, Evans SE (1995) Albanerpetontid amphibians from the Cretaceous of Spain. Nature 373:143–145.

Mendelson I, Joseph R, da Silva HR, Maglia AM (2000) Phylogenetic relationships among marsupial frog genera (Anura: Hylidae: Hemiphractinae) based on evidence from morphology and natural history. Zool J Linn Soc Lond 128:125–148.

Milner AR (1988) The relationships and origin of living amphibians. In: Benton MJ (ed) The Phylogeny and Classification of the Tetrapods 1. Amphibians, Reptiles, Birds. Oxford: Oxford University Press, pp. 59–102.

Milner AR (1993) The Paleozoic relatives of lissamphibians. Herp Monogr 7:8–27.

Milner AR (2000) Mesozoic and Tertiary Caudata and Albanerpetontidae. In: Heatwole H, Carroll RL (eds) Amphibian Biology Volume 4. Chipping Norton, UK: Surrey Beatty, pp. 1412–1444.

Mueller RL, Macey JR, Jaekel M, Wake DB, Boore JL (2004) Morphological homoplasy, life history evolution, and historical biogeography of plethodontid salamanders inferred from complete mitochondrial genomes. Proc Natl Acad Sci USA 101:13820–13825.

Nussbaum RA (1977) Rhinatrematidae: A new family of caecilians (Amphibia: Gymnophiona). Occ Pap Mus Zool Univ Michigan 682:1–30.

Nussbaum RA, Wilkinson M (1989) On the classification and phylogeny of caecilians (Amphibia: Gymnophiona), a critical review. Herp Monogr 3:1–42.

Özeti N, Wake DB (1969) The morphology and evolution of the tongue and associated structures in salamanders and newts (family Salamandridae). Copeia 1969:205–215.

Pagel M (1999) Inferring the historical patterns of biological evolution. Nature 401:877–884.

Parsons TS, Williams EE (1962) The teeth of Amphibia and their relation to amphibian phylogeny. J Morph 110:375–383.

Parsons TS, Williams EE (1963) The relationships of the modern Amphibia: a re-examination. Q Rev Biol 38:26–53.
Pauly GB, Hillis DM, Cannatella DC (2004) The history of a Nearctic colonization: Molecular phylogenetics and biogeography of the Nearctic toads (*Bufo*). Evolution 58:2517–2535.
Púgener LA, Maglia AM, Trueb L (2003) Revisiting the contribution of larval characters to an analysis of phylogenetic relationships of basal anurans. Zool J Linn Soc Lond 139:129–155.
Purgue AP (1997) Tympanic sound radiation in the bullfrog *Rana catesbeiana*. J Comp Physiol A 181:438–445.
Rage J-C, Rocek Z (1989) Redescription of *Triadobatrachus massinoti* (Piveteau, 1936) an anuran amphibian from the early Triassic. Palaeontographica Abt A 206:1–16.
Rand AS, Dudley R (1993) Frogs in helium: The anuran vocal sac is not a cavity resonator. Physiol Zool 66:793–806.
Reiss JO (1996) Palatal metamorphosis in basal caecilians (Amphibia: Gymnophiona) as evidence for lissamphibian monophyly. J Herp 30:27–39.
Richards C, Moore WS (1996) A phylogeny for the African treefrog family Hyperoliidae based on mitochondrial DNA. Mol Phylogenet Evol 5:522–532.
Richards CM, Moore WS (1998) A molecular phylogenetic study of the Old World treefrog family Rhacophoridae. Herp Journal 8:41–46.
Rocek Z (2000) Mesozoic anurans. *in* Amphibian Biology. In: Heatwole H, Carroll RL (eds) Amphibian Biology Volume 4. Chipping Norton, UK: Surrey Beatty, pp. 1295–1331.
Rocek Z, Rage J-C (2000) Proanuran stages (*Triadobatrachus*, *Czatkobatrachus*). In: Heatwole H, Carroll RL (eds) Amphibian Biology Volume 4. Chipping Norton, UK: Surrey Beatty, pp. 1283–1294.
Roelants K, Bossuyt F (2005) Archaeobatrachian paraphyly and Pangaean diversification of crown-group frogs. Syst Biol 54:111–126.
Ruta M, Coates MI, Quicke DLJ (2003) Early tetrapod relationships revisited. Biol Rev Camb Phil Soc 78:251–345.
Ruvinsky I, Maxson L (1996) Phylogenetic relationships among bufonoid frogs (Anura:Neobatrachia) inferred from mitochondrial DNA sequences. Mol Phylogenet Evol 5:533–547.
Ryan MJ (1986) Neuroanatomy influences speciation rates among anurans. Proc Natl Acad Sci USA 83:1379–1382.
Ryan MJ (1988) Constraints and patterns in the evolution of anuran acoustic communication. In: Fritzsch B, Ryan MJ, Wilczynski W, Hetherington TE, Walkowiak W (eds) The Evolution of the Amphibian Auditory System. New York: Wiley Interscience, pp. 637–677.
Ryan MJ, Rand AS (1995) Female responses to ancestral advertisement calls in túngara frogs. Science 269:390–392.
San Mauro D, García-París M, Zardoya R (2004a) Phylogenetic relationships of discoglossid frogs (Amphibia:Anura:Discoglossidae) based on complete mitochondrial genomes and nuclear genes. Gene 343:357–366.
San Mauro D, Gower DJ, Oommen OV, Wilkinson M, Zardoya R (2004b) Phylogeny of caecilian amphibians (Gymnophiona) based on complete mitochondrial genomes and nuclear RAG1. Mol Phylogenet Evol 33:413–427.
San Mauro D, Vences M, Alcobendas M, Zardoya R, Meyer A (2005) Initial diversification of living amphibians predated the breakup of Pangaea. Am Nat 165:590–599.

Sanchíz B (1998) Salientia. Encyclopedia of paleoherpetology. Volume 4. München: Verlag Dr. Friedrich Pfeil.

Santos JC, Coloma LA, Cannatella DC (2003) Multiple, recurring origins of aposematism and diet specialization in poison frogs. Proc Natl Acad Sci USA 100:12792–12797.

Shaffer HB (1984) Evolution in a paedomorphic lineage. I. An electrophoretic analysis of the Mexican ambystomatid salamanders. Evolution 38:1194–1206.

Shaffer HB, Clark JM, Kraus F (1991) When molecules and morphology clash: a phylogenetic analysis of the north american ambystomatid salamanders Caudata: Ambystomatidae. Syst Zool 40:284–303.

Shubin NH, Jenkins FA (1995) An Early Jurassic jumping frog. Nature 377:49–52.

Sokol OM (1975) The phylogeny of anuran larvae: A new look. Copeia 1975:1–24.

Spinar ZV (1972) Tertiary Frogs from Central Europe. The Hague: W. Junk.

Starrett PH (1973) Evolutionary patterns in larval morphology. In: Vial JL (ed) Evolutionary Biology of the Anurans: Contemporary Research on Major Problems. Columbia: University of Missouri Press, pp. 251–271.

Taylor EH (1968) The Caecilians of the World. Lawrence: University of Kansas Press.

Titus TA, Larson A (1995) A molecular phylogenetic perspective on the evolutionary radiation of the salamander family Salamandridae. Syst Biol 44:125–151.

Trueb L (1973) Bones, frogs, and evolution. In: Vial JL (ed) Evolutionary Biology of the Anurans: Contemporary Research on Major Problems. Columbia: University of Missouri Press, pp. 65–132.

Trueb L (1996) Historical constraints and morphological novelties in the evolution of the skeletal system of pipid frogs (Anura: Pipidae). In: Tinsley RC, Kobel HR (eds) The Biology of *Xenopus*. Oxford: Clarendon, pp. 349–377.

Trueb L, Alberch P (1985) Miniaturization and the anuran skull: A case study of heterochrony. In: Duncker H-R, Fleischer G (eds) Vertebrate Morphology. Stuttgart: Gustav Fischer Verlag, pp. 113–121.

Trueb L, Cloutier R (1991a) A phylogenetic investigation of the inter- and intrarelationships of the Lissamphibia (Amphibia: Temnospondyli). In: Schultze H-P, Trueb L (eds) Origins of the Higher Groups of Tetrapods: Controversy and Consensus. Ithaca, NY: Cornell University Press, pp. 233–313.

Trueb L, Cloutier R (1991b) Toward an understanding of the amphibians: two centuries of systematic history. In: Schultze H-P, Trueb L (eds) Origins of the Higher Groups of Tetrapods: Controversy and Consensus. Ithaca, NY: Cornell University Press, pp. 175–193.

van der Meijden A, Vences M, Meyer A (2004) Novel phylogenetic relationships of the enigmatic brevicipitine and scaphiophrynine toads as revealed by sequences from the nuclear *Rag-1* gene. Proc R Soc Lond B (Biol Letters Suppl) 271:378–381.

Vences M, Glaw F (2003) When molecules claim for taxonomic change: New proposals on the classification of Old World treefrogs. Spixiana 24:85–92.

Vences M, Kosuch J, Boistel R, Haddad CF, La Marca E, Lötters S, Veith M (2003) Convergent evolution of aposematic coloration in Neotropical poison frogs: A molecular phylogenetic perspective. Organisms Divers Evol 3:215–226.

Wake DB (1966) Comparative osteology and evolution of the lungless salamanders, family Plethodontidae. Mem So Calif Acad Sci 4:1–111.

Wake DB, Özeti N (1969) Evolutionary relationships in the family Salamandridae. Copeia 1969:124–137.

Wake MH (1993) Non-traditional characters in the assessment of caecilian phylogenetic relationships. Herp Monogr 7:42–55.

Wake TA, Wake MH, Lesure R (1999) A Mexican archaeological site includes the first Quaternary fossil of caecilians. Quat Res 52:138–140.
Wiens JJ, Bonett RM, and Chippindale PT (2005) Ontogeny discomcobulates phylogeny: Paedomorphosis and higher-level salamander relationships. Syst Biol 54: 91–110.
Wilkinson J, Drewes RC, Tatum OL (2002a) A molecular phylogenetic analysis of the family Rhacophoridae with an emphasis on the Asian and African genera. Mol Phylogenet Evol 24:265–273.
Wilkinson M (1989) On the status of *Nectocaecilia fasciata* Taylor, with a discussion of the phylogeny of the Typhlonectidae (Amphibia: Gymnophiona). Herpetologica 45:23–36.
Wilkinson M (1997) Characters, congruence, and quality. A study of neuroanatomical and traditional data in caecilian phylogeny. Biol Rev 72:423–470.
Wilkinson M, Nussbaum RA (1996) On the phylogenetic position of the Uraeotyphlidae (Amphibia: Gymnophiona). Copeia 1996:550–562.
Wilkinson M, Nussbaum RA (1999) Evolutionary relationships of the lungless caecilian *Atretochoana eiselti* (Amphibia: Gymnophiona: Typhlonectidae). Zool J Linn Soc Lond 126:191–223.
Wilkinson M, Sheps JA, Oommen OV, Cohen BL (2002b) Phylogenetic relationships of Indian caecilians (Amphibia: Gymnophiona) inferred from mitochondrial rRNA sequences. Mol Phylogenet Evol 23:401–407.
Wu S-H (1994) Phylogenetic relationships, higher classification, and historical biogeography of the microhyloid frogs (Lissamphibia: Anura: Brevicipitidae and Microhylidae). (Doctoral dissertation) Ann Arbor: University of Michigan.
Zardoya R, Meyer A (2001) On the origin of and phylogenetic relationships among living amphibians. Proc Natl Acad Sci USA 98:7380–7383.
Zhang P, Chen YQ, Zhou H, Wang XL, Qu LH (2003) The complete mitochondrial genome of a relic salamander, *Ranodon sibiricus* (Amphibia: Caudata) and implications for amphibian phylogeny. Mol Phylogenet Evol 28:620–626.
Zug GR, Vitt LJ, Caldwell JP (2001) Herpetology. An Introductory Biology of Amphibians and Reptiles. 2nd edition. San Diego: Academic.

3
The Behavioral Ecology of Anuran Communication

KENTWOOD D. WELLS AND JOSHUA J. SCHWARTZ

1. Introduction

As the last rays of sunlight disappear from the evening sky, a shallow marsh in Panama begins to come alive with the calls of frogs and toads. Among these is a small yellow and brown hourglass treefrog, *Hyla ebraccata* (Fig. 3.1C). First, a single male begins giving a tentative series of single-note, buzzlike advertisement calls. Soon other males join the first one, and a chorus begins to develop. The first male responds to the calls of his neighbors by placing his own calls immediately after their calls, and he soon increases his calling rate and begins to add clicklike secondary notes to his calls in an attempt to outsignal his rivals. Suddenly another male calls only a few centimeters away, and the first male responds by modifying the introductory notes of his calls, producing aggressive notes with a pulse repetition rate about three times that of his advertisement calls. As the two males approach each other, they gradually increase the duration of their aggressive calls and eventually stop giving secondary click notes as a short wrestling bout ensues. After a few seconds, the intruding male withdraws, and the first male returns to advertisement calling.

Having sorted out spacing within the chorus, most of the males soon settle into a regular rhythm of advertisment calling, punctuated by occasional aggressive calls. Periodically they stop calling as their calls are overpowered by bursts of calling from groups of males of another frog, the small-headed treefrog (*Hyla microcephala*) (Fig. 3.1D). The males of *H. ebraccata* have difficulty making their calls audible when surrounded by the other species, and they attempt to place their calls in the silent periods between bursts of *H. microcephala* calling activity. After two hours of calling, the first male detects the movement of a noncalling frog nearby. Sensing that a female may be approaching, he immediately switches to a rapid series of repeated introductory advertisement call notes. The female turns toward the male, and with a few zigzag hops, approaches his calling site and allows him to clasp her in amplexus. The pair then moves off to find a suitable leaf on which to lay their eggs, positioned a half meter or so above the shallow water where the tadpoles will complete their development. After mating,

FIGURE 3.1. Calling males of some anurans in which vocal communication has been studied in detail. (A) *Pseudacris crucifer* (Hylidae). (B) *Hyla versicolor* (Hylidae). (C) *Hyla ebraccata* (Hylidae). (D) *Hyla microcephala* (Leptodactylidae). (E) *Eleutherodactylus coqui* (Leptodactylidae). (F) *Physalaemus pustulosus* (Leptodactylidae). Photos by Kentwood D. Wells.

the male takes no further interest in his offspring, but moves to a nearby calling perch and resumes calling.

These scenes, which can be repeated dozens of times each night in a single breeding assemblage in Panama (Schwartz and Wells 1984a; Wells and Schwartz 1984a,b), illustrate the complexity of vocal interactions that can occur in a chorus of frogs. Many anurans have repertoires consisting of several distinct types of calls, and they often exhibit considerable plasticity in their use of different call elements, responding to changes in local chorus density, the presence of nearby conspecific and heterospecific callers, and to approaching females by modifying their vocal signals (Wells 1988; Gerhardt and Huber 2002). All male frogs have the same ultimate goal: to outsignal their competitors and attract females, eventually fertilizing their eggs to contribute their genes to the next generation of frogs. The ways in which they accomplish this goal vary among species, however. Some anurans have relatively simple calls, whereas others have exceedingly complex calls. Males of some species are very aggressive toward other males in a chorus, whereas males of other species seldom react to their neighbors.

This chapter reviews the behavioral ecology of anuran vocal communication. First, the influence of sexual selection on the production and energetic cost of calls is briefly reviewed. Next, the major types of calls produced by male anurans, as well as the less common vocal signals of females are discussed. The interactions among males in choruses and the ways in which these interactions affect the ability of males to attract mates are covered in some detail, followed by a brief discussion of ways in which features of the anuran auditory system contribute to communication within a chorus setting.

2. Sexual Selection, Energetic Constraints, and Signaling System Evolution

When Charles Darwin originally outlined his theory of sexual selection in his book, he had relatively little to say about sexual selection in amphibians. He did suggest that the calls of frogs are analogous to the songs of birds and probably were shaped by sexual selection. Indeed, subsequent research has shown that sexual selection is the main driving force in the evolution of anuran acoustic communication (Gerhardt and Huber 2002). Many features of anuran calls can be shaped by sexual selection, including call intensity, calling rate, call duration, call pitch, and the temporal pattern of interaction among competing males. In addition, all of the morphological, physiological, and biochemical machinery involved in call production is molded by sexual selection. Because the energetic cost of calling in many species is quite high, selection should favor mechanisms to increase the efficiency of sound production and transmission, thereby enabling a calling male to conserve energy reserves while maximizing the transmission of signals to receivers, especially females.

2.1 Morphology and Physiology of Call-Producing Muscles

The basic mechanisms of sound production during calling are discussed in the next chapter (see Walkowiak, Chapter 4). The power for sound production is provided mainly by the sexually dimorphic trunk muscles, and sexual selection has produced a number of morphological and biochemical adaptations for call production by males, including highly aerobic muscle fibers, high concentrations of mitochondria, high activities of aerobic enzymes, heavy vascularization, and ample supplies of lipid and carbohydrates to fuel call production. There is a strong interspecific correlation between muscle structure and biochemistry and typical calling rates, with the most aerobic muscles being characteristic of species with high calling rates (Wells 2001).

The repeated contraction of the trunk muscles to produce calls can be energetically expensive in species with high calling rates (Wells 2001). The North American spring peeper (*Pseudacris crucifer*; Fig. 3.1A) produces single-note calls (Fig. 3.2A), each representing one contraction of the trunk muscles, and can produce up to 100 call notes per minute. Similarly, the tiny Neotropical small-headed treefrog (*Hyla microcephala*) (Fig. 3.1D) produces long trains of notes grouped into multinote calls (Fig. 3.2E) and also can produce up to 100 notes per minute (Wells and Taigen 1989). Males can call at these levels for several hours each night. Sustaining such high calling effort requires a high aerobic capacity, because anaerobic metabolism plays little role in call production. For these small treefrogs, rates of oxygen consumption during calling can be up to 25 times resting rates. This results in a significant drain on stored energy reserves. Short-term performance probably is limited mainly by carbohydrate reserves in the form of glycogen stored in the muscle tissue (Bevier 1997b), whereas lipid reserves are depleted over longer time intervals (Ressel 2001). The high energetic cost of calling probably explains the relatively short average chorus tenure of many male frogs (Murphy 1994), which exerts strong selective pressures on males to outsignal their competitors and attract females as rapidly as possible. Some frogs, however, invest much less effort in calling each night, but can remain in a chorus for several months (Bevier 1997a; Wells 2001). For these species, the ability to remain active for long periods of time probably is a more important determinant of mating success than nightly calling performance.

2.2 Vocal Sacs as Sound Radiators and Visual Signals

Because the metabolic cost of calling is high for many anurans, any adaptation to increase the efficiency of sound transmission will be favored by selection. Most anurans that call in air have inflatable vocal sacs that radiate sound to the external environment, providing greater energetic efficiency than would be possible with the larynx alone, because the vocal cords are much smaller than the wavelength of the call. Even so, the efficiency with which frogs convert metabolic energy into radiated sound energy is quite low (Prestwich 1994; McLister 2001). This problem is most acute for species with very low-pitched calls,

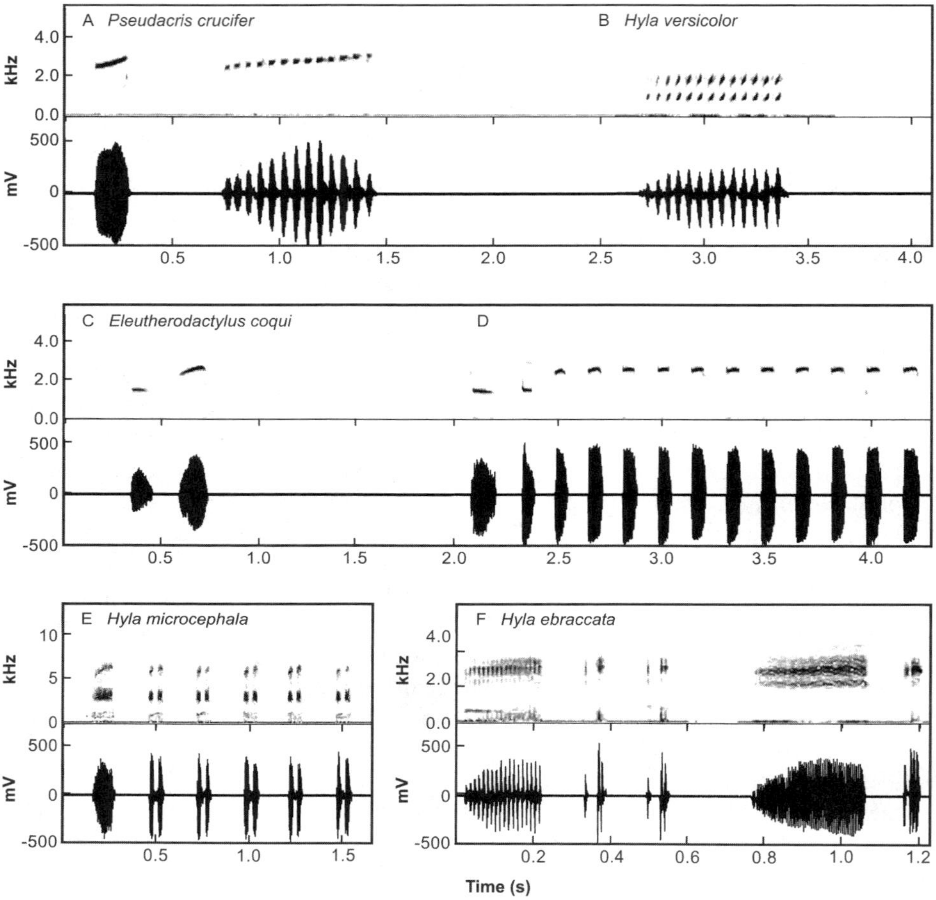

FIGURE 3.2. Representative anuran calls. The sound spectrograms at the top of each part show changes in frequency (kHz) over time. The oscillograms at the bottom of each part show changes in amplitude (mV), over time. (A) Tonelike peep of *Pseudacris crucifer*. (B) Amplitude-modulated aggressive call of *P. crucifer*. (B) Amplitude-modulated advertisement call of *Hyla versicolor*. (C) Tonelike "co" note and frequency-modulated "qui" note of *Eleutherodactylus coqui* advertisement call. (D) Multinote aggressive call of *E. coqui*. (E) Multinote advertisement call of *Hyla microcephala*, composed of an introductory note followed by several biphasic secondary click notes. (F) Three-note advertisement call of *Hyla ebraccata* (left) and two-note aggressive call of *H. ebraccata* (right), with much higher pulse rate in the introductory note. Recordings by Kentwood D. Wells and Joshua J. Schwartz.

because low-pitched sounds have long wavelengths. Small species with high-pitched calls and large vocal sacs relative to the size of the head (and larynx) can be expected to be more efficient sound radiators and have unusually loud calls for their body size. For example, both *Pseudacris crucifer* and *Hyla micro-*

cephala, which have proportionately large vocal sacs, weigh only about 1 g, but can produce calls as loud as those of songbirds that are 10 to 100 times heavier (Pough et al., 1992).

In addition to radiating sound, vocal sacs also can serve as visual signals, which can increase the effectiveness of acoustic signals by making them more detectable by receivers, especially in noisy environments. In many frogs, the throat region and vocal sac are conspicuously colored. Usually these are highly reflective colors such as white or yellow, but some frogs have black vocal sacs that also make the males conspicuous (see Hödl and Amezquita 2001 for a review of visual signaling in frogs). Experimental studies using a mechanical frog model showed that a combination of an acoustic signal and a visual signal of a moving vocal sac was most effective in eliciting aggressive responses from males of a South American dendrobatid frog, the brilliant-thighed poison frog (*Allobates* [*Epipedobates*] *femoralis*; Narins et al. 2003). In another dendrobatid frog, the palm rocket frog (*Colostethus palmatus*) from Colombia, females were attracted to the moving vocal sac of a model male frog, but not to a model with a deflated vocal sac (Lüddecke 1999). Males of a leptodactylid frog, the Túngara frog (*Physalaemus pustulosus*), have unsually large and conspicuous vocal sacs (Fig. 3.1F). Video playback of a calling male with a moving vocal sac enhanced the attractiveness of an acoustic stimulus to females (Rosenthal et al. 2004).

2.3 Other Sound Radiators

Although vocal sacs probably are the main sound-radiating organs in most frogs, they are not the only ones. For example, much of the sound energy produced by a calling male North American bullfrog (*Rana catesbeiana*) is radiated not from the vocal sacs, which sit in the water, but from the very large tympanic membranes (Purgue 1997). Male bullfrogs, and males of a number of other ranid frogs, have tympana up to 50% larger than those of females. These enlarged tympana have a thickened central patch that increases the mass of the eardrum and apparently serves to decouple the auditory and sound-broadcasting functions of the eardrum. Males of a West African frog, Parker's water frog (*Petropedetes parkeri*), have a conspicuous spongy papilla projecting from the tympanum, offset from the center of the membrane. There is some evidence that the ears of these frogs are used for both sound reception and sound transmission, as in North American bullfrogs (Narins et al. 2001).

3. The Vocal Repertoires of Frogs and Toads

Some anurans have relatively limited repertoires of call types, whereas others have a diverse array of calls used in different social contexts. Most anurans have advertisement calls that are given by males to advertise their positions to females and to other males, although some species have secondarily lost the advertisement call (Wells 1977a). Many species also have release calls, produced by both males and females, which are given when an unreceptive individual is clasped by a male. Some species also have distinct courtship calls, given by males when they detect females nearby. Less common are courtship calls given by females, often in response to the calls of males. Aggressive calls, used during agonistic interactions among males, are common, although anurans in some clades typically lack distinct aggressive calls (e.g., many toads in the genus *Bufo*). Some anurans also produce distress calls when seized by predators, although there is little evidence to suggest that these function in intraspecific communication. We focus our discussion on the three categories of calls used most commonly in social interactions in choruses: advertisement calls, courtship calls, and aggressive calls.

3.1 Advertisement Calls

The advertisement calls of anurans convey the same sorts of messages as do advertising signals of many other animals: they signal the species identity, sexual receptivity, position, size, and in some cases, the individual identity of males in a chorus. Hundreds of playback experiments with scores of species have shown that female frogs will approach conspecific calls presented alone or in choice tests with heterospecific calls (Gerhardt and Huber 2002). In species in which males call over long periods of time and females are in the same habitat, advertisement calls could stimulate hormone production in females and maintain reproductive condition in females, although this has rarely been demonstrated experimentally (Lea et al. 2001). Advertisement calls also advertise a male's position to other males and help to maintain spacing between calling individuals, with perceived call intensity providing information about the spatial proximity of competing males (Brenowitz et al. 1984; Wilczynski and Brenowitz 1988; Brenowitz, 1989; Gerhardt et al. 1989).

Males of some species can use the pitch of other males' advertisement calls to assess the body size of competitors (Davies and Halliday 1978; Arak, 1983b; Robertson 1984; Given 1987; Wagner 1989c). Males are more likely to approach or attack speakers playing high-pitched calls of small males, but retreat from the low-pitched calls of large males. Males of some species alter the pitch of their calls in response to those of neighboring males (e.g., *Rana catesbeiana*; Bee and Bowling 2002), but it is not always clear that this provides more accurate information about male body size to opponents (Bee et al. 2000; Bee 2002; see further discussion of aggressive interactions below).

Although the advertisement calls of most anurans consist of a single note, a series of identical repeated notes, or a long trill, some have complex advertisement calls with more than one kind of note (Wells 1988). Most frogs have only a few kinds of notes in their calls, but some rhacophorid and mantellid treefrogs have extraordinarily complex calls, with a dozen or more distinct kinds of notes. The functions of these very complex calls are not fully understood, but some call components appear to be used in aggressive interactions among males (Narins et al. 2000, Christensen-Dalsgaard et al. 2002; Feng et al. 2002).

In some species, different types of notes apparently convey separate messages to males and females. For example, in the Puerto Rican common coquí (*Eleutherodactylus coqui*) (Figs. 3.1E, 3.2C), the beginning "co" note elicits calling and aggressive responses from males, but they show little response to the secondary "qui" note (Narins and Capranica 1978). Females are attracted to the "qui" note, but show little response to the "co" note alone (Narins and Capranica 1976). In the Australian eastern smooth frog (*Geocrinia victoriana*), the long introductory note conveys an aggressive message to males, whereas the shorter secondary notes are attractive to females (Littlejohn and Harrison 1985). A somewhat similar system is found in the short-legged spiny reed frog (*Afrixalus brachycnemis*) from southern Africa, which has a rapidly pulsed note that serves as an aggressive signal and a longer trill that is attractive to females (Backwell 1988). In Fornasini's spiny reed frog (*Afrixalus fornasinii*), males give long trains of very short pulses (trills) that often grade into a series of repeated pulsed notes that are given mainly in response to other males (Schneichel and Schneider 1988). Some frogs add secondary notes to their calls during chorusing interactions (Figs. 3.2E,F), including *Hyla ebraccata* (Fig. 3.1C) and *Hyla microcephala* (Fig. 3.1D) from Panama; these are discussed in a later section.

3.2 Male Courtship Calls

Male frogs often alter their vocal behavior when females are nearby, producing calls that are likely to increase the signal-to-noise ratio of the male's calls or provide directional cues to females (Wells 1977b, 1988). Male spring peepers (*Pseudacris crucifer*) give longer peeps when females are nearby (Rosen and Lemon 1974). Male gray treefrogs (*Hyla versicolor*) respond to approaching females by giving trills that can be several times the length of normal advertisement calls (Wells and Taigen 1986; Klump and Gerhardt 1987). Similar behavior is seen in the Trinidad poison frog (*Mannophyrne trinitatis*; Wells 1980b). Males normally give two-note advertisement calls, but combine these into a continuous trill when females are approaching, producing a call with 50% more notes than the normal advertisement call (Fig. 3.3).

Often males simply increase calling rates in response to approaching females (reviewed by Wells 1988 and Gerhardt and Huber 2002). Others produce distinctive courtship calls, especially species in which the male leads the female to a concealed oviposition site during courtship (Wells 1977b, 1988; Townsend and Stewart 1986; Hoskin 2004). Distinctive courtship calls also occur in some species in which the male calls from a fixed location to attract the female (Greer and Wells 1980; Kluge, 1981; Robertson 1986). In midwife toads (*Alytes obstetricans*), males give courtship calls while moving toward females and females sometimes respond with calls of their own (Bush 1997). Courtship calls sometimes are given at lower intensity than advertisement calls, perhaps to avoid alerting other males to the presence of a female (see discussion of "eavesdropping" below). Ovaska and Caldbeck (1997b) showed that males of the Antilles robber frog (*Eleutherodactylus antillensis*) respond to playbacks of courtship calls by

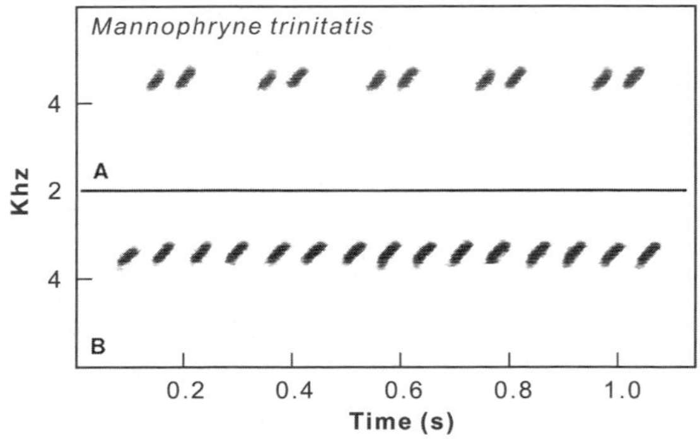

FIGURE 3.3. Sound spectrograms of (A) the advertisement call and (B) a courtship trill of *Mannophryne trinitatis*. Recordings by Kentwood D. Wells.

approaching the speaker playing the call, a behavior not seen during playbacks of advertisement or aggressive calls.

Some frogs have more than one type of courtship call. In Fleischmann's glass frog (*Hyalinobatrachium fleischmanni*) from Panama, a male gives long, frequency-modulated calls when a moving frog is first detected nearby. This appears to serve as an aggressive call when directed at males, but probably also provides direction cues to females. Once a female begins to approach the male, he switches to a series of short chirps (Greer and Wells 1980). In *Mannophryne trinitatis*, a male gives a long trill when courting a female at a distance, but switches to quiet chirps as he leads the female to a hidden oviposition site (Wells 1980b). Males of the Australian ornate frog (*Cophixalus ornatus*) give long courtship calls while leading females to nest sites, but give shorter calls when in a nest with a female (Hoskin 2004). Both types of calls are delivered at a high rate, but low intensity. Short-range courtship calls also have been reported in several dendrobatid frogs and some species of *Eleutherodactylus* with concealed oviposition sites (Ovaska and Hunte 1992; Bourne 1997; Ovaska and Caldbeck 1997a, 1999).

3.3 Female Courtship Calls

Some female frogs vocalize in response to the calls of males (Emerson and Boyd 1999), although all female frogs lack vocal sacs. The best-studied species are midwife toads in the genus *Alytes*. Female midwife toads call in response to male calls, and these calls elicit soft courtship calls from males (Bush 1997; Bosch and Márquez 2001). The calls given by females probably enhance the ability of males and females to find each other. Male midwife toads often call from hidden locations in rock crevices or burrows, but sometimes move toward females and engage in vocal duets with them (Bush et al. 1996; Bush 1997). Duetting between males and females also occurs in the African common platanna (*Xenopus laevis*), which often calls in muddy water where males may not be visible to females

(Tobias et al. 1998). Males of Serro Utyum robber frog (*Eleutherodactylus podiciferus*) from Costa Rica switch from the normal advertisement call trill to a series of squeak calls after hearing similar squeaks given by females (Schlaepfer and Figeroa-Sandí 1998). Low-intensity courtship calls have been reported in a number of ranid frogs, including the North American carpenter frog (*Rana virgatipes*) (Given 1987), bullfrog (*R. catesbeiana*; Judge et al. 2000), fanged frogs of the genus *Limnonectes* from Borneo and Southeast Asia (Emerson 1992; Orlov 1997), the Asian rice frog (*Limnonectes limnocharis*), water skipping frog (*Euphlyctis cyanophlyctis*), and red-eared frog (*Rana erythraea*), and the Levuka wrinkled ground frog (*Platymantis vitiensis*) from Fiji (Roy et al. 1995; Boistel and Sueur 1997). At least two nonexclusive functions of female courtship calling have been proposed: (1) facilitating mate location by eliciting more calling from nearby males, enabling females to distinguish territorial from satellite males, and (2) identifying of females as potential mates rather than territorial competitors (Emerson and Boyd 1999). To date, however, very few experimental studies of male responses to female calls have been done (Given 1993a; Bush et al. 1996; Bush 1997; Tobias et al. 1998; Bosch 2001, 2002), and both of these remain viable hypotheses.

3.4 Aggressive Calls

Many male frogs defend their calling sites against conspecifics and often have distinctive aggressive vocalizations. Aggressive and advertisement calls usually have similar dominant frequencies, but differ in temporal structure, but there is no unique temporal structure common to all aggressive calls. Certain constraints on call production probably limit divergence between aggressive calls and advertisement calls within species. Frogs that produce wideband calls typically produce relatively short pulses of sound and probably are incapable of producing notes of long duration, whereas frogs that produce narrowband calls typically have longer notes and probably cannot produce very short calls (Gerhardt and Huber, 2002). For example, in *Pseudacris crucifer*, the advertisement call is a tonelike peep, whereas the aggressive call is a long trill consisting of a series of slightly shorter pulses (Fig. 3.2A). In other chorus frogs in the genus *Pseudacris*, the advertisement call consists of trains of extremely short pulses, whereas aggressive calls are longer trains of the same type of pulses, sometimes delivered at faster rates (Owen 2003).

In a dendrobatid frog from Panama, *Colostethus panamensis*, the advertisement call is a short trill and the aggressive call is a long tonelike peep that resembles a trill with the notes merged together (Wells 1980a). These calls sometimes grade into each other as a male makes the transition from aggressive to advertisement calling (Fig. 3.4). Males of another species in the same genus, the Bogata rocket frog (*C. subpunctatus*), sometimes respond to calls of other males by grouping call notes into bouts of two or three notes. This does not appear to enhance the attractiveness of males to females, but does function as an aggressive signal (Lüddecke 2002). The Santo Andre snouted treefrog (*Scinax rizibilis*)

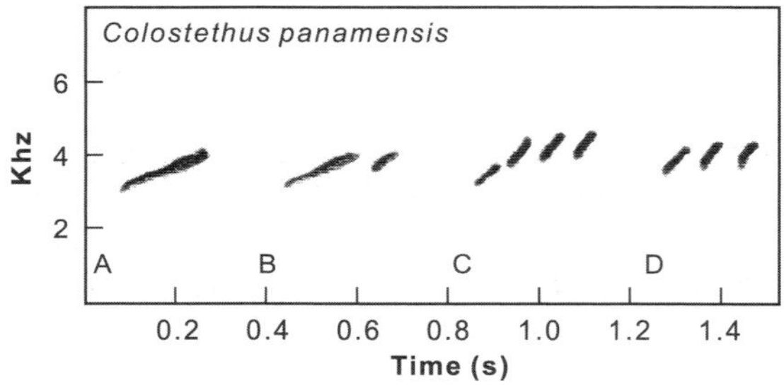

FIGURE 3.4. Sound spectrograms of the calls of *Colostethus panamensis*. (A) Single-note aggressive call. (B) Two-note aggressive call. (C) Transition from the aggressive call to an advertisement call. (D) Three-note advertisement call. Recordings by Kentwood D. Wells.

from Brazil has an advertisement call consisting of a series of repeated notes, whereas aggressive calls consist of irregular trains of short pulses. This species also has a close-range aggressive call consisting of short bursts of pulses (Bastos and Haddad 2002). In the Lesser Antilles robber frog (*Eleutherodactylus urichi*) from Trinidad, the advertisement call is a short, tonelike peep, the aggressive call an even shorter click (Wells 1981). In contrast, the aggressive call of *Eleutherodactylus coqui* is a long train of notes (Fig. 3.2D). The advertisement calls of the Angola forest treefrog (*Leptopelis viridis*) are short clicks, whereas the aggressive calls are about twice as long and have a slightly lower dominant frequency (Grafe et al. 2000).

A number of hylid treefrogs from South and Central America, including the hourglass treefrog (*Hyla ebraccata*), small-headed treefrog (*H. microcephala*), and veined treefrog (*H. phlebodes*) produce aggressive calls with a structure similar to that of advertisement calls (Fig. 3.2F), but with a much higher pulse repetition rate (Schwartz and Wells 1984a,b, 1985; Wells and Schwartz 1984b). Because aggressive calls do not function in species recognition, one might expect such calls to be less stereotyped than are advertisement calls. Indeed, in these and other anuran species, temporal features such as pulse repetition rate and number of pulses are much more variable in aggressive calls than in advertisement calls (Schwartz and Wells 1984a; Littlejohn 2001; Owen 2003).

Some frogs change the dominant frequency of their advertisement calls when responding to the calls of other males. This type of behavior has been reported in the white-lipped frog (*Leptodactylus albilabris*) from Puerto Rico (Lopez et al. 1988) and in several North American anurans, including Northern cricket frogs (*Acris crepitans*; Wagner 1989b, 1992), green frogs (*Rana clamitans*; Bee and Perrill 1996; Bee et al. 1999, 2000), carpenter frogs (*R. virgatipes*; Given 1999), bullfrogs (*R. catesbeiana*; Bee and Bowling 2002), and American toads (*Bufo americanus*; Howard and Young 1998). In all cases except *L. albilabris*, males lower the dominant frequency of their calls. This generally has been interpreted as an aggressive response, perhaps a means of conveying information about the size of the caller, although in bullfrogs, such changes do not appear to be correlated with fighting ability (Bee 2002).

Many anurans have graded signaling systems in which long- and short-range aggressive calls represent two ends of a continuum. This has been studied in most detail in *Hyla ebraccata* from Panama (Wells and Schwartz 1984b; Wells and Bard 1987; Wells 1989). This species has advertisement calls consisting of an introductory note and a series of shorter secondary notes with the same dominant frequency and pulse repetition rate. Aggressive calls have introductory notes with much higher pulse rates and are more variable in duration. As males approach each other, they lengthen the introductory notes of their aggressive calls while gradually dropping secondary click notes (Fig. 3.5). The secondary notes make the calls more attractive to females, which are not strongly attracted to aggressive calls with high pulse repetition rate (Wells and Bard 1987). Hence, males gradually adjust the relative aggressiveness and attractiveness of their calls, depending on the proximity of their opponents. Similar behavior has been described in two other Panamanian treefrogs, *H. microcephala* (Schwartz and Wells 1985) and *H. phlebodes* (Schwartz and Wells 1984b). As in *H. ebraccata*, females of *H. microcephala* prefer the lower pulse rates of advertisement calls to the higher pulse rates of aggressive calls (Schwartz 1987a).

Some North American hylid frogs also have graded aggressive calls. Male spring peepers (*Pseudacris crucifer*) increase the duration of their trilled aggressive calls in response to increases in the intensity and duration of an aggressive call stimulus (Schwartz 1989). Several other species in the genus *Pseudacris* also have trilled aggressive calls, and some of these show evidence of graded variation similar to that seen in spring peepers (Owen 2003). Northern cricket frogs (*Acris crepitans*) have a somewhat simpler system. Males produce calls with progressively more pulses as they approach each other, but they do not have structurally distinct aggressive calls (Wagner 1989a,c, 1992; Burmeister et al. 1999, 2002). In contrast to many other species, these changes in call structure actually make the calls more attractive to females, rather than less attractive (Kime et al. 2004). Hence, these calls may convey aggressive messages to males, or they may be a form of escalated competition among males for the attention of females, or both. Multinote aggressive calls are characteristic of several species of Caribbean robber frogs in the genus *Eleutherodactylus*, although the extent to which these calls are graded is not clear (Stewart and Rand 1991; Stewart and Bishop 1994; Michael 1997; Ovaska and Caldbeck 1997b; O'Brien 2002).

In the Old World, graded aggressive calls have been described in several clades of frogs, although most species have not been studied in as much detail as the New World species. Examples include Australian ground froglets (*Geocrinia*) and crowned toadlets (*Pseudophryne*; Pengilley 1971; Littlejohn and Harrison 1985), African reed frogs (*Hyperolius*) and spiny reed Frogs (*Afrixalus*; Backwell 1988; Grafe 1995), and Asian Bubble-nest Frogs (*Philautus*; Arak 1983a). Some frogs exhibit graded variation in both advertisement and aggressive calls. Males of the

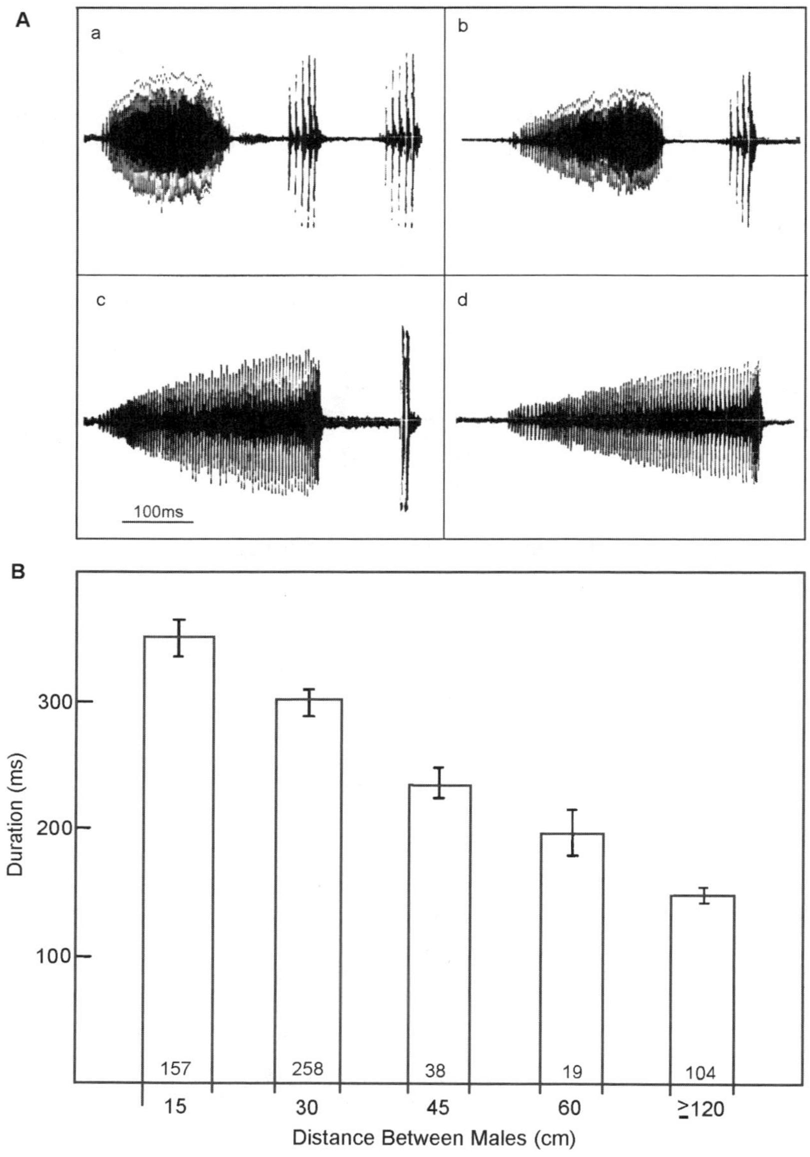

FIGURE 3.5. Graded aggressive calls of the Panamanian treefrog *Hyla ebraccata*. (A) Oscillograms of four aggressive calls recorded from the same male, showing a gradual increase in the duration of the introductory note and reduction in the duration and number of secondary click notes. Call (a) was given at the longest distance between males; call (d) was given at the shortest distance. (B) Duration of introductory notes of aggressive calls as a function of the distance between interacting males. Numbers at the bottom of each column are sample sizes. Recordings by Kentwood D. Wells. Data from Wells and Schwartz 1984b.

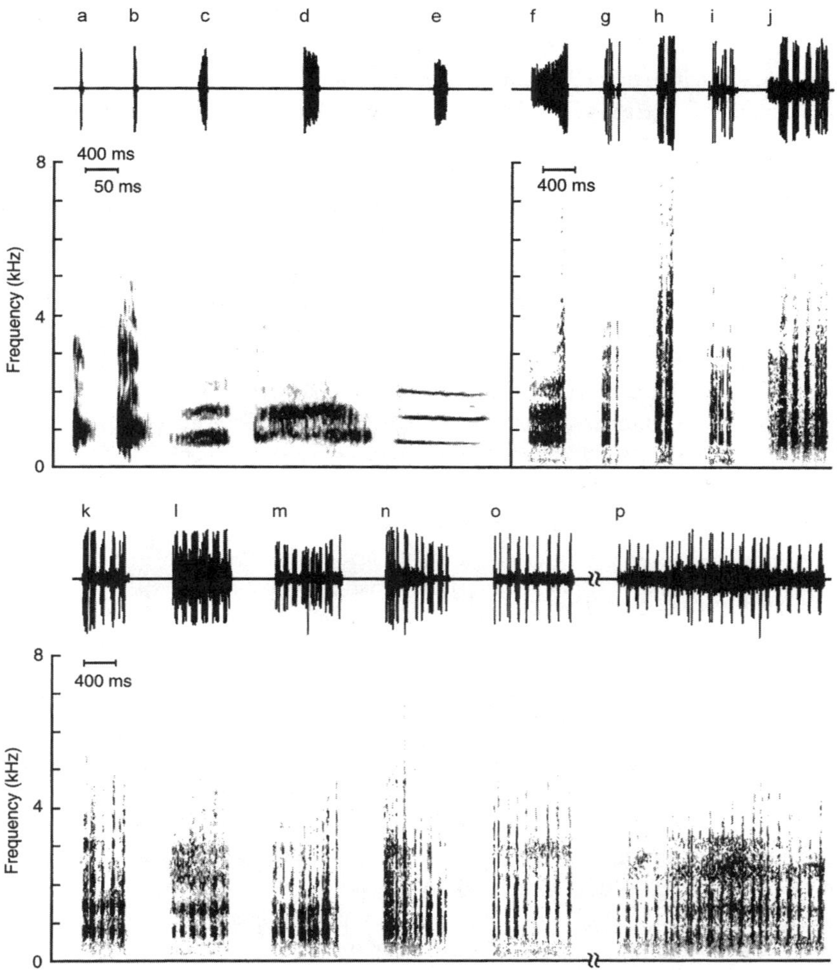

FIGURE 3.6. Complex vocal repertoire of *Boophis madagascariensis*. Oscillograms are shown on top of each part, sound spectrograms on the bottom. For the first set of calls (a)–(e), the time scale on the sound spectrograms has been magnified to show details of call structure. Time scales are the same for oscillograms and sound spectrograms for all other calls: (a) toc note; (b) short click note; (c) short rip note; (d) loud click note; (e) tonelike note; (f) long rip note; (g) creak note; (h)–(p) iambic notes with increasing number of pulses. Males give iambic notes more frequently in response to playbacks of conspecific calls, and these may represent a graded aggressive call system. Reprinted from Narins et al. (2000), Fig. 3, p. 287 with the permission of Cambridge University Press.

Nicobar frog (*Fejervarya nicobariensis*) from Malaysia produce multinote advertisement calls with 1 to 6 click notes, but when males are calling in close proximity, they produce up to 25 notes (Jehle and Arak 1998). Males in dense choruses give distinctly different squawklike aggressive calls, and these sometimes are followed by a series of clicklike notes similar to those in the advertisement calls. These compound calls apparently represent transition calls that convey an aggressive message to other males while retaining notes attractive to females. A rhacophorid treefrog from Thailand, the Javan whipping frog (*Polypedates leucomystax*), has a repertoire of at least 12 distinct call types, many consisting of trains of pulses or clicks that appear to function as aggressive signals (Christensen-Dalsgaard et al. 2002). The Madagascar bright-eyed frog (*Boophis madagascariensis*) has an even more variable vocal repertoire of at least 28 different types of calls, although most of these appear to be variants of a single call type that differ in the number of notes and pulses produced (Narins et al. 2000). The most variable call types often were given in response to playbacks of similar notes at high intensities and have many of the characteristics of the graded aggressive calls seen in other species (Fig. 3.6).

4. Chorusing Behavior

Many frogs and toads form aggregations in which males call to attract mates (Zelick et al. 1999; Gerhardt and Huber 2002). The term "chorus" is used here to describe any group of signaling animals (Gerhardt and Huber 2002), without specifying the spatial distribution or call timing relationships among individual males (Brush and Narins 1989). Choruses can both facilitate and impede communication between males and females. The acoustic environment of a chorus can be complex because of the spatial distribution of males, intense competition for mates, high levels of background noise, and temporal overlap among calls of neighboring males. The close proximity of calling males allows females to quickly assess multiple mates and may promote vocal competition among males (Wiley and Poston 1996). Males can acquire information about the capabilities of rivals that can be used when adopting perch sites, mating tactics (Humfeld 2003), or calling tactics (Wells 1988). However, a loud chorus also can make signal detection, localization, discrimination, and interpretation difficult (Wollerman and Wiley 2002a,b). For example, the call preferences of females in the field or in experiments using multispeaker designs that mimic the sonic complexity of natural choruses often differ from those in simple two-choice laboratory experiments (Gerhardt 1982; Telford et al. 1989; Márquez and Bosch 1997; Schwartz et al. 2001). Within a chorus, however, some females may be better able than others to discriminate among males because of differences in the local acoustic environment (Gerhardt and Klump 1988; Schwartz and Gerhardt 1989). The presence of calling heterospecific anurans may also create opportunities for mismatings and wasted reproductive effort (Gerhardt 1994; Pfennig et al. 2000; Gerhardt and Schwartz 2001).

4.1 Choruses as Venues for Communication Networks and Competition

Anuran choruses allow for the exchange of information between numerous signalers and receivers, and choruses can be considered "communication networks" (McGregor and Peake 2000; Grafe 2005), rich in "public information" (Valone and Templeton 2002; Danchin et al. 2004). A network of signalers can affect the temporal dynamics of male calling behavior and the elaboration of male vocalizations. For example, playbacks of calls typically increase the calling effort of male receivers (see Table 9.1 in Gerhardt and Huber 2002), as manifested in increased calling rate, call complexity, or call duration. The sounds produced by a chorus can stimulate males to call (Schwartz 1991), with some minimum number of males being necessary to initiate and maintain chorusing activity (Brooke et al. 2000). Where many males can hear one another, calling efforts of individual males may rapidly escalate as a chain reaction occurs among signalers. This rapid escalation in male calling effort should facilitate comparison of potential mates by females (Wiley and Poston 1996), possibly outweighing any disadvantages imposed by masking and degradation of signals in the chorus.

Eavesdropping refers to "extracting information from signaling interactions between others" (McGregor and Peake 2000), and the network environment clearly makes possible eavesdropping by both males and females (Grafe 2005). Eavesdropping has not been studied experimentally in anurans, so we can only speculate on how it might affect their behavior. Males can acquire information on the capabilities of their competitors (Johnstone 2001), and they may detect nearby females as a result of a change in the calling behavior of a neighbor. For example, in *Hyla versicolor*, a male dramatically increases both call duration and calling effort when it detects a female. This change sometimes triggers similar changes in the calling behavior of neighbors (Schwartz et al. unpubl. data). Controlled experiments are needed to exclude the possibility that such neighbors are responding to the calls of their neighbors and not to cues produced by the female. Female-induced elevation in calling also may be exploited by satellite males, who may move towards females or even to begin to call (Grafe 2005). Nonsatellites also could take advantage of acoustic cues by moving towards individuals about to mate. Therefore, the low amplitude of courtship calls of some species could reduce the likelihood that satellite males will intercept females (Given 1993a). Satellites or less capable calling males may move to sites where more vigorous callers are likely to attract females (Pfennig et al. 2000, Gerhardt and Huber 2002, Humfeld 2003), as proposed in "hot-shot" models of lek evolution (Höglund and Alatalo 1995).

Females that eavesdrop on aggressive exchanges between males could utilize public information to select winners (McGregor and Peake 2000; Danchin et al. 2004) and so augment information supplied in advertisement calls with that transmitted in signals that are generally unattractive to females (the aggressive call).

4.2 Agonistic Interactions and Male Spacing

The use of aggressive vocalizations, including graded aggressive calls, was discussed in an earlier section. In most frogs, males seek to maintain some mininum distance between themselves and their nearest neighbors. Competition for calling sites can be mediated by both aggressive calling and actual fighting (Wells 1988; Gerhardt and Huber 2002). Male frogs apparently assess the proximity of competitors by the intensity of their calls (Wilczynski and Brenowitz 1988; Brenowitz 1989; Gerhardt et al. 1989; Marshall et al. 2003). Murphy and Floyd (2005) found that male barking treefrogs (*Hyla gratiosa*) entering low-density choruses spaced themselves farther from the louder of two speakers broadcasting advertisement calls than they did in high chorus densities. This difference could be because males have more difficulty gauging relative call intensities in high-density situations (perhaps due to noise-induced masking and threshold shifts; Schwartz and Gerhardt 1998). The increased proximity of calling males also could have made it hard for males to distance themselves from a speaker (Murphy and Floyd, 2005).

In many species, aggressive interactions tend to occur early in the evening as males sort out spatial relationships in the chorus (e.g., Wells and Bard 1987; Backwell 1988). Aggressive interactions probably are less costly at this time because females typically arrive later at night (Backwell 1988; Murphy 1999; Gerhardt and Huber 2002). This is important, because aggressive calls often are less attractive to females than are advertisement calls (Oldham and Gerhardt 1975; Schwartz and Wells 1985; Wells and Bard 1987; Backwell 1988; Grafe 1995; Brenowitz and Rose 1999; Marshall et al. 2003).

During the course of an evening, males may habituate to the calls of near neighbors, making them less likely to engage in costly aggressive interactions (Brenowitz and Rose 1999). Marshall et al. (2003) found that after just 10 min of advertisement call broadcasts (at 4 and 8 dB above the prestimulus aggressive threshold) to male spring peepers (*Pseudacris crucifer*), aggressive call thresholds were elevated nearly 10 dB. Qualitatively similar data were obtained for the Pacific treefrog (*P. regilla*; Brenowitz and Rose 1994; Rose and Brenowitz 1997). One important consequence of such plasticity is that on nights when large numbers of males enter the chorus, the percapita frequency of agonistic interactions and intermale distances may be relatively low. Thus, there is not only a synergistic relationship but also a dynamic interaction between spacing and aggression that largely explains the shifting spatial distribution of males in choruses over time. Stable choruses may develop not only when male attendance at a breeding site is low but also when it is high (Rose and Brenowitz 2002). At a proximate level, it appears that short-term habituation to specific callers, rather than adaptation of the auditory system, is sufficient to explain the experimental results with *P. crucifer* and *P. regilla*, as males responded with aggressive calls when the advertisement call stimulus was changed to aggressive calls. However, it seems reasonable that neural threshold shifts (Narins and Zelick 1988; Schwartz and Gerhardt 1998) also contribute to lower aggressive thresholds when background noise levels in the chorus are high.

Discrimination between the calls of familiar neighbors and unfamiliar "strangers" has been demonstrated in the green frog (*Rana clamitans*) (Owen and Perrill 1998), bullfrog (*R. catesbeiana*) (Davis 1987), and Beebe's rocket frog (*Colostethus beebei*) (Bourne et al. 2001). Bullfrogs habituate to the calls originating from a particular location, as reflected by a reduced tendency to give aggressive calls or approach the caller. Habituation also could explain why bullfrogs are more likely to answer calls of distant males than those of near neighbors (Boatright-Horowitz et al. 2000). Nevertheless, males become disinihibited if the spectral characteristics of the vocalization are altered by 10% during playback tests, or if the source of the calls is moved (Bee and Gerhardt 2001). Potential recognition cues also include fine temporal call features and even the pattern of variation of call features within males (Bee 2004).

4.3 Advertisement Call Plasticity

Competition among males for the attention of females often results in considerable plasticity in advertisement calling, with males modifying their calls in ways that increase the signal-to-noise ratio of their vocalizations or make their calls more attractive to females. For example, males often respond to the calls of others by increasing the energy content of their signals by elevating calling rate, call duration, or call complexity (Wells 1988, 2001; see Table 9.1 in Gerhardt and Huber 2002 for examples and exceptions). Males of some species alter call dominant frequency or the distribution of spectral energy (Lopez et al. 1988; Wagner 1989a, 1992; Bee and Perrill 1996; Howard and Young 1998; Given 1999) and adjust call amplitude (Lopez et al. 1988). Many of these changes are presumed to increase a male's relative attractiveness to females (Ryan and Keddy-Hector 1992; Andersson 1994; Halliday and Tejedo 1995; Sullivan et al. 1995). This hypothesis has been supported by phonotaxis experiments in which gravid females were presented with acoustic alternatives broadcast from speakers in a laboratory arena (e.g., Ryan 1980) or in the field (e.g., Schwartz et al. 2001). Additional support comes from observations of mate choice in nature (e.g., Passmore et al. 1992; Schwartz et al. 1995; Grafe 1997) or artificial choruses with real males (Schwartz et al. 2001). For example, computer-based monitoring of choruses of male *Hyla microcephala* confirmed that males with the highest rates of note production were the first to attract females (Schwartz et al. 1995). Males of this species tend to match the number of notes in their calls during pairwise interactions (Schwartz 1986).

Approximate note matching has been reported in other species as well (Arak 1983a; Pallett and Passmore 1988; Jehle and Arak 1998; Gerhardt et al. 2000a) and may be a way for males to fine-tune calling effort to match that of their closest competitors. Males are expected to expend only the minimum energy necessary to nullify another caller's advantage (Arak 1983a; Jehle and Arak 1998; Benedix and Narins 1999). Such behavior also could reduce a male's risk of predation (Tuttle and Ryan 1981; Zuk and Kolluro 1998; Gerhardt and Huber 2002, page 2004). In the Australian red-legged froglet (*Crinia georgiana*), males responded to playbacks of calls from two speakers as if they were summing the notes from these different sources (Gerhardt et al. 2000a). Therefore males appeared to expend more energy than was necessary to surpass the calling performance of individual rivals. This behavior probably reflects erroneous auditory grouping (see Farris et al. 2002) and thus a failure of test males to take thorough advantage of available spatial information. In *C. georgiana*, males that produce more notes per call do not necessarily gain a mating advantage (Smith and Roberts 2003). There also are different signaling routes to mating success in this species: large males did best when they produced more pulses in the first notes of their calls, whereas smaller males did best when they called at high rates.

Broadcasts of low-frequency calls that resemble those produced by large males can induce males to move away from a speaker (e.g., Arak 1983b), and the frequency shifts observed in the calls of some species may be an attempt by males to dupe rivals into misjudging their size. For example, the playback tests of Bee et al. (2000) indicate that such a bluffing strategy is conditional in green frogs, *Rana clamitans*, depending on the relative size of the interactants. Alternatively, spectral changes may honestly communicate size or size-independent fighting ability (Wagner 1992).

In some species, a subset of males may reduce or stop calling when exposed to the calls of another male (Gerhardt and Huber 2002; Humfeld 2003; Tobias et al. 2004). Males also may adopt satellite tactics and attempt to parasitize the calling efforts of other males (Halliday and Tejedo 1995). Such behavior probably represents an attempt by individuals to minimize energy expenditure for either mate attraction or aggression in the face of superior competition. In explosive breeders, very high chorus densities may cause males to cease vocal activity altogether and actively search for females (Wells 1977a; Halliday and Tejedo 1995).

Socially mediated changes in calls or calling behavior may render signals inherently more attractive to females (Wells 1988; Ryan and Keddy-Hector 1992; Sullivan et al. 1995; Schwartz 2001; Gerhardt and Huber 2002), but could these alterations improve a male's odds of mating in other ways? One possibility is that such changes modify the redundancy of signals and so improve signal detection and localizability, and reduce recognition errors by receivers under noisy conditions (Wiley 1983; Bradbury and Verhencamp 1998; Ronacher 2000; Narins et al. 2000). Kime (2001) tested and rejected the hypothesis that call complexity reduces masking vulnerability in the northern cricket frog (*Acris crepitans*), and the Túngara frog (*Physalaemus pustulosus*). Males of the former species cluster their calls within "call groups" and typically add calls to these groups, as well as the number of pulses per call, in response to the calls of other males (Wagner 1989b; Burmeister et al. 1999). Male *P. pustulosus* produce FM "whines" to which they append a variable number of chuck notes following acoustic stimulation by neighbors (Ryan 1980). Although females of both species find calls with greater complexity more attractive (Ryan 1980; Wagner 1991), these changes did not enhance signal efficacy in noise (Kime 2001).

Schwartz et al. (2001, 2002) hypothesized that call-induced increases in call duration and accompanying reductions in calling rate in *Hyla versicolor* are related to the increased threat of acoustic interference in larger choruses. In this species, pulse shape and the duration of pulses and interpulse intervals have a strong impact on the relative attractiveness of calls (Gerhardt 2001; Schul and Bush 2002) and can easily be obscured during call overlap. Females also exhibit strong discrimination against very short calls (Gerhardt et al. 2000b; Schwartz et al. 2001), and they prefer long calls delivered at a low rate to short calls delivered at a high rate. Therefore, by giving long calls, even at a low rate, in an acoustically cluttered environment a male may increase the chances that there will be a sufficient number of call pulses and interpulse intervals clear of call overlap to attract a female. Preliminary data on male call overlap (Schwartz et al. 2001, 2002) were consistent with the hypothesis, and more focused experiments to test the idea are in progress. In addition, experiments to test whether longer calls are more easily detected in chorus noise are underway. The threat of call overlap also could explain why males of *E. coqui* increase call duration (albeit to a much smaller degree than *H. versicolor*) in response to the vocalizations of conspecifics. Benedix and Narins (1999) suggested that by shifting to longer calls, a male compensates for constraints on calling rate imposed in choruses by the reduced number of available quiet intervals into which a male could insert his calls without interference. Male frogs also could increase the detectability and attractiveness of vocalizations under noisy conditions by increasing signal amplitude. This has been reported for Puerto Rican white-lipped frogs (*Leptodactylus albilabris*) (Lopez et al. 1988), but whether this is a general response to background noise is not known.

4.4 Patterns of Call Timing

Call interaction between males is a dynamic process and the timing relationships between males typically are fluid and change in response to the ambient acoustics or the level of male–male competition. Accordingly, leader–follower relationships may shift during chorusing (Moore et al. 1989; Bosch and Marquez 2001; Gerhardt and Huber 2002; Grafe 2003), yielding timing patterns that temporarily are perceived as alternating, synchronized, or partially overlapping (Fig. 3.7). Nevertheless, at particular spatial and temporal scales (Schwartz and Wells 1985; Given 1993b; Boatright-Horowitz et al. 2000), certain call timing patterns may dominate and a variety of hypotheses is available to explain such behavior at both proximate and ultimate levels (Greenfield 2002; Gerhardt and Huber 2002).

At a coarse temporal scale, call-timing shifts may occur in response to the calls of other species of frogs. For example, Littlejohn and Martin (1969) reported that males of one species of myobatrachid frog with especially long calls inhibited calling by another species with shorter calls. Schwartz and Wells (1983a,b) reported similar behavior in Panamanian tree frogs. Calling by males of *Hyla ebraccata* was inhibited by chorusing of groups of nearby *H. microcephala* or *H. phlebodes*. For *H. ebraccata*, these two species are especially potent sources of interference. *Hyla microcephala* calls in dense aggregations and employs calls with many notes. Even pairs of *H. phlebodes* can produce rapid-fire sequences

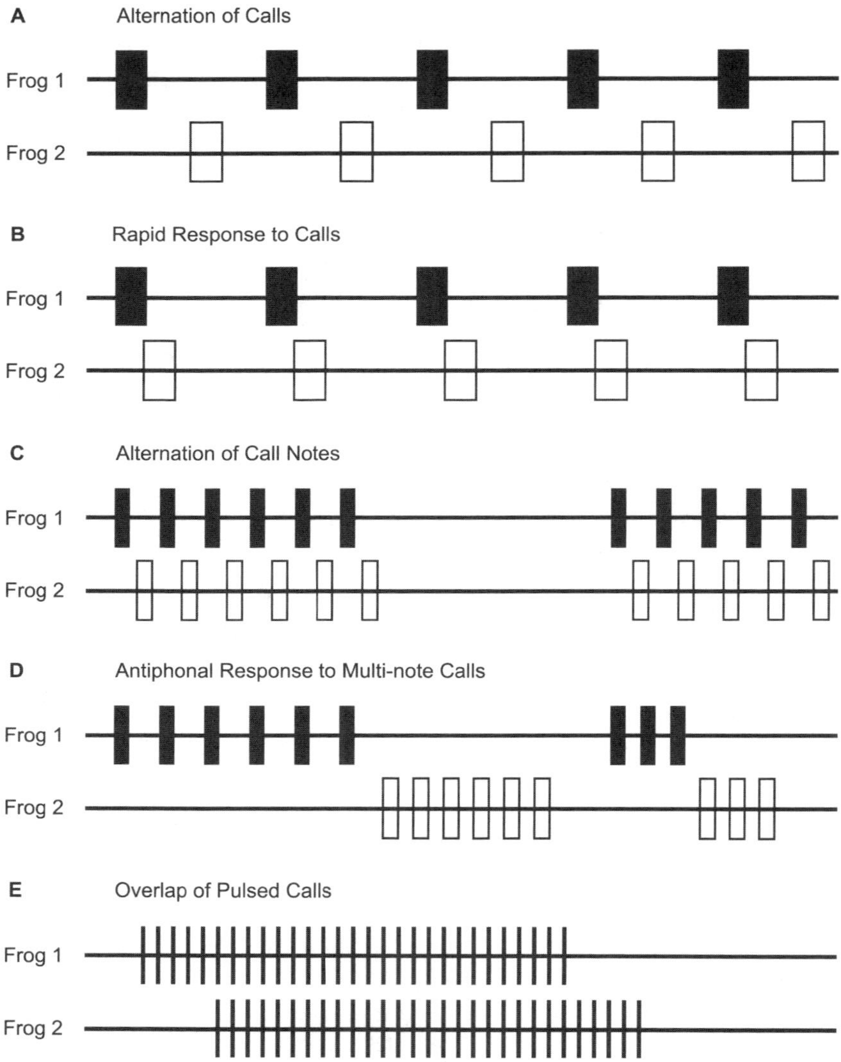

FIGURE 3.7. Diagram of several possible types of vocal interaction between neighboring males in a frog chorus. (A) Calls consist of relatively long notes given at regular intervals, with the calls of the second male precisely alternated with those of the first. (B) Calls consist of relatively long notes given at regular intervals, with the calls of the second male starting immediately after the end of the first male's calls. (C) Calls consist of a variable number of closely spaced short notes, with individual notes of the second male's calls alternating with those of the first male. The result is minimal acoustic interference and relatively precise matching of the number of call notes. (D) Calls consist of a variable number of short notes. Calls of the second male are given immediately after the entire sequence of notes of the first male has ended, with fairly precise matching of the number of call notes. (E) Calls are trills made up of a rapid series of short pulses. Calls are overlapped with no attempt to avoid acoustic interference.

of alternating multinote calls that briefly monopolize the "air-time" for vocal advertisement. Males of *H. ebraccata* exploit silent periods when calling by these species subsides. This is most obvious during interactions with *H. microcephala*, because males of this species chorus in a roughly cyclical pattern in which bouts of vocal activity lasting about 5 to 25 seconds alternate with periods of inactivity of comparable duration.

The roughly on–off or cyclical pattern of chorusing observed in *H. microcephala*, sometimes referred to as unison bout singing (Greenfield and Shaw 1983; Schwartz 1991), also has been reported in other species (Duellman and Trueb 1966; Rosen and Lemon 1974; Whitney and Krebs 1975; Schneider 1977; Zimmerman and Bogart 1984; Ibañez 1991). The cycles last from several seconds to a few minutes, and calling bouts are initiated when the calls of one male stimulate others to join the chorus. Although males of some species of anurans appear oblivious to the calls of others (e.g., the American toad, *Bufo americanus*, and the southern toad, *B. terrestris*; Gerhardt and Huber 2002), male frogs of many species adjust the timing of their calls or call elements relative to the individual calls of conspecifics and heterospecifics. Thus these changes can occur extremely rapidly (e.g., Narins 1982b; Schwartz and Wells 1985; Narins and Zelick 1988; Grafe, 2003) and typically involve either abbreviating or elongating the call period in response to specific calls or call elements of neighbors (Klump and Gerhardt 1992). The outcome of the behavior is often called alternation and males of some species may even interleave notes of multipart calls (or calls of call groups) with those of other males (Schwartz and Wells 1984a; Schwartz and Wells 1985; Grafe 2003; Fig. 3.8). However, alternation between pairs of males may be inconsistent and so result in some acoustic interference (e.g., Schwartz et al. 2002; Gerhardt and Huber 2002). Moreover, in vocal dyads among heterospecifics, there may be species asymmetries in responsiveness (e.g., between the two species of gray treefrogs, *H. versicolor* and *H. chrysoscelis*, Marshall 2004).

Hyla microcephala males provide an excellent example of how selection has acted at different levels to shape call timing in a noisy assemblage of calling males. Males produce multinote calls and are stimulated to call and add notes to their calls in response to vocalizations. Chorusing by even a small number of males can be quite noisy, and the calls of males frequently overlap, but when overlap occurs, the constituent notes of neighboring males usually do not. Rather, notes of each interacting male are timed so as to fall within the internote intervals of the other male (Schwartz and Wells 1985; Schwartz 1993). The resulting pattern of note alternation is facilitated by mutual inhibition of note production by each note of the neighbor (Fig. 3.8B). Accordingly, during call overlap between two males, each male will lengthen an inter-note interval when the note of the other male falls with the interval. Conversely, during call overlap, the drop in sound intensity accompanying the end of each interrupting note triggers a male to produce his next note. The ability to rapidly interleave notes is also present in *Hyla phlebodes*, although this occurs without concomitant changes in internote intervals (Schwartz and Wells 1984b; Fig. 3.8A). Precise note alternation may be

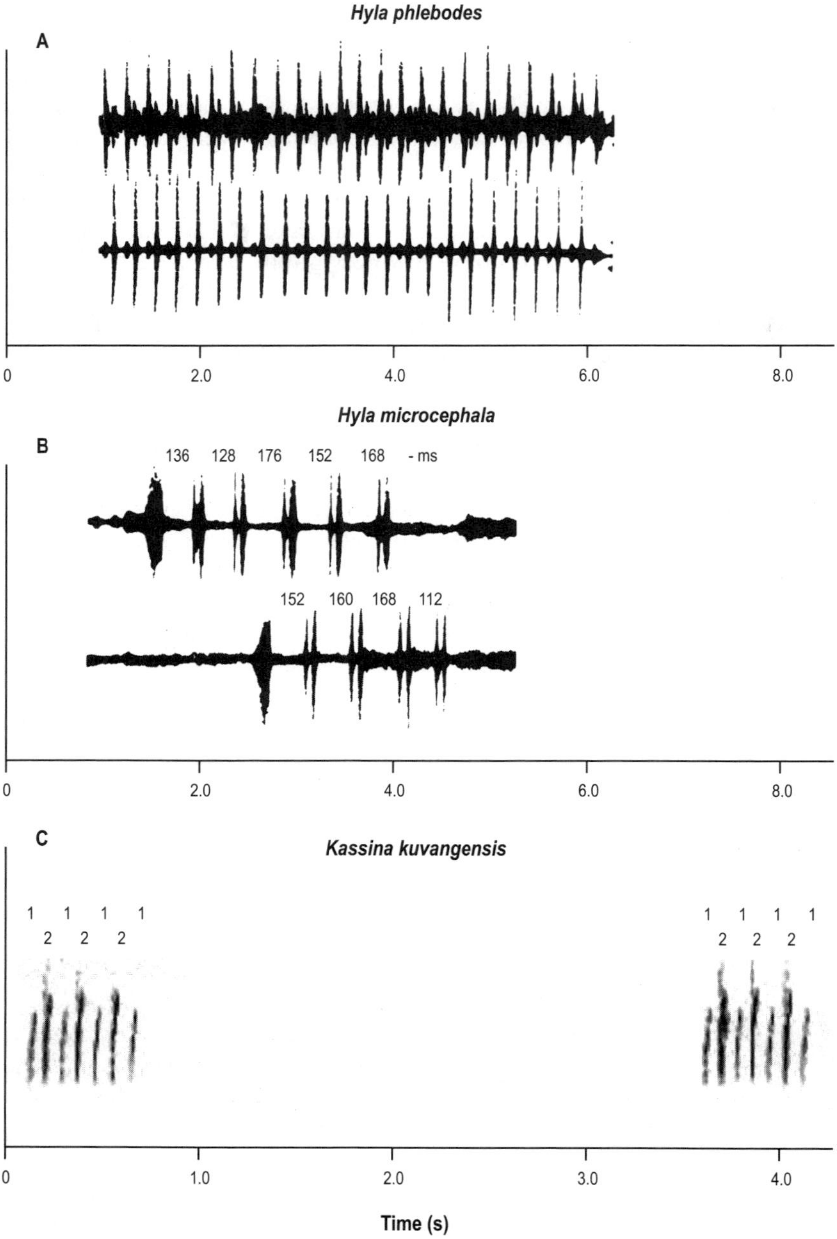

FIGURE 3.8. Two males of (a) *Hyla phlebodes* alternating notes within multinote calls (recordings by Kentwood D. Wells and Joshua J. Schwartz). (b) Same for *Hyla microcephala*, showing an increase in internote intervals (in ms) in overlapped portions of the calls (modified from Schwartz and Wells 1985; reproduced by permission of the American Society of Ichthyologists and Herpetologists). (c) Pair of *Kassina kuvangensis* males alternating calls within "call groups" (modified from Grafe 2005, Fig. 13.1, p. 281; reprinted with the permission of Cambridge University Press). Males of *H. microcephala* and *K. kuvangensis* also alter the spacing between their notes or calls, respectively, in response to the signal elements of alternating competitors.

difficult to achieve, as there are males of other species that produce multinote calls that fail to exhibit note alternation (reviewed in Grafe 2003).

Obviously, in a chorus with many males, pairwise note-by-note timing would not be an effective means to reduce acoustic interference. The solution adopted by *H. microcephala* is to adjust note timing with respect to only a male's loudest (and nearest) one or two neighbors in the chorus while ignoring (for note-timing purposes) the notes of more distant individuals (Schwartz 1993). This behavior has been referred to as "selective attention" and was first reported in frogs by Brush and Narins (1989) in their study of *Eleutherodactylus coqui*. Greenfield and Rand (2000) described similar behavior in *Physalaemus pustulosus* and further demonstrated that the "rules" frogs use to delimit their zone of selective attention are flexible enough to accommodate the dynamic nature of frog choruses. A combination of chorus monitoring and playback tests with interrupting stimuli indicated that such flexibility also characterizes selective attention in *Hyla microcephala*. Nevertheless, additional work on the relative importance of spatial and intensity cues are clearly needed.

The gray treefrog (*H. versicolor*) does not exhibit a comparable pattern of selective attention. In pairwise interactions, males significantly reduced call overlap, but this was not so in groups of three to eight males (Schwartz et al. 2001). Moreover, adjacent males overlapped calls more than did more widely separated individuals. It is possible that these findings were an artifact of the testing environment: an artificial pond with males equally spaced around the pond perimeter. With the additional spatial cues and more pronounced intensity differences present in a natural chorus, male behavior might be similar to that of the aforementioned species. Another possible explanation is that males of *H. versicolor* are not as severely penalized when calls overlap as are some other species. Schwartz and Gerhardt (1995) found that spatial separation of interfering call sources mitigated the effects of acoustic interference. This was not the case with the smaller species *H. microcephala* (Schwartz 1993), however. As discussed above, an intriguing possibility is that males rely on changes in call duration and rate, rather than selective attention, to compensate for the increased risk of call overlap in dense choruses.

Some species of frogs appear to time their calls so that they are more likely to overlap than to alternate (e.g., Ryan 1986; Ibañez 1993; Grafe 1999). At a proximate level, such (approximate) synchrony on a fine-scale may occur via call-period changes induced by a neighbor's call that falls in a certain time-window after the subject's call (Gerhardt and Huber 2002; Greenfield 2002). Alternatively, signal detection may trigger a short-latency vocal response that may or may not occur before the stimulating call has ended (Fig. 3.9A). For example, calling by males of the brown running frog (*Kassina fusca*) from West Africa is triggered by the onset of conspecific calls, leading to overlap (Fig. 3.9B), and the offset of some heterospecific calls, leading to alternation (Grafe 1999). In *Hyla ebraccata*, signals with a rapid rise time are especially effective in eliciting short-latency vocal responses (Schwartz and Wells 1984a). In *H. microcephala*, such sound-induced stimulation evidently occurs in conjunction with sound-induced

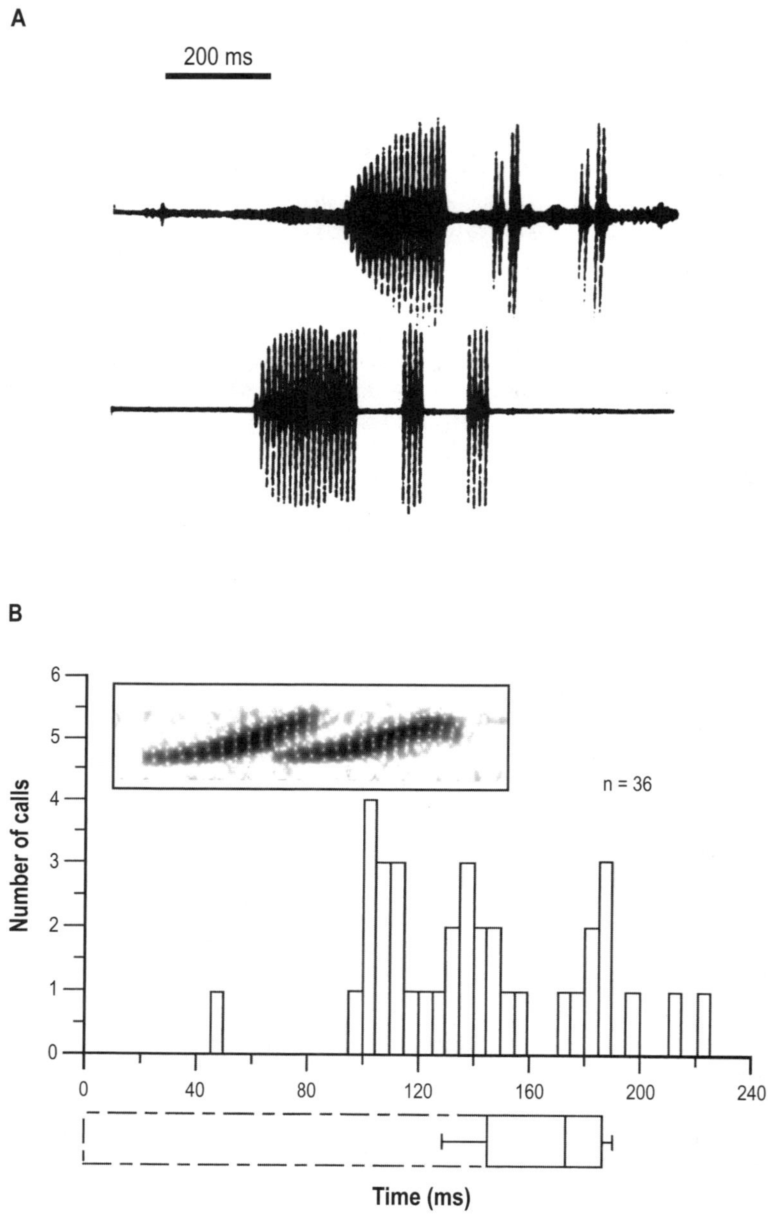

FIGURE 3.9. (A) Call timing between two males of *Hyla ebraccata*, resulting in the primary note of the following male overlapping a secondary note of the leading male. Recordings by Kentwood D. Wells and Joshua J. Schwartz. (B) Call overlap between two males of *Kassina fusca*. The histogram shows the distribution of call latencies of one male to the calls of the other male. The dashed box gives the duration of male calls and encloses a box plot giving the median, interquartile range, and 10th and 90th percentiles of response call latencies. Modified from Grafe 1999, Fig. 1, p. 2333; reprinted with the permission of the Royal Society of London.

inhibition. As mentioned previously, note production also is stimulated by a drop in sound intensity following the end of a neighbor's note (Schwartz 1993; also see Zelick and Narins 1983 for another report of this phenomenon). Thus interacting males overlap calls but rapidly alternate the notes within their calls. Elucidating the neural mechanisms controlling this behavior will be both challenging and fascinating.

4.5 Ultimate Explanations for Call-Timing Adjustments

The relationships between the timing of male vocalizations within anuran choruses can have a profound impact on mating success that is comparable to those associated with call structure and rate (Schwartz 1987b; Klump and Gerhardt 1992; Grafe 1999; Schwartz et al. 2001). Background noise levels within choruses often can be very high, and the problem of call overlap and masking interference can be ameliorated if males adjust the timing of their calls relative to those of other individuals. As described above, the time scale of these adjustments may be flexible and relate to the nature of the source of acoustic interference. For example, males of *Hyla ebraccata* may adjust the fine-scale timing of their calls in a way that reduces overlap with the individual calls of males of *H. microcephala* or alternate with groups of chorusing *H. microcephala* on a coarse scale. Phonotaxis experiments showed that males of *H. ebraccata* improved their chances of attracting females by avoiding call overlap with neighboring males (Schwartz and Wells 1984a) and by concentrating calling during quiet periods (Schwartz and Wells 1983b).

Broadcasts of chorus noise did not support the hypothesis that males of *H. microcephala* periodically quiet down during unison bout singing because of an increased threat of masking and acoustic interference (Schwartz 1991). Analysis of muscle glycogen reserves and calling rates suggested that males periodically stop calling to save energy and increase total calling time (Schwartz et al. 1995). Additional factors also may be relevant. For example, cyclical patterns of activity may emerge as a result of the intrinsic auditory sensitivities and response properties of individuals when grouped, but have no functional basis per se (Schwartz 2001). An intriguing possibility is that cyclical calling reduces individual risk of predation, but this hypothesis has yet to be tested.

Both competition and cooperation can occur simultaneously in choruses, and these interactions have been invoked to explain both call synchrony and call alternation (Greenfield 2002; Gerhardt and Huber 2002; Grafe 2005). Males reduce the chances that their signals overlap by alternating calls, part of a general strategy to exploit brief periods of relative quiet (Grafe 2003). On the other hand, males may synchronize calls because the resulting overlap amplifies their signals. This form of cooperation could be advantageous for individuals that call in areas with chronic high background noise, such as streams (Marshall and Gerhardt, unpublished data on canyon treefrogs, *H. arenicolor*). Whether males of some anuran species gain a per capita mating advantage by elevating the signal amplitude in this manner, or by concentrating calls in time, is unknown, but deserves further study. However, the risk of degrading important fine-temporal information (Schwartz 1987b) within the calls might outweigh any advantages of improved detectability. Reduced risk of predation, perhaps when coupled with cyclical or unpredictable bouts of chorusing, may benefit synchronizing callers. Tuttle and Ryan (1982) presented evidence consistent with this hypothesis in their study of the Panama cross-banded treefrog (*Smilisca sila*) and the frog-eating bat *Trachops cirrhosus*.

Rather than being a manifestation of male–male cooperation, synchrony sometimes may result from male–male competition for females (Greenfield 2002). In *Hyla ebraccata*, rapid male vocal responses result in masking of shorter secondary notes of a leading conspecific male with the longer primary note of a following male (Fig. 3.9A). Tests with females demonstrated that following males are favored under such circumstances (Wells and Schwartz 1984a). In *Kassina fusca*, females also discriminate in favor of overlapping follower calls under some relative call timing arrangements but leader calls under others (Fig. 3.9B), a result that may be due to backward masking or a precedence effect, respectively (Grafe 1999). During interactions and playback tests, Grafe found that responding males timed their overlapping calls to fall in a time window preferred by females.

Greenfield (2002, 2005) proposed that both alternation and synchrony result from a neural process that resets a male's call-timing following perception of another male's call. This call-timing change can increase the likelihood that a male's calls will occupy a leading position relative to those of his neighbor's. Males that are successful in this regard may gain a mating advantage because of an inherent response property of the auditory system of many species known as the precedence effect (for reviews see Zurek 1987, Litovskya and Colburn 1999). In fact, computer modeling has demonstrated that "inhibitory-resetting" of call-timing and also selective attention may be favored by selection when female mate choice is biased by a precedence effect (Greenfield, 2005). Although the term "precedence effect" has been applied when there is a preference for a leading call, in auditory psychophysics the application of the term is more restricted. Under appropriate conditions of signal duration and timing, lagging sounds will be localized at the source of a leading sound. If this phenomenon occurs in female frogs, the advantage to a leading male is obvious. Unfortunately, data are not yet available to conclusively demonstrate a precedence effect in the restrictive sense in anurans. Whatever the mechanistic explanation (e.g., precedence effect, forward masking; Grafe 1996; Gerhardt and Huber 2002), there is growing evidence for preferences by females for leading calls (Dyson and Passmore 1988a,b; Klump and Gerhardt 1992; Grafe 1996; Greenfield et al. 1997; Bosch and Marquez 2002; Marshall 2004; Schwartz unpublished data), although species are known in which females show a follower or no order preference (Wells and Schwartz 1984a; Ibañez 1993; Bosch and Marquez 2001; Gerhardt and Huber 2002; Grafe 2003). In fact, in some cases, leader preferences may be sufficiently strong to counteract or reverse other preferences. For example, in the spring peeper (*Pseudacris crucifer*) females show a leader preference that can tolerate a 6 to 9 dB relative intensity disadvantage (Schwartz unpublished). If calls do not overlap, a preference is absent. Thus, the call-timing relationships that might confer an advantage to a male are quite circumscribed in this species. Moreover, whether some males are sufficiently consistent as leaders to achieve an advantage and how putative female call-timing preferences respond to increasing levels of acoustic complexity within natural choruses are currently unknown. In the gray treefrog, *Hyla versicolor*, Marshall (2004) demonstrated that females prefer leading calls only when calls, and their component pulses, overlap. Under these circumstances, the preference is so strong that it can reverse the female aversion to calls of *H. chrysoscelis* (Marshall et al., in press). Thus in mixed-species choruses of gray treefrogs, call overlap and a leader preference could result in mismatings and loss of fitness. In Fischer's dwarf frog (*Physalaemus fischeri* [*enesefae*]), call order can reverse the bias of females favoring calls with lower dominant frequencies (Tárano and Herrera 2003).

Schwartz (1987b) and Schwartz and Rand (1991) tested three hypotheses, using four species, for why males alternate calls. Hypothesis 1 proposed that alternation allows interacting males to more easily hear one another. This could be advantageous because (a) call intensity cues are used to mediate intermale spacing and/or (b) call detection enables males to adjust their signal attractiveness to match or exceed that of competitors. Interactive playback experiments (Schwartz 2001) supported Hypothesis 1. Hypothesis 2 proposed that alternation helps preserve the fine temporal structure within calls that might otherwise be obscured or degraded by call overlap among males. This hypothesis also was supported. Females of *Hyla versicolor* and *H. microcephala*, species with pulsatile calls or call notes, respectively, discriminated in favor of alternating relative to out-of-phase overlapping calls in four-speaker choice tests. *Pseudacris crucifer* and *Physalaemus pustulosus* females failed to discriminate between calls in the same circumstances. Both of these species lack calls consisting of pulses. *Physalaemus pustulosus* has a frequency-modulated introductory "whine" in its call that contributes to call recognition by females (Rose et al. 1988; Wilczynski et al. 1995). Schwartz and Rand (1991) speculated that the spectral filtering characteristics of the auditory system enable females to sufficiently discern the downward frequency sweep of the whine, even when calls partially overlap. Hypothesis 3 proposed that alternation facilitates the localization of call sources. If this were the case, females of all species should have discriminated against overlapped calls when these were presented precisely in phase. This did not occur. Results from some other studies are also inconsistent with Hypothesis 3 (Passmore and Telford 1981; Backwell and Passmore 1991; Grafe 1996; Marquez and Bosch 2001). Nevertheless, certain call-timing relationships (e.g., overlapped calls with leading versus following pulses, Marshall 2004) may have an impact on localization in a way that was not detected using the stimulus arrangements in the aforementioned experiments.

In addition to the advantages described above for Hypothesis 1, call-timing shifts may have an additional role during male–male interactions. Based on field observations and results of playbacks to calling *Polypedates leucomystax*, Christensen-Dalsgaard et al. (2002) recently proposed that short-latency responses are a way for males to direct their calls to a particular individual.

5. Auditory System Features: Contributions to Communication in Choruses

Features of the auditory system of anurans may facilitate the task of detecting, discriminating, and localizing relevant communication signals within the often noisy and spatially complex "real-world" acoustic environment within choruses (Feng and Ratnam 2000). Most of these characteristics are not qualitatively unique to members of this taxon, although evolution may have fine-tuned the relevant attributes in ways that improve their effectiveness under biologically relevant circumstances.

5.1 Signal Detection and Discrimination

The tuning of the peripheral auditory system of frogs tends to be well (but not perfectly) matched to the dominant frequency of the species-specific advertisement call and often more complex spectral patterns of call energy distribution (Gerhardt and Schwartz 2001). In fact, the role of the anuran auditory system as a matched filter that can improve the detection of biologically relevant signals in the presence of background noise has long been appreciated (Capranica and Rose 1983). This filtering potential is reflected not only in audiograms (obtained at threshold) but also in critical ratios (e.g., Narins 1982a; Moss and Simmons 1986; Simmons 1988). Certain characteristics of the acoustic milieu of choruses may also be exploited by central neuronal processes and so facilitate call detection. For example, the amplitude envelope of natural background noise can be dramatically modulated with this temporal structure correlated across sound frequencies (Nelkan et al. 1999). In some taxa (e.g., Klump and Langemann 1995), tone detection thresholds are reduced when embedded in noise with such structure as compared to detection thresholds in noise lacking modulations. The actual contribution of this "comodulation masking release (CMR)" to communication of frogs is currently poorly understood (Goense and Feng 2003), but could be significant in situations with considerable abiotic noise or in multispecies assemblages. However, for most chorusing species the most potent source of background noise is that produced by conspecifics rather than hetersopecifics with call spectra different from their own. Thus solutions other than matched filtering or CMR must play a part in reducing the potentially serious problems for males and female anurans imposed by masking and call overlap. If males cannot detect the individual calls of neighboring males they may not be able to accurately assess the nature and intensity of competition in their vicinity and may fail to adjust their spacing appropriately. Masking of conspecific calls may impede a female's ability to not only find a mate, but assess relative performance within a group of males and so possibly more effectively improve her fitness (e.g., Welch et al. 1998).

In spite of their relatively small interaural distances (e.g., <2.5 cm, Gerhardt and Huber 2002, p. 230), available data indicate that some anurans are able to exploit directional cues to extract signals from the background din of a chorus or calls of overlapping males. In *Hyla cinerea*, separation of speakers broadcasting calls from those broadcasting noise facilitated both detection of advertisement calls and discrimination of advertisement calls from aggressive calls by females during phonotaxis experiments (Schwartz and Gerhardt 1989). Schwartz and Gerhardt (1995) also found that separation of speakers (by 120 degrees) broadcasting overlapping calls of *Hyla versicolor* elicited discrimination in their favor relative to speakers that were not separated. The timing of the overlapped calls was such that call interference rendered the resulting pulse pattern unattractive (Schwartz 1987b). At each ear, separation of call sources may reduce the strength of the auditory input contributed by one of the overlapping calls and so facilitate encoding of an effective pulse pattern (Schwartz and Gerhardt 1995). Interestingly, an earlier experiment with *H. microcephala* failed to reveal such discrimination when speakers were separated by 120 degrees, perhaps because the interaural separation of females in this species is less than half that of *H. versicolor* (Schwartz 1993). Discrepancies between the note-timing behavior of males during natural interactions as compared to those in response to overlapping notes broadcast from a single speaker suggest that angular separation of callers may contribute to the ability of *H. microcephala* males to selectively time their call notes with respect to a subset of chorus members (Schwartz 1993).

The data of Schwartz and Gerhardt (1989) on green treefrogs (*Hyla cinerea*) are consistent with the notion that signal discrimination is a more difficult task than signal detection. However, under some circumstances low to moderate noise levels within choruses may actually enhance the ability of females to discriminate among males. Schwartz and Gerhardt (1998) found that females of *Pseudacris crucifer* preferred synthetic advertisement calls of 3500 Hz to those of 2600 Hz only in the presence of background noise (filtered to resemble that produced by a natural chorus). Multiunit recordings from the auditory midbrain of females suggested a likely explanation. The noise induced a desensitization of the auditory system (this phenomenon is quite familiar to anyone who has been to a loud rock concert) that, in turn, increased the stimulus level at which auditory neurons would reach saturation in their firing rate. This threshold shift was reflected not only in right-shifts in plots of neural activity versus stimulus amplitude but also in plots of neural activity versus stimulus frequency. Relatively flat isointensity response plots obtained at high call intensities became peaked after exposure to noise and so resembled those obtained at low to moderate call intensities. Perhaps most significantly, only in the presence of noise was there a significant relationship between the frequency eliciting the maximum multiunit neural activity and the frequency preference of individual females.

Schwartz and Gerhardt (1998) speculated that noise-induced threshold shifts may have their greatest impact on discrimination not for females within aggregations of conspecifics but for females attempting to discriminate males of their own species within mixed-species assemblages (where spectral differences would be more profound than between conspecifics). The peripheral auditory system of anurans is particularly adept at detecting the amplitude–time envelope of sounds and thus, through phase-locking, neurons of the eighth nerve encode modulations in amplitude (e.g., waveform periodicity, pulses) present in the calls of many species (Feng and Shellart 1999; Gerhardt and Huber 2002). High levels of background noise impair this process (Simmons et al. 1992). In mammals low-intensity noise may improve phase-locking (Rhode et al. 1978; Lewis and Henry 1995), but whether noise might augment signal detection or discrimination in anurans through such a mechanism is largely unknown (but see Narins et al. 1997).

5.2 Signal Restoration

In humans (Warren 1970; Samuel 1981) and starlings (Braaten and Leary 1999), the brain can fill in signal elements that are missing or inaudible due to the presence of masking noise. This process, known as phonemic restoration or temporal induction, provides an illusion of signal continuity and could potentially be useful to anurans within loud choruses. For example, a female *Hyla versicolor* might more rapidly and effectively compare the call duration of adjacent males if she could interpolate between inaudible or obscured sections of calls that might be overlapped by the calls of other males or the background noise of the assemblage. This hypothesis was recently tested using phonotaxis tests using calls containing silent gaps, portions masked by filtered noise, or interrupted by overlapping calls (Schwartz et al. 2004). Results failed to support the presence of a significant restorative process. Females did not "fill in" missing information when large gaps were present, although obscuring pulses within a call with other signals appeared less detrimental than removing pulses.

6. Summary

Anuran amphibians are unique among ectothermic tetrapods in the degree to which they depend on acoustic communication to attract mates, advertise territory ownership, or otherwise communicate with conspecifics. The sound production mechanism of most frogs also is unique in that the trunk muscles involved in forcing air out of the lungs and through the vocal chords are not used for normal respiration. Hence, the hypertrophied muscles of male frogs can be considered a sexually selected trait, driven by competition among males for access to females. Anurans with very high calling rates have highly aerobic muscles with high mitochondrial and capillary densities and often large reserves of lipids that are not present in leg muscles. These morphological and physiological traits support the high metabolic demands of calling. Our knowledge of the physiological, biochemical, and morphological basis of call production is limited, however, to a relatively small sample of anuran species and is not necessarily representative of all anurans. There also has been relatively little research on the structure and function of vocal sacs and how variation in vocal sac structure relates to differences in calling behavior, or the use of vocal sacs as visual signals that complement or amplify acoustic signals.

Many anurans produce several types of calls with different functions, although some have relatively simply vocal repertoires. Most anurans have advertisement calls given spontaneously by males to advertise their species identity, sexual receptivity, and spatial location to females and to other males. Males of many species also have distinctive aggressive calls, which sometimes are graded in a way that allows males to modify the intensity of their aggressive message or trade off female-attracting and male-repelling functions of their calls. Males of some species modify advertisement calls during close-range courtship interactions with females, and in some species, females respond with calls of their own. Both male and female courtship calls are poorly studied and probably are much more common than currently recognized.

Although a frog chorus often seems to be a disorganized cacophony, closer examination often reveals complex networks of interactions among males in a chorus. Males of many species probably attend to the calls of only a few near neighbors, with the remaining males simply contributing to background noise. Males respond to calls of near neighbors in ways that enable them to minimize acoustic interference and maximize the signal-to-noise ratio of their calls. Male frogs become habituated to particular levels of calling activity, so the acoustic threshold for responding to other calling males changes with chorus density. Males of some species that maintain long-term territories are able to recognize neighbors individually and respond more aggressively toward intruding strangers than toward familiar neighbors. Many of the acoustic interactions in choruses can be seen as products of intense sexual selection, with males competing to outsignal their competitors for the attention of females, often increasing rates of signaling as chorus density increases.

Characteristics of the anuran auditory system facilitate the detection and discrimination of biologically relevant signals. The peripheral auditory system of both males and females is selectively tuned to the frequencies of conspecific calls, allowing the frogs to filter out heterospecific calls broadcast on other frequency bands. Upward shifts in auditory response thresholds and directional cues aid communication by some species within noisy chorus environments. Although background noise can interfere with call discrimination, males of some species exhibit enhanced signal discrimination in the presence of low to moderate levels of background noise. This may be particularly important in mixed-species choruses in which heterospecific calls elevate levels of background noise, but are mostly broadcast on different frequency bands from conspecific calls.

References

Andersson M (1994) Sexual Selection. Princeton, NJ: Princeton University Press.

Arak A (1983a) Vocal interactions, call matching and territoriality in a Sri Lankan treefrog, *Philautus leucorhinus* (Rhacophoridae). Anim Behav 31:292–302.

Arak A (1983b) Sexual selection by male-male competition in natterjack toad choruses. Nature 306:261–262.

Backwell PRY (1988) Functional partitioning in the two-part call of the leaf-folding frog, *Afrixalus brachycnemis*. Herpetologica 44:1–7.

Backwell PRY, Passmore NI (1991) Sonic complexity and mate localization in the leaf-folding frog, *Afrixalus delicates*. Herpetologica 47:226–229.

Bastos RP, Haddad CFB (2002) Acoustic and aggressive interactions in *Scinax rizibilis* (Anura: Hylidae) during the reproductive activity in southeastern Brazil. Amphibia-Reptilia 23:97–104.

Bee MA (2002) Territorial male bullfrogs (*Rana catesbeiana*) do not assess fighting ability based on size-related variation in acoustic signals. Behav Ecol 13:109–124.

Bee MA (2004) Within-individual variation in bullfrog vocalizations: Implications for a vocally mediated social recognition system. J Acoust Soc Am 116:3770–3781.

Bee MA, Bowling AC (2002) Socially mediated pitch alteration by territorial male bullfrogs, *Rana catesbeiana*. J Herpetol 36:140–143.

Bee MA, Gerhardt HC (2001) Neighbour–stranger discrimination by territorial male bullfrogs (*Rana catesbeiana*): II. Perceptual basis. Anim Behav 62:1141–1150.

Bee MA, Perrill SA (1996) Responses to conspecific advertisement calls in the green frog (*Rana clamitans*) and their role in male–male communication. Behaviour 133:283–301.

Bee MA, Perrill SA, Owen PC (1999) Size assessment in simulated territorial encounters between male green frogs (*Rana clamitans*). Behav Ecol Sociobiol 45:177–184.

Bee MA, Perrill SA, Owen PC (2000) Male green frogs lower the pitch of acoustic signals in defense of territories: A possible dishonest signal of size? Behav Ecol 11:168–177.

Benedix JH Jr, Narins PM (1999) Competitive calling behavior by male treefrogs, *Eleutherodactylus coqui* (Anura: Leptodactylidae). Copeia 1999:1118–1122.

Bevier CR (1997a) Breeding activity and chorus tenure of two Neotropical hylid frogs. Herpetologica 53:297–311.

Bevier CR (1997b) Utilization of energy substrates during calling activity in tropical frogs. Behav Ecol Sociobiol 41:343–352.

Boatright-Horowitz SL, Horowitz SS, Simmons AM (2000) Patterns of vocal interactions in a bullfrog (*Rana catesbeiana*) chorus: Preferential responding to far neighbors. Ethology 106:701–712.

Boistel R, Sueur J (1997) Comportement sonore de la femelle de *Platymantis vitiensis* (Amphibia, Anura) en l'absence du mâle. C R Acad Sci Paris, Sci Vie 320:933–941.

Bosch J (2001) Female reciprocal calling in the Iberian midwife toad (*Alytes cisternasii*) varies with male call rate and dominant frequency: Implications for sexual selection. Naturwissenschaften 88:434–437.

Bosch J (2002) The function of female reciprocal calls in the Iberian midwife toad (*Alytes cisternasii*): Female–female acoustic competition? Naturwissenschaften 89:575–578.

Bosch J, Márquez R (2001) Female courtship call of the Iberian midwife toad (*Alytes cisternasii*). J Herpetol 35:647–652.

Bosch J, Márquez R (2002) Female preference function related to precedence effect in an amphibian anuran (*Alytes cisternasii*): tests with non-overlapping calls. Behav Ecol 13:149–153.

Bourne GR (1997) Reproductive behavior of terrestrial breeding frogs *Eleutherodactylus johnstonei* in Guyana. J Herpetol 31:221–229.

Bourne GR, Collins AC, Holder AM, McCarthy CL (2001) Vocal communication and reproductive behavior of the frog *Colostethus beebei* in Guyana. J Herp 35:272–281.

Boyd SK, Wissing KD, Heinsz JE, Prins GS (1999) Androgen receptors and sexual dimorphisms in the larynx of the bullfrog. Gen Comp Endocrinol 113:59–68.

Braaten RF, Leary JC (1999) Temporal induction of missing birdsong segments in European starlings. Psychol Sci 10:162–166.

Bradbury JH, Verhencamp SL (1998) Principles of Animal Communication. Sunderland, MA: Sinauer.

Brenowitz EA (1989) Neighbor call amplitude influences aggressive behavior and intermale spacing in choruses of the Pacific treefrog (*Hyla regilla*). Ethology 83:69–79.

Brenowitz EA, Rose GJ (1994) Behavioural plasticity mediates aggression in choruses of the Pacific treefrog. Anim Behav 47:633–641.

Brenowitz EA, Rose GJ (1999) Female choice and plasticity of male calling behaviour in the Pacific treefrog. Anim Behav 57:1337–1342.

Brenowitz EA, Wilczynski W, Zakon HH (1984) Acoustic communication in spring peepers: environmental and behavioral aspects. J Comp Physiol A 155:585–592.

Brooke PN, Alford RA, Schwarzkopf L (2000) Environmental and social factors influence chorusing behaviour in a tropical frog: examining various temporal and spatial scales. Behav Ecol Sociobiol 49:79–87.

Brush JS, Narins PM (1989) Chorus dynamics of a Neotropical amphibian assemblage: comparison of computer simulation and natural behaviour. Anim Behav 37:33–44.

Burmeister S, Wilczynski W, Ryan MJ (1999) Temporal call changes and prior experience affect graded signalling in the cricket frog. Anim Behav 57:611–618.

Burmeister S, Ophir AG, Ryan MJ, Wilczynski W (2002) Information transfer during cricket frog contests. Anim Behav 64:715–725.

Bush SL (1997) Vocal behavior of males and females in the Majorcan midwife toad. J Herpetol 31:251–257.

Bush SL, Dyson ML, Halliday TR (1996) Selective phonotaxis by males in the Majorcan midwife toad. Proc R Soc Lond B 263:913–917.

Capranica RR, Rose G (1983) Frequency and temporal processing in the auditory system of anurans. In: Huber F, Markl H (eds) Neuroethology and Behavioral Physiology. Berlin: Springer-Verlag, pp. 136–152.

Christensen-Dalsgaard J, Ludwig TA, Narins PM (2002) Call diversity in an Old World treefrog: level dependence and latency of acoustic responses. Bioacoustics 13:21–35.

Danchin E, Giraldeau L, Valone TJ, Wagner RH (2004). Public information: From nosy neighbors to cultural evolution. Science 305:487–491.

Davies NB, Halliday TR (1978) Deep croaks and fighting assessment in toads, *Bufo bufo*. Nature 274:683–685.

Davis MS (1987) Acoustically mediated neighbor recognition in the North American bullfrog, *Rana catesbeiana*. Behav Ecol Sociobiol 21:185–190.

Dyson ML, Passmore NI (1988a) Two-choice phonotaxis in *Hyperolius marmoratus*: The effect of temporal variation in presented stimuli. Anim Behav 36:648–652.

Dyson ML, Passmore NI (1988b) The combined effect of intensity and the temporal relationship of stimuli on the phonotactic responses of female painted reed frogs (*Hyperolius marmoratus*). Anim Behav 36:1555–1556.

Duellman WE, Trueb L (1966) Neotropical hylid frogs, genus *Smilisca*. Univ Kansas Publ Mus Nat Hist 17:281–375.

Emerson SB (1992) Courtship and nest-building behavior of a Bornean frog, *Rana blythi*. Copeia 1992:1123–1127.

Emerson SB, Boyd SK (1999) Mating vocalizations of female frogs: Control and evolutionary mechanisms. Brain Behav Evol 53:187–197.

Farris EH, Rand AS, Ryan MJ (2002) The effects of spatially separated call components on phonotaxis in Túngara frogs: Evidence for auditory grouping. Brain Behav Evol 60: 181–188.

Feng AS, Ratnam R (2000) Neural basis of hearing in real-world situations. Ann Rev Psychol 51:699–725.

Feng AS, Shellart NAM (1999) Central auditory processing in fish and amphibians. In: Fay RR, Popper AN (eds) Comparative Hearing: Fish and Amphibians. New York: Springer-Verlag, pp. 218–268.

Feng AS, Narins PM, Xu C-H (2002) Vocal acrobatics in a Chinese frog, *Amolops tormotus*. Naturwissenschaften 89:352–356.

Gerhardt HC (1982) Sound pattern recognition in some North American treefrogs (Anura: Hylidae): Implications for mate choice. Am Zool 22:581–595.

Gerhardt HC (1998) Acoustic signals of animals: Field measurements, recording, analysis and description. In: Hopp S, Owren M, Evans C (eds) Techniques in the Study of Animal Communication by Sound. New York: Springer-Verlag, pp. 1–25.

Gerhardt HC (2001) Acoustic communication in two groups of closely related treefrogs. Adv Study Behav 30:99–167.

Gerhardt HC, Huber F (2002) Acoustic Communication in Insects and Frogs: Common Problems and Diverse Solutions. Chicago: University of Chicago Press.

Gerhardt HC, Klump GM (1988) Masking of acoustic signals by the chorus background noise in the green treefrog: A limitation on mate choice. Anim Behav 36:1247–1249.

Gerhardt HC, Schwartz JJ (2001) Auditory and frequency preferences in anurans. In: Ryan MJ (ed) Anuran Communication. Washington DC: Smithsonian Institution Press, pp. 73–85.

Gerhardt HC, Diekamp B, Ptacek M (1989) Inter-male spacing in choruses of the spring peeper, *Pseudacris* (*Hyla*) *crucifer*. Anim Behav 38:1012–1024.

Gerhardt HC, Ptacek MB, Barnett L, Torke K (1994) Hybridization in the diploid-tetraploid treefrogs, *Hyla chrysoscelis* and *Hyla versicolor*. Copeia 1994:51–59.

Gerhardt HC, Roberts JD, Bee MA, Schwartz JJ (2000a) Call matching in the quacking frog (*Crinia georgiana*). Behav Ecol Sociobiol 48:243–251.

Gerhardt HC, Tanner SD, Corrigan CM, Walton HC (2000b) Female preference functions based on call duration in the gray tree frog (*Hyla versicolor*). Behav Ecol 11:663–669.

Girgenrath M, Marsh RL (1997) In vivo performance of trunk muscles in tree frogs during calling. J Exp Biol 200:3101–3108.

Given MF (1987) Vocalizations and acoustic interactions of the carpenter frog, *Rana virgatipes*. Herpetologica 43:467–481.

Given MF (1993a) Male response to female vocalizations in the carpenter frog, *Rana virgatipes*. Anim Behav 46:1139–1149.

Given MF (1993b) Vocal interactions in *Bufo woodhousii fowleri*. J Herp 27:447–452.

Given MF (1999) Frequency alteration of the advertisement call in the carpenter frog, *Rana virgatipes*. Herpetologica 55:304–317.

Goense JBM, Feng AS (2003) Effects of Noise Bandwidth and Modulation on Signal Detection for Single Neurons in the Frog Auditory Midbrain. Abstr 26th Ann Midw Meet Assoc Res in Otolaryngol.

Grafe TU (1995) Graded aggressive calls in the African painted reed frog *Hyperolius marmoratus* (Hyperoliidae). Ethology 101:67–81.
Grafe TU (1996) The function of call alternation in the African reed frog *Hyperolius marmoratus*: precise call timing prevents auditory masking. Behav Ecol Sociobiol 38:149–158.
Grafe TU (1997) Costs and benefits of mate choice in the lek-breeding reed frog, *Hyperolius marmortus*. Anim Behav 53:1103–1117.
Grafe TU (1999) A function of synchronous chorusing and a novel female preference shift in an anuran. Proc R Soc London B 266:2331–2336.
Grafe TU (2003) Synchronised interdigitated calling in the Kuvangu running frog (*Kassina kuvangensis*). Anim Behav 66:127–136.
Grafe TU (2005) Anuran choruses as communication networks. In: McGregor PK (ed) Communication Networks. Cambridge, UK: Cambridge University Press, pp. 277–299.
Grafe TU, Steffan JO, Stoll C (2000) Vocal repertoire and effect of advertisement call intensity on calling behaviour in the West African tree frog, *Leptopelis viridis*. Amphibia-Reptilia 21:13–23.
Greenfield MD (1994). Cooperation and conflict in the evolution of signal interactions. Ann Rev Ecol Syst 25:97–126.
Greenfield MD (2002) Signalers and Receivers: Mechanisms and Evolution of Arthropod Communication. Oxford: Oxford University Press.
Greenfield MD (2005) Mechanisms and evolution of communal sexual displays in arthropods and anurans. Adv Stud Behav 35:1–62.
Greenfield MD, Rand AS (2000) Frogs have rules: Selective attention algorithms regulate chorusing in *Physalaemus pustulosus* (Leptodactylidae). Ethology 106:331–347.
Greenfield MD, Shaw KC (1983) Adaptive significance of chorusing with special reference to the orthoptera. In: Morris GK, Boulder DT (eds) Orthopteran Mating Systems: Sexual Competition in a Diverse Group of Insects. Boulder, CO: Westview, pp. 1–27.
Greenfield MD, Tourtellot MK, Snedden WA (1997) Precedence effects and the evolution of chorusing. Proc Roy Soc London B 264:1355–1361.
Greer BJ, Wells KD (1980) Territorial and reproductive behavior of the tropical American frog *Centrolenella fleischmanni*. Herpetologica 36:318–326.
Halliday TR, Tejedo M (1995) Intrasexual selection and alternative mating behavior. In: Heatwole H, Sullivan BK (eds) Amphibian Biology: Vol 2: Social Behaviour. Chipping Norton, UK: Surrey Beatty, pp. 419–468.
Hödl W, Amezquita A (2001) Visual signaling in anuran amphibians, pp. 121–141. In: Ryan MJ (ed) Anuran Communication. Washington, DC: Smithsonian Institution Press.
Höglund J, Alatalo RV (1995) Leks. Princeton, NJ: Princeton University Press.
Hoskin CJ (2004) Australian microhylid frogs (*Cophixalus* and *Austrochaperina*): phylogeny, taxonomy, calls, distribution and breeding biology. Aust J Zool 52:237–269.
Howard RD, Young JR (1998) Individual variation in male vocal traits and female mating preferences in *Bufo americanus*. Anim Behav 55:1165–1179.
Humfeld SC (2003) Signaling, intersexual dynamics and the adoption of alternative male mating behaviors in green treefrogs, *Hyla cinerea*. Unpubl. PhD Dissertation, Univ. of Missouri, Columbia.
Ibañez R (1991) Synchronized calling in *Centrolenella granulosa* and *Smilisca sila* (Amphibia, Anura). Unpubl. PhD Dissertation, Univ. of Connecticut, Storrs.
Ibañez R (1993) Female phonotaxis and call overlap in the Neotropical glassfrog, *Centrolenella granulosa*. Copeia 1993:846–850.

Jehle R, Arak A (1998) Graded call variation in the Asian cricket frog *Rana nicobariensis*. Bioacoustics 9:35–48.

Johnstone RA (2001) Eavesdropping and animal conflict. Proc Natl Acad Sci USA 98: 9177–9180.

Judge KA, Swanson SJ, Brooks RJ (2000) *Rana catesbeiana* (bullfrog). Female vocalization. Herp Rev 31:236–237.

Kelley DB (1996) Sexuanl differentiation in *Xenopus laevis*. In: Tinsley RC, Kobel HR (eds) The Biology of Xenopus. New York: Oxford University Press, pp. 143–176.

Kime NM (2001) Female Mate Choice for Socially-Variable Advertisement Calls in the Cricket Frog, *Acris crepitans*. Unpublished PhD Dissertation, Univ. of Texas, Austin.

Kime NM, Burmeister SS, Ryan MJ (2004) Female preferences for socially variable call characters in the cricket frog, *Acris crepitans*. Anim Behav 68:1391–1399.

Kluge AG (1981) The life history, social organization, and parental behavior of *Hyla rosenbergi* Boulenger, a nest-building gladiator frog. Misc Publ Mus Zool Univ Michigan 160:1–170.

Klump GM, Gerhardt HC (1987) Use of non-arbitrary acoustic criteria in mate choice by female gray tree frogs. Nature 326:286–288.

Klump GM, Gerhardt HC (1992) Mechanisms and function of call-timing in male–male interactions in frogs. In: McGregor PK (ed) Playback and Studies of Animal Communication. New York: Plenum, pp. 153–174.

Klump GM, Langemann U (1995) Comodulation masking release in a songbird. Hear Res 87:157–164.

Lea J, Dyson M, Halliday T (2001) Calling by male midwife toads stimulates females to maintain reproductive condition. Anim Behav 61:373–377.

Lewis ER, Henry KR (1995) Nonlinear effects of noise on phase-locked cochlear-nerve responses to sinusoidal stimuli. Hear Res 92:1–16.

Litovskya RY, Colburn HS (1999) The precedence effect. J Acoust Soc Am 106: 1633–1654.

Littlejohn MJ (2001) Patterns of differentiation in temporal properties of acoustic signals of anurans. In: Ryan MJ (ed) Anuran Communcation. Washington, DC: Smithsonian Institution Press, pp. 102–120.

Littlejohn MJ, Harrison PA (1985) The functional significance of the diphasic advertisement call of *Geocrinia victoriana* (Anura: Leptodactylidae). Behav Ecol Sociobiol 16: 363–373.

Littlejohn MJ, Martin AA (1969) Acoustic interaction between two species of leptodactylid frogs. Anim Behav 17:785–791.

Lopez PT, Narins PM, Lewis ED, Moore SW (1988) Acoustically induced call modification in the white-lipped frog, *Leptodactylus albilabris*. Anim Behav 36:1295–1308.

Lüddecke H (1999) Behavioral aspects of the reproductive biology of the Andean frog *Colostethus palmatus* (Amphibia: Dendrobatidae). Rev Acad Colombiana Cienc 23 (Suplemento especial):303–316.

Lüddecke, H (2002) Male and female responses to call playback in the Andean frog *Colostethus subpunctatus*. Amphibia-Reptilia 23:141–150.

Márquez R, Bosch J (1997) Female preference in complex acoustical environments in midwife toads *Alytes obstetricans* and *Alytes cisternasii*. Behav Ecol 8:588–594.

Márquez R, Bosch J (2001) Communication and mating in the midwife toads (*Alytes obstetricans* and *Alytes cisternasii*). In: Ryan MJ (ed) Anuran Communication. Washington DC: Smithsonian Institution Press, pp. 220–231.

Marshall VT (2004) Social aspects of communication in gray treefrogs: Intraspecific and interspecific interactions. Unpubl. PhD Dissertation, Univ. of Missouri, Columbia.

Marshall VT, Humfeld SC, Bee MA (2003) Plasticity of aggressive signalling and its evolution in male spring peepers, *Pseudacris crucifer*. Anim Behav 65:1223–1234.

Marshall VT, Schwartz JJ, Gerhardt HC (2006) The effects of heterospecific call overlap on the phonotactic behaviour of grey treefrogs. Anim Behav, in press.

Martin WF (1972) Evolution of vocalization in the genus *Bufo*. In: Blair WF (ed) Evolution in the Genus *Bufo*. Austin: University of Texas Press, pp. 279–309.

McGregor PK, Peake TM (2000) Communication networks: Social environments for receiving and signalling behaviour. Acta Etholog 2:71–81.

McLister JD (2001) Physical factors affecting the cost and efficiency of sound production in the treefrog *Hyla versicolor*. J Exp Biol 203:69–80.

Michael SF (1997) Vocalization and diurnal retreat defense in the Puerto Rican frog *Eleutherodactylus cochranae*. J Herpetol 31:453–456.

Moore SW, Lewis ER, Narins PM, Lopez P (1989) The call-timing algorithm of the white-lipped frog, *Leptodactylus albilabris*. J Comp Physiol 164A:309–319.

Moss CF, Simmons AM (1986) Frequency selectivity of hearing in the green treefrog, *Hyla cinerea*. J Comp Physiol A 159:257–266.

Murphy CG (1994) Chorus tenure of male barking treefrogs, *Hyla gratiosa*. Anim Behav 48:763–777.

Murphy CG (1999) Nightly timing of chorusing by male barking treefrogs (*Hyla gratiosa*): The influence of female arrival and energy. Copeia 1999:333–347.

Murphy CG, Floyd SB (2005) The effect of call intensity on male spacing in choruses of barking treefrogs (*Hyla gratiosa*). Anim Behav 69:419–426.

Narins PM (1982a) Effects of masking noise on evoked calling in the Puerto Rican coqui (Anura: Leptodactylidae). J Comp Physiol 147:439–446.

Narins PM (1982b) Behavioral refractory period in Neotropical treefrogs. J Comp Physiol 148:337–344.

Narins PM, Capranica RR (1976) Sexual differences in the auditory system of the tree frog *Eleutherodactylus coqui*. Science 192:378–380.

Narins PM, Capranica RR (1978) Communicative significance of the two-note call of the treefrog *Eleutherodactylus coqui*. J Comp Physiol 127:1–9.

Narins PM, Zelick R (1988) The effects of noise on auditory processing and behavior in amphibians. In: Fritszch B, Wilczynski W, Ryan MJ, Hetherington T, Walkowiak W (eds) The Evolution of the Amphibian Auditory System. New York: Wiley, pp. 511–536.

Narins PM, Benedix JH, Moss F (1997) Does stochastic resonance play a role in hearing? In: Lewis ER, Lyon R, Long GR, Narins PM, Steele CR (eds) Diversity in Auditory Mechanics. Singapore: World Scientific, pp. 83–90.

Narins PM, Lewis ER, McClelland BE (2000) Hyperextended call note repertoire of the endemic Madagascar treefrog *Boophis madagascariensis* (Rhacophoridae). J Zool, London 250:283–298.

Narins PM, Hödl W, Grabul DS (2003) Biomodal signal requisite for agonistic behavior in a dart-poison frog, *Epipedobates femoralis*. Proc Natl Acad Sci USA 100:577–580.

Narins PM, Lewis ER, Purgue AP, Bishop PJ, Minter LR, Lawson DP (2001) Functional consequences of a novel middle ear adaptation in the central African frog, *Petropedetes parkeri* (Ranidae). J Exp Biol 204:1223–1232.

Nelken I, Rotman Y, Bar Yosef O (1999) Responses of auditory-cortex neurons to structural features of natural sounds. Nature 397:154–157.

O'Brien JA (2002) Territory acquisition and maintenance in a Neotropical frog, *Eleutherodactylus coqui*. Unpublished PhD dissertation, University of Connecticut, Storrs.

Oldham RS, Gerhardt HC (1975) Behavioral isolating mechanisms of the treefrogs *H. cinerea* and *H. gratiosa*. Copeia 1975:223–231.

Orlov N (1997) Breeding behavior and nest construction in a Vietnam frog related to *Rana blythi*. Copeia 1997:464–465.

Ovaska KE, Caldbeck J (1997a) Courtship behavior and vocalizations of the frogs *Eleutherodactylus antillensis* and *E. cochranae* on the British Virgin Islands. J Herpetol 31:149–155.

Ovaska KE, Caldbeck J (1997b) Vocal behaviour of the frog *Eleutherodactylus antillensis* on the British Virgin Islands. Anim Behav 54:181–188.

Ovaska KE, Caldbeck J (1999) Courtship call of the frog *Eleutherodactylus schwartzi* from the British Virgin Islands. J Herpetol 33:501–504.

Ovaska K, Hunte W (1992) Male mating behavior of the frog *Eleutherodactylus johnstonei* (Leptodactylidae) in Barbados, West Indies. Herpetologica 48:40–49.

Owen PC (2003) The structure, function, and evolution of aggressive signals in anuran amphibians. Unpublished PhD dissertation, University of Connecticut, Storrs.

Owen PC, Perrill SA (1998) Habituation in the green frog, *Rana clamitans*. Behav Ecol Sociobiol 44:209–213.

Page R (2004) The costs of túngara calling behavior. Animal Behavior Society: Abstracts of 41st Animal Behavior Meeting.

Pallett JR, Passmore NI (1988) The significance of multi-note advertisement calls in a reed frog, *Hyperolius tuberilinguis*. Bioacoustics 1:13–23.

Passmore NI, Telford SR (1981) The effect of chorus organization on mate localization in the painted reed frog (*Hyperolius marmoratus*). Behav Ecol Sociobiol 9:291–293.

Passmore NI, Bishop PJ, Caithness N (1992) Calling behaviour influences mating success in male painted reed frogs, *Hyperolius marmoratus*. Ethology 92:227–241.

Pengilley RK (1971) Calling and associated behavior of some species of *Pseudophryne* (Anura: Leptodactylidae). J Zool, London 163:73–92.

Pfennig KS, Rapa K, McNatt R (2000) Evolution of male mating behavior: Male spadefoot toads preferentially associate with conspecific males. Behav Ecol Sociobiol 48: 69–74.

Pough FH, Magnusson WE, Ryan MJ, Wells KD, Taigen TL (1992) Behavioral energetics. In: Feder ME, Burggren WW (eds) Environmental Physiology of the Amphibians. Chicago: University of Chicago Press, pp. 395–436.

Prestwich KN (1994) The energetics of acoustic signaling in anurans and insects. Amer Zool 34:625–643.

Purgue, AP (1997) Tympanic sound radiation in the bullfrog *Rana catesbeiana*. J Comp Physiol A 181:438–445.

Ressel SJ (2001) Ultrastructural design of anuran muscles used for call production in relation to the thermal environment of a species. J Exp Biol 204:1445–1457.

Rhode WS, Geisler CD, Kennedy DT (1978) Auditory nerve fiber responses to wide-band noise and tone combinations. J Neurophysiol 41:692–704.

Robertson JGM (1984) Acoustic spacing by breeding males of *Uperoleia rugosa* (Anura: Leptodactylidae). Z Tierpsychol 64:283–297.

Robertson JGM (1986) Female choice, male strategies and the role of vocalizations in the Australian frog *Uperoleia rugosa*. Anim Behav 34:773–784.

Ronacher B, Krahe B, Hennig RM (2000) Effects of signal duration on the recognition of masked communication signals by the grasshopper *Chorthippus biguttulus*. J Comp Physiol A 186:1065–1072.

Rose GJ, Brenowitz EA (1997) Plasticity of aggressive thresholds in *Hyla regilla*: Discrete accommodation to encounter calls. Anim Behav 53:353–361.

Rose GJ, Brenowitz EA (2002) Pacific treefrogs use temporal integration to differentiate advertisement from encounter calls. Anim Behav 63:1183–1190.

Rose GJ, Zelick R, Rand AS (1988) Auditory processing of temporal information in a neotropical frog is independent of signal intensity. Ethology 77:330–336.

Rosen M, Lemon RE (1974) The vocal behavior of spring peepers, *Hyla crucifer*. Copeia 1974:940–950.

Rosenthal GG, Rand AS, Ryan MJ (2004) The vocal sac as a visual cue in anuran communication: an experimental analysis using video playback. Anim Behav 68:55–58.

Roy D, Borah B, Sarma A (1995) Analysis and significance of female reciprocal call in frogs. Curr Sci 69:265–270.

Ryan MJ (1980) Female mate choice in a Neotropical frog. Science 209:523–525.

Ryan MJ (1986) Synchronized calling in a treefrog (*Smilisca sila*). Brain Behav Evol 29:196–206.

Ryan MJ, Keddy-Hector A (1992) Directional patterns of female mate choice and the role of sensory biases. Amer Nat 139:S4–S35.

Samuel AG (1981) Phonemic restoration: Insights from a new methodology. J Exp Psychol General 110:474–494.

Schlaepfer MA, Figeroa-Sandí R (1998) Female reciprocal calling in a Costa Rican leaf-litter frog, *Eleutherodactylus podiciferus*. Copeia 1998:1076–1080.

Schneichel W, Schneider H (1988) Hearing and calls of the banana frog, *Afrixalus fornasinii* (Bianconi) (Anura: Rhacophoridae). Amphibia-Reptilia 9:251–264.

Schneider H (1977) Acoustic behavior and physiology of vocalization in the European tree frog, *Hyla arborea* (L.). In: Taylor DH, Guttman SI (eds) The Reproductive Biology of Amphibians. New York: Plenum, pp. 295–335.

Schul J, Bush SL (2002) Non-parallel coevolution of sender and receiver in the acoustic communication system of treefrogs. Proc R Soc Lond B 269:1847–1852.

Schwartz JJ (1986) Male calling behavior and female choice in the Neotropical treefrog *Hyla microcephala*. Ethology 73:116–127.

Schwartz JJ (1987a) The importance of spectral and temporal properties in species and call recognition in a Neotropical treefrog with a complex vocal repertoire. Anim Behav 35:340–347.

Schwartz JJ (1987b) The function of call alternation in anuran amphibians: A test of three hypotheses. Evolution 41:461–471.

Schwartz JJ (1989) Graded aggressive calls of the spring peeper, *Pseudacris crucifer*. Herpetologica 45:172–181.

Schwartz JJ (1991) Why stop calling? A study of unison bout singing in a Neotropical treefrog. Anim Behav 42:565–577.

Schwartz JJ (1993) Male calling behavior, female discrimination and acoustic interference in the Neotropical treefrog *Hyla microcephala* under realistic acoustic conditions. Behav Ecol Sociobiol 32:401–414.

Schwartz JJ (2001) Call monitoring and interactive playback systems in the study of acoustic interactions among male anurans. In: Ryan MJ (ed) Anuran Communication. Washington DC: Smithsonian Institution Press, pp. 183–204.

Schwartz JJ, Gerhardt HC (1989) Spatially-mediated release from masking in an anuran amphibian. J Comp Physiol A 166:37–41.
Schwartz JJ, Gerhardt HC (1995) Directionality of the auditory system and call pattern recognition during acoustic interference in the gray treefrog, *Hyla versicolor*. Aud Neurosci 1:195–206.
Schwartz JJ, Gerhardt HC (1998) The neuroethology of frequency preferences in the spring peeper. Anim Behav 56:55–69.
Schwartz JJ, Rand AS (1991) The consequences for communication of call overlap in the tungara frog, a neotropical anuran with a frequency-modulated call. Ethology 89:73–83.
Schwartz JJ, Wells KD (1983a) The influence of background noise on the behavior of a Neotropical treefrog, *Hyla ebraccata*. Herpetologica 39:121–129.
Schwartz JJ, Wells KD (1983b) An experimental study of acoustic interference between two species of Neotropical treefrogs. Anim Behav 31:181–190.
Schwartz JJ, Wells KD (1984a) Interspecific acoustic interactions of the Neotropical treefrog *Hyla ebraccata*. Behav Ecol Sociobiol 14:211–224.
Schwartz JJ, Wells KD (1984b) Vocal behavior of the Neotropical treefrog *Hyla phlebodes*. Herpetologica 40:452–463.
Schwartz JJ, Wells KD (1985) Intra- and interspecific vocal behavior of the Neotropical treefrog *Hyla microcephala*. Copeia 1985:27–38.
Schwartz JJ, Buchanan B, Gerhardt HC (2001) Female mate choice in the gray treefrog (*Hyla versicolor*) in three experimental environments. Behav Ecol Sociobiol 49: 443–455.
Schwartz JJ, Buchanan B, Gerhardt HC (2002) Acoustic interactions among male gray treefrogs (*Hyla versicolor*) in a chorus setting. Behav Ecol Sociobiol 53:9–19.
Schwartz JJ, Huth K, Lasker J (2004) Impact of the chorus environment on temporal processing of advertisement calls by gray treefrogs. Abstracts, 147th Meeting Acoustical Society of America: JASA 115:2374.
Schwartz JJ, Ressel S, Bevier CR (1995) Carbohydate and calling: Depletion of muscle glycogen and the chorusing dynamics of the Neotropical treefrog *Hyla microcephala*. Behav Ecol Sociobiol 37:125–135.
Simmons AM (1988) Selectivity for harmonic structure in complex sounds by the green treefrog (*Hyla cinerea*). J Comp Physiol A 162:397–403.
Simmons AM, Schwartz JJ, Ferragamo M (1992) Auditory-nerve representation of a complex communication sound in background noise. J Acoust Soc Am 91:2831–2844.
Smith MJ, Roberts JD (2003) Call structure may affect male mating success in the quacking frog *Crinia georgiana* (Anura: Myobatrachidae). Behav Ecol Sociobiol 53:221–226.
Stewart MM, Bishop PJ (1994) Effects of increased sound level of advertisement calls on calling male frogs, *Eleutherodactylus coqui*. J Herpetol 28:46–53.
Stewart MM, Rand AS (1991) Vocalizations and the defense of retreat sites by male and female frogs, *Eleutherodactylus coqui*. Copeia 1991:1013–1024.
Sullivan BK, Ryan MJ, Verrill PA (1995) Female choice and mating system structure. In: Heatwole H, Sullivan BK (eds) Amphibian Biology: Vol 2: Social Behaviour. Chipping Norton, UK: Surrey Beatty, pp. 469–517.
Tárano Z, Herrera EA (2003) Female preferences for call traits and mating success in the Neotropical frog, *Physalaemus enesefae*. Ethology 109:121–134.
Telford SD, Dyson ML, Passmore NI (1989) Mate choice occurs only in small choruses of painted reed frogs *Hyperolius marmoratus*. Bioacoustics 2:47–53.
Tobias ML, Barnard C, O'Hagan R, Horng SH, Rand M, Kelley DB (2004) Vocal communication between male *Xenopus laevis*. Anim Behav 67:353–365.

Tobias ML, Viswanathan SS, Kelley DB (1998) Rapping, a female receptive call, initiates male–female duets in the South African clawed frog. Proc Natl Acad Sci USA 95: 1870–1875.

Townsend DS, Stewart MM (1986) Courtship and mating behavior of a Puerto Rican frog, *Eleutherodactylus coqui*. Herpetologica 42:165–170.

Tuttle MD, Ryan MJ (1981) Bat predation and the evolution of frog vocalizations in the Neotropics. Science 214:677–678.

Tuttle MD, Ryan MJ (1982) The role of synchronized calling, ambient light, and ambient noise, in anti-bat predator behavior of a treefrog. Behav Ecol Sociobiol 11:125–131.

Valone TJ, Templeton JJ (2002) Public information for the assessment of quality: a widespread social phenomenon. Phil Trans R Soc Lond B 357:1549–1557.

Wagner WE Jr (1989a) Fighting, assessment, and frequency alteration in Blanchard's cricket frog. Behav Ecol Sociobiol 25:429–436.

Wagner WE Jr (1989b) Social correlates of variation in male calling behavior in Blanchard's cricket frog, *Acris crepitans blanchardi*. Ethology 82:27–45.

Wagner WE Jr (1989c) Graded aggressive signals in Blanchard's cricket frog: Vocal responses to opponent proximity and size. Anim Behav 38:1025–1038.

Wagner WE Jr (1991) Social selection on male calling behavior in Blanchard's cricket frog. PhD thesis, University of Texas, Austin.

Wagner WE Jr (1992) Deceptive or honest signalling of fighting ability? A test of alternative hypotheses for the function of changes in call dominant frequency by male cricket frogs. Anim Behav 44:449–462.

Warren RM (1970) Perceptual restoration of missing speech sounds. Science 167:392–393.

Welch AM, Semlitsch RD, Gerhardt HC (1998) Call duration as an indicator of genetic qualtity in male gray treefrogs. Science 280:1928–1930.

Wells KD (1977a) The social behaviour of anuran amphibians. Anim Behav 25:666–693.

Wells KD (1977b) The courtship of frogs. In: Taylor D, Guttman S (eds) The Reproductive Biology of Amphibians. New York: Plenum, pp. 233–262.

Wells KD (1980a) Behavioral ecology and social organization of a dendrobatid frog (*Colostethus inguinalis*). Behav Ecol Sociobiol 6:199–209.

Wells KD (1980b) Social behavior and communication of a dendrobatid frog (*Colostethus trinitatis*). Herpetologica 36:189–199.

Wells KD (1981) Territorial behavior of the frog *Eleutherodactylus urichi* in Trinidad. Copeia 1981:726–728.

Wells KD (1988) The effects of social interactions on anuran vocal behavior. In: Fritszch B, Wilczynski W, Ryan MJ, Hetherington T, Walkowiak W (eds) The Evolution of the Amphibian Auditory System. New York: Wiley, pp. 433–454.

Wells KD (1989) Vocal communication in a Neotropical treefrog, *Hyla ebraccata*: Responses of males to graded aggressive calls. Copeia 1989:461–466.

Wells KD (2001) The energetics of calling in frogs. In: Ryan MJ (ed) Anuran Communication. Washington, DC: Smithsonian Institution Press, pp. 45–60.

Wells KD, Bard KM (1987) Vocal communication in a Neotropical treefrog, *Hyla ebraccata*: Responses of females to advertisement and aggressive calls. Behaviour 101: 200–210.

Wells KD, Schwartz JJ (1984a) Vocal communication in a Neotropical treefrog, *Hyla ebraccata*: Advertisement calls. Anim Behav 32:405–420.

Wells KD, Schwartz JJ (1984b) Vocal communication in a Neotropical treefrog, *Hyla ebraccata*: Aggressive calls. Behaviour 91:128–145.

Wells KD, Taigen TL (1986) The effect of social interactions on calling energetics in the gray treefrog (*Hyla versicolor*). Behav Ecol Sociobiol 19:9–18.

Wells KD, Taigen TL (1989) Calling energetics of a Neotropical treefrog, *Hyla microcephala*. Behav Ecol Sociobiol 25:13–22.

Whitney CL, Krebs JR (1975) Mate selection in Pacific treefrogs. Nature 255:325–326.

Wilczynski W, Brenowitz EA (1988) Acoustic cues mediate inter-male spacing in a Neotropical frog. Anim Behav 36:1054–1063.

Wilczynski W, Rand AS, Ryan MJ (1995) The processing of spectral cues by the call analysis system of the tungara frog, *Physalaemus pustulosus*. Anim Behav 49:911–929.

Wiley RH (1983) The evolution of communication: Information and manipulation. In: Halliday TR, Slater PJB (eds) Animal Behaviour 2. Communication. Oxford: Blackwell, pp. 156–189.

Wiley RH, Poston J (1996) Perspective: Indirect mate choice, competition for mates, and coevolution of the sexes. Evolution 50:1371–1381.

Wollerman L, Wiley RH (2002a) Background noise from a natural chorus alters female discrimination of male calls in a Neotropical frog. Anim Behav 63:15–22.

Wollerman L, Wiley RH (2002b) Possibilities for error during communication by Neotropical frogs in a complex acoustic environment. Behav Ecol Sociobiol 52:465–473.

Yager DD (1996) Sound production and acoustic communication in *Xenopus borealis*. In: Tinsley RC, Kobel HR (eds) The Biology of Xenopus. New York: Oxford University Press, pp. 121–141.

Zelick R, Mann DA, Popper AN (1999) Acoustic communication in fishes and frogs. In: Fay RR, Popper AN (eds) Comparative Hearing: Fish and Amphibians. New York: Springer-Verlag, pp. 363–411.

Zelick RD, Narins PM (1983) Intensity discrimination and the precision of call timing in two species of Neotropical treefrogs. J Comp Physiol A 153:403–412.

Zimmerman BL, Bogart JP (1984) Vocalizations of primary forest frog species in the Central Amazon. Acta Amazonica 14:473–519.

Zuk M, Kolluru GR (1998) Exploitation of signals by predators and parasitoids. Quart Rev Biol 73:415–438.

Zurek PM (1987) The precedence effect. In: Yost WA, Gourevitch G (eds) Directional Hearing. New York: Springer-Verlag, pp. 85–105.

4
Call Production and Neural Basis of Vocalization

W. Walkowiak

1. Introduction

A common feature of most anuran species is the use of calls in intraspecific communication. Vocalizations play an essential role in various behavioral contexts and particularly in the reproductive behavior of anuran amphibians. Several approaches have been used to study anuran acoustic communication. Most early studies focused on cataloguing and classifying calls. In general, frogs utter a limited number of call types; these are classifiable either according to the behavioral contexts in which calls are produced as advertisement calls, courtship calls, aggressive calls (i.e., close-range and long-distance territorial calls), release calls, warning calls, distress calls, and so on, or according to their physical attributes as trills, chuckle calls, rapping, and others (see Wells and Schwartz, Chapter 3). Spectral composition and temporal patterns of these call-types are distinct, and show differences in one or more acoustic parameters (Walkowiak 1988a).

In the early studies, it was generally assumed that call patterns are relatively fixed (i.e., genetically determined) and that vocal learning or individual experience plays little or no role in vocal production. This view has been revised over the years and it is now believed that the social context, learning, and the inner state of the animal (e.g., motivation) can modify the frog's calling.

In this chapter, different levels of organization of the vocal production system are presented, starting with the peripheral structures and mechanisms of call production, to neural control of muscles involved in calling, and finally to the central pattern generators and their control by higher vocal centers. Because internal as well as external factors influence calling behavior, the role of forebrain centers (e.g., the striatum and the limbic system) is also described. The emphasis of this chapter is placed on audio–motor integration.

2. Peripheral Structures and Mechanisms of Call Production

In spite of the small call repertoire for most species, the acoustical characteristics of anuran calls may differ markedly from one species to the next. Some calls

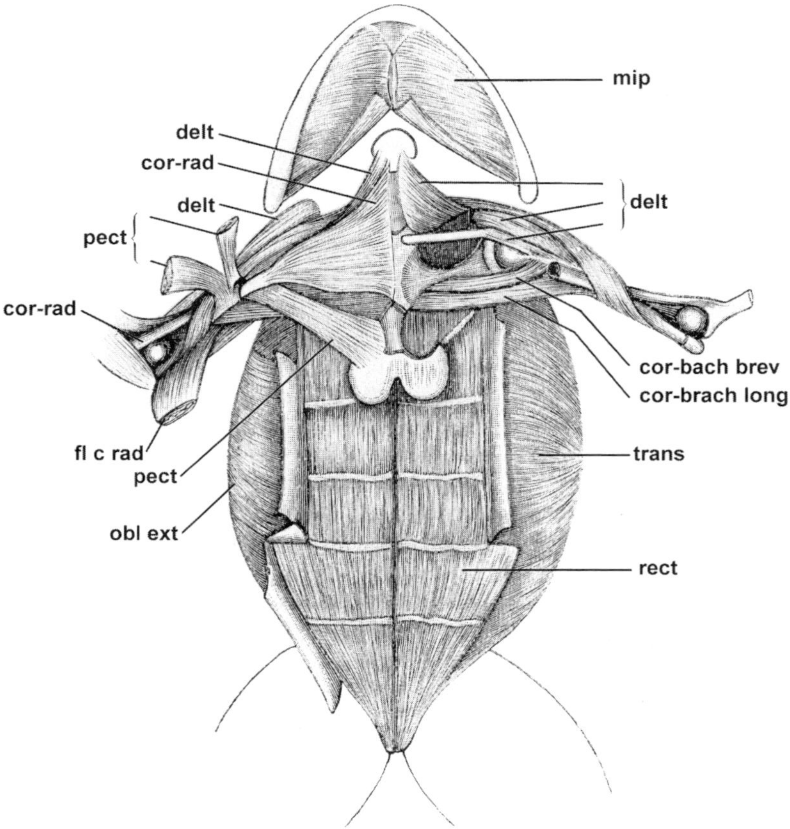

FIGURE 4.1. Ventral view of the body musculature of *Rana esculenta.* Modified from Gaupp (1896).

consist of a single-tone peep (i.e., a short note with a constant frequency), others contain a train of sound pulses with complex temporal structures as well as spectral characteristics (see Wells and Schwartz, Chapter 3). This raises the question about the control of the different call parameters, that is, which parameters are controlled by the peripheral structures (e.g., the larynx or vocal sacs), and which are controlled by the CNS?

Among anuran amphibians, various calling mechanisms have evolved. Neobatrachian species generally use the expiratory air stream to generate sound. For them, the driving power is produced by contraction of the trunk muscles (M. rectus abdominis, M. transverses, M. obliquus externus, and internus; Fig. 4.1). The contraction forces the air to leave the lungs and pass through the larynx, and the air is then pressed into the buccal cavity and to the paired or unpaired vocal sacs. The vocal sacs are diverticula of the mouth cavity and may serve as resonators, sound couplers, or acoustic radiators. Contraction of the mouth floor and vocal sac muscles presses the air back through the open glottis into the lungs silently. The trunk muscles are sexually dimorphic; that is, male muscles exhibit morphological as well as biochemical adaptations for call production (for review see Wells 2001).

The archaeobatrachian genus *Bombina* utilizes an inspiratory call production mechanism (Lörcher 1969). Male specimens take up air into the mouth cavity and force the air volume through the larynx into the lungs, thus producing whistle-like advertisement calls. All species of the genus *Discoglossus* use both the inspiratory as well as the expiratory air stream to produce calls (Weber 1974); their calls are more complex than *Bombina* calls.

Frogs in the family Pipidae call underwater, and thus sound production cannot be based on movement of the air between the two compartments mentioned above (Yager 1992, 1996). Instead, sound pulses are produced by two cartilaginous discs in the larynx which are in tight contact during silent phases. Fast separation of the discs produces the acoustic clicks. Call production in the genus *Xenopus* is highly specialized and involves intrinsic laryngeal muscles exclusively.

2.1 The Larynx

The anuran larynx is a complex structure with a cartilaginous skeleton. A ring-shaped crico-tracheal cartilage supports two arytenoid cartilages. When the glottis is closed the medial edges of the arytenoids touch each other. The larynx is connected to the hyoid in the mouth floor via the posterior–medial processes (Fig. 4.2). The glottis is opened by the dilator muscles of the larynx which pull the arytenoids aside. Two pairs of muscles serve to close the larynx, the hyolaryngeal muscles, and the anterior sphincter muscles. The posterior sphincter muscles pull the vocal cords together in the midline (Schneider 1970; Schmidt 1972; for other variants see Wahl 1969; Schmid 1977, 1978).

Vocal cords are attached to the posterior end of the arytenoids. Schmid (1978) examined their structures in a number of European species. The vocal cords of Pelobatidae, Hylidae, and Ranidae are T-shaped; that is, the lateral parts are connected to the laryngeal skeleton and the medial parts of the vocal cords extend cranially and caudally. In Bufonidae, the vocal cords are L-shaped, whereas Bombinatoridae and Discoglossidae possess only a tissue pad at the inner side of the arytenoids. Only in *Pelobates fuscus* and *Bombina orientalis* the laryngeal muscles (part of the dilator muscle) project into the vocal cords (Schmid 1978), thus controlling the tension of the vocal cords. In pipid frogs, vocal cords are missing. Like the trunk muscles, the laryngeal muscles show pronounced morphological and biochemical specializations (Eichelberg and Schneider 1973) as well as sexual dimorphism (Wells 2001).

Compared to the auditory system, relatively fewer studies have been directed towards understanding of the muscular mechanisms and neural control of calling. In an isolated preparation in three ranid species (*Rana temporaria*, *R. esculenta*, *R. ridibunda*), Paulsen (1965, 1967) investigated the effects of expiratory air streams passing through the larynx. A low-pressure air stream (<100 mm H_2O)

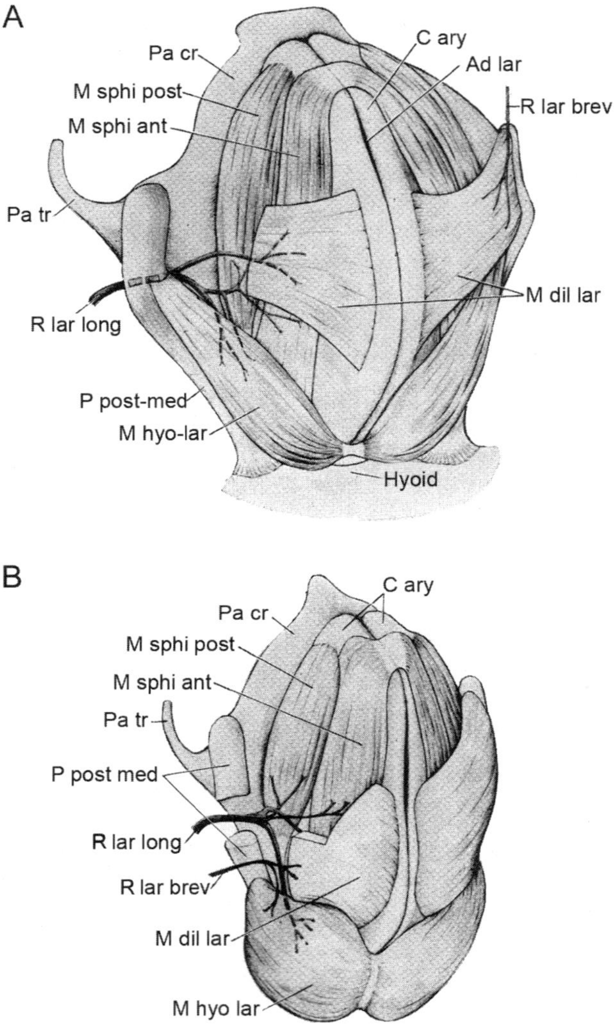

FIGURE 4.2. Laryngeal musculature and nerves of *Hyla a. arbora*. Modified from Schneider (1970). With kind permission of Springer Science and Business Media LLC.

elicits no vibrations of the vocal cords when the glottis is open. In contrast, higher pressure levels elicit a valvelike behavior: the caudal pouches of the vocal cords are filled with air thus moving the medial part of the vocal cords towards each other and occluding the opening. When the air pressure rises, the vocal cords are forced apart. The oscillations produced by this mechanism are responsible for the pulse structure of the calls (Fig. 4.3). These data indicate that at least some parameters of frog calls, such as their fundamental frequency and pulse shape, are

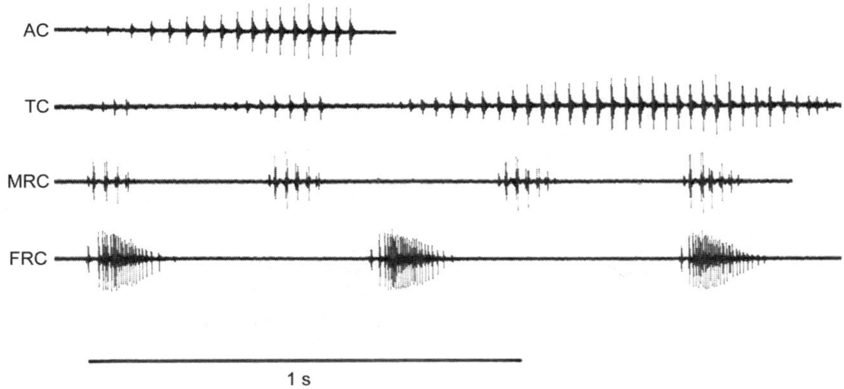

FIGURE 4.3. Oscillograms of an advertisement call (AC), aggressive calls (TC), male release calls (MRC), and female release calls (FRC) of *Rana temporaria*. Modified from Walkowiak (1980). With kind permission of Springer Science and Business Media LLC.

determined by the physical properties of the larynx skeleton and laryngeal muscles. Recently, similar results have been obtained from four species of North American treefrogs (Gridi-Papp 2003). The mass of the vocal cords generally correlates with the overall body mass, and so larger frogs typically produce lower-pitched calls, both within and between species (Gerhard and Huber 2002), although there are some notable exceptions (Ryan and Drewes 1990). Schmidt (1965, 1968a, 1972) examined the activity of laryngeal muscles in *Rana pipiens*, *Hyla versicolor*, and *H. cinerea* during advertisement and release calling with cinematographic and electromyographic techniques and showed that glottal opening is caused by contraction of the dilator muscles (m. dilatator larynges). The external and anterior constrictors (m. hyolaryngis and m. sphincter anterior) act as alternating constrictor muscles of the glottis. The posterior constrictor (m. sphincter posterior) brings the vocal cords into opposition.

2.2 Expiratory Call Generation

In expiratory calling, the pressure for forcing the air volume out of the lungs through the larynx into the buccal cavity is produced by contraction of the flank musculature. Two modes of muscle contraction are possible (Martin 1971, 1972). A continuous contraction produces an uninterrupted air stream which in turn leads to passive oscillations of the intrinsic laryngeal structures such as vocal cords or the arytenoids. Another mode involves repetitive contractions of the trunk muscles that directly produce pulsed vocalizations. Martin and Gans (1972) recorded the air pressure, sound output, and electromyograms of the trunk musculature during release calling in *Bufo valliceps*. They found that the trunk muscles contract in a pulsatile fashion, with each contraction producing a pulse in pulmonary pressure and therefore a sound pulse. Girgenrath and Marsh (1997)

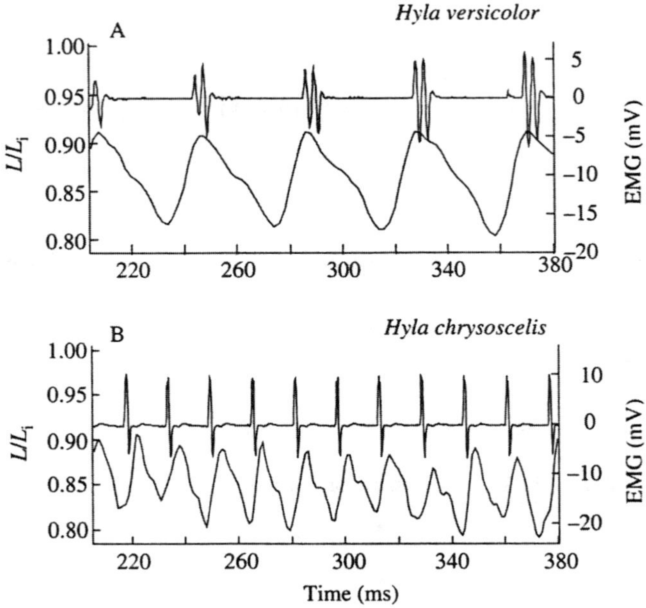

FIGURE 4.4. Electromyograms and muscle length traces associated with calls of *Hyla versicolor* (A) and *H. chrysoscelis* (B). L/L_i is the ratio of the length of the external oblique muscle to its initial length. From Giergenrath and Marsh (1997). Reprinted from J Exp Biol, 200, M Girgenrath and RL Marsh In vivo performance of trunk muscles in tree frogs during calling 3101–3108, 1997, with permission from Company of Biologists Ltd.

used high-speed video and electromyography to measure in vivo performance of the trunk muscles (m. obliquus externus) in *Hyla versicolor* and *H. chrysoscelis*, two closely related species. The pulse rates of the advertisement calls of these species differ by a factor of two; that is, at 25°C the pulse rates for *H. versicolor* and *H. chrysoscelis* are 25 Hz and ~50 Hz, respectively. In both species, during calling, the external oblique muscle shortens progressively over the first pulses; subsequently there are single strain cycles that correspond to the individual pulses in the call. In *H. versicolor* two biphasic electromyographic potentials correspond to each strain cycle, whereas in *H. chrysoscelis* there is only one biphasic potential for each strain cycle (Fig. 4.4). In general, the trunk musculature and the laryngeal muscles operate in a well-coordinated manner to generate the pulsed call structure.

2.3 Inspiratory Call Generation

Inspiratory call generation is closely tied to respiration. De Jongh and Gans (1969) analyzed the mechanisms of respiration in the American bullfrog, *Rana catesbeiana*, with a variety of techniques. They found three phenomena that could be attributed to breathing. (1) Exchange of fresh air in the buccal cavity is managed by periodical elevation and lowering of the mouth floor when the nares are open. (2) Pulmonary gas exchange during a ventilatory cycle involves opening and closing of the nares and the glottis. (3) A series of ventilatory cycles builds up inflation cycles that are interrupted by an apneic pause. In each case it is the buccal force pump that produces the air stream.

Mechanisms of respiration and inspiratory call generation have been studied extensively in *Bombina orientalis* and *Discoglossus pictus* (Weirich et al. 1989; Walkowiak 1992; Strake 1995). Male *Bombina* utter their calls while floating on the water surface. Advertisement calls have durations of ~120 ms at 25°C (Akef and Schneider 1985; Schneider et al. 1986). To call, they take up a large volume of air into the lungs and the mouth cavity, via several buccal oscillation and lung inflation cycles (Fig. 4.5). The driving force for the inspiratory air stream is generated by the musculature in the mouth floor (Fig. 4.6). The most ventral layer is composed of three muscles: (1) the anterior intermandibular muscle (*mia*) which is the smallest muscle and spans the most rostral parts of the dental bones; (2) the paired posterior mandibular muscles (*mip*) which are the largest, attach to the dental bones, meet medially, and are connected via a tendon (*raphe*); and (3) the interhyoid muscles (*mih*) which are the most caudal muscles, originate from the hyoid and also meet in the midline of the mouth floor. Contraction of the *mia*

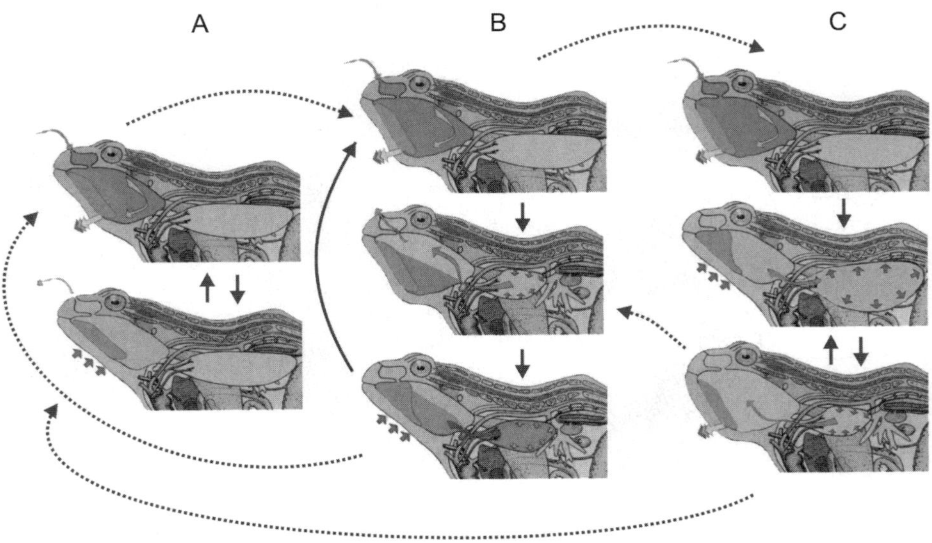

FIGURE 4.5. Schematic description of buccal oscillation (A), lung inflation (B), and calling (C) with respect to airflow in *Bombina orientalis*. Solid arrows show the sequence of movements of single cycles and dotted arrows indicate the transition of different respiratory and calling movements.

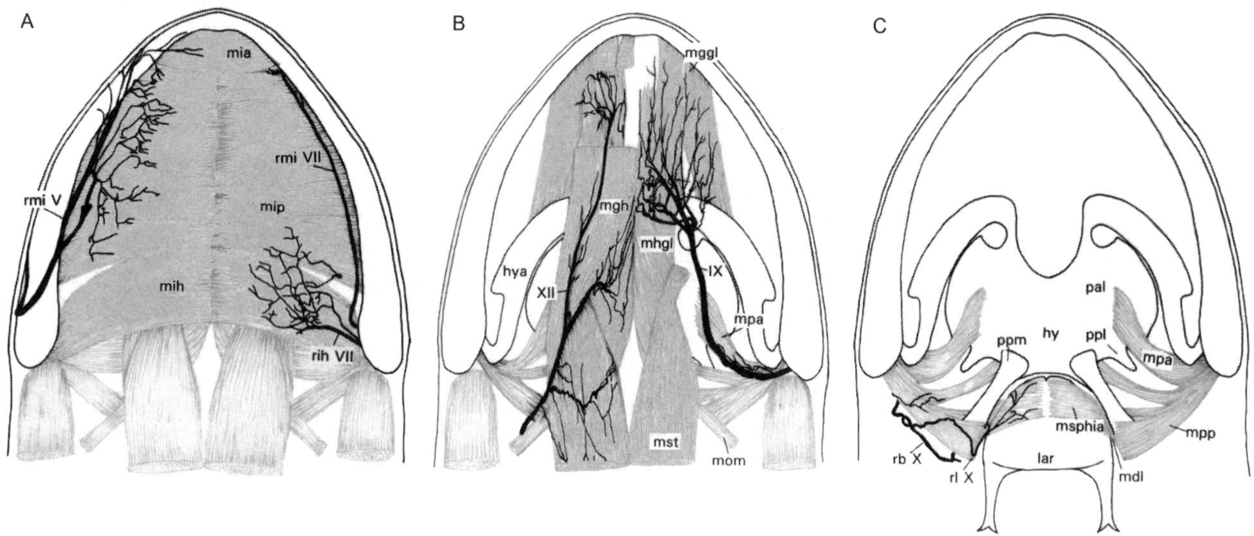

FIGURE 4.6. Ventral view of the muscles of the mouth floor and the larynx, and their innervations in *Discoglossus pictus* after removal of the skin (A), the ventralmost muscle layer (B), and the longitudinal muscles (C). From Strake (1995).

leads to closure of the nares. A detailed description of this complicated mechanism is given by Trueb and Cannatella (1982). Contraction of the *mip* and *mih* lifts the mouth floor upwards thus narrowing the volume of the buccal cavity. These muscles of the mouth floor produce the main power for elevation of the mouth floor. Whereas these muscles are oriented perpendicularly to the head axis, the next dorsal group of muscles stretches in the longitudinal direction. One pair of these muscles, the geniohyoid muscles (*mgh*), also participates in narrowing of the buccal cavity. The *mgh* are attached to the posterior process of the hyoid and to the chin. Their contraction moves the anterior part of the hyoid upwards. Occluding the nares, contraction of the mouth floor levators, and opening of the glottis permit the air to be forced into the lungs. Depression of the mouth floor is achieved by contraction of the only antagonists, the sternohyoid muscles (*msh*), which are attached to the anterior part of the hyoid and the sternum. Further longitudinally oriented muscles are the protractor of the tongue [i.e., the hyoglossal muscle (*mhgl*)], and the retractors of the tongue [i.e., the genioglossal muscle (*mggl*)]. The hyoid is supported by the petrohyoid muscles which narrow the posterior part of the buccal cavity by their contraction.

Weirich et al. (1989) and Strake (1995) were the first to record electromyograms from corresponding muscles during respiration, advertisement call generation, and release call generation in freely moving specimens of *B. bombina* and *D. pictus*. The sequence and duration of the muscular activation during lung inflation are depicted in Figure 4.7a. The first step involves contraction of the *msh* that depresses the mouth floor. Because the nares are open, this contraction draws fresh air into the buccal cavity. In the following sequence, the first muscle that is excited is the *mia*, which closes the nares. Then, with a high degree of synchronization, the *mip* and the *mih* contract, followed by the activation of the *mgh* after a short delay. These muscle activities move the mouth floor upwards. The synchronized muscle activity forces the air volume to move from the oral cavity through the open glottis into the lungs. Surprisingly, both tongue muscles (i.e., the protractor and, following a brief delay, the retractor) contract during elevation of the mouth floor. This forms a tissue block that acts as a valve to occlude the nares from the inside of the mouth.

In principle, the sequence of muscular activation is similar during advertisement call generation (Fig. 4.7b). In *B. orientalis*, the onset of muscle activation precedes the audible vocalization by approximately 40 ms. This is noteworthy for audio–motor integration because the time window for call suppression by other acoustic signals (e.g., conspecific calls) closes approximately 110 ms before the expected call in *B. bombina* (Walkowiak 1988b) and ca. 80 ms in *B. orientalis* (Mohr and Schneider 1993). As call suppression is an important mechanism for establishing antiphonal calling, the maximum time for neural processing of auditory input and generating the vocal response is only 40 to 70 ms in this genus (Walkowiak 1992), and a remarkable 26 to 37 ms in *Pseudacris streckeri* (Loftus-Hills 1974). Differences between muscular activity during lung inflation and advertisement call generation relate mainly to the average duration of the muscle activity of the levator muscles. Major differences are obvious concerning the

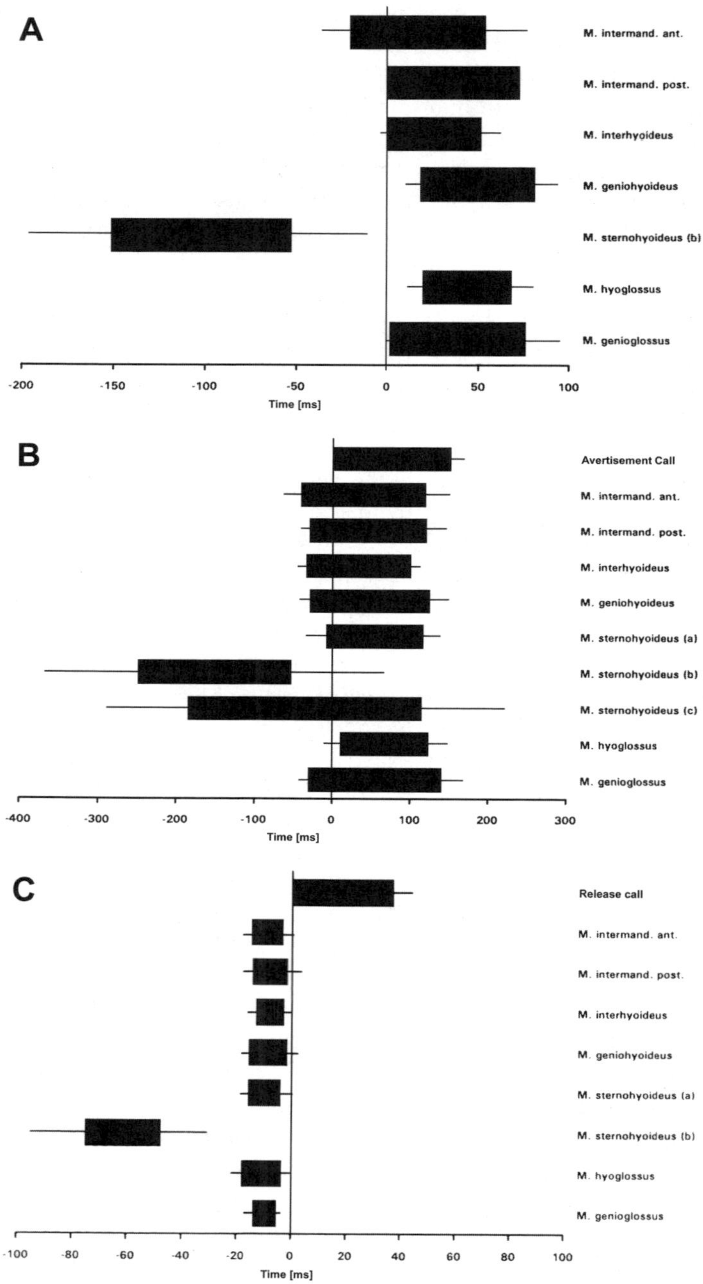

FIGURE 4.7. Electromyographical bar diagram of different muscles involved in generation of lung inflation (A), advertisement calls (B), and release calls (C) in *Bombina orientalis*. Length of bars represents the mean duration of activity; mean relative onsets are related to the onset of *mip* (A) and calls (B, C). Standard deviations of the onset of the muscle activity are indicated to the left; standard deviations of the duration of muscle activities are indicated on the right. Note the different activity patterns of *msh*. From Strake et al. (1994) and Strake (1995).

activity pattern of the *msh*. The depressor of the mouth floor is either active in the interval between two elevations of the mouth floor, in parallel with the elevator activities, or active during both phases of buccal movements. This variability of *msh* motor patterns can be observed from one call to the next. It may be regulated by sensory feedback from hyoid position sensors, lung or body wall stretch receptors, and so on, but this has not been validated experimentally. Coactivation of the *mgh* and *msh* may stabilize the position of the hyoid, and thus the position of the larynx in the air stream. Although the activity of the *msh* may vary from one call to the other, the call properties, that is, call duration or fundamental frequency, are not affected. The trunk muscles are not active during advertisement call generation, even in the expiratory phase. Taken together, the results from archaeobatrachian species support the hypothesis that calling in terrestrial species or species that call on the water surface have evolved from respiration.

Release calls in *Bombina* are uttered in a series of short ca. 40 ms calls that are generated as are the advertisement calls by an inspiratory air stream (Strake 1995). In contrast to these, the myograms show an extremely short and highly synchronized activation of the elevators of the mouth floor (Fig. 4.7c). It is not known whether the difference is caused by recruitment of a different set of motoneurons or neuronal network, or whether different sensory stimuli (i.e., somatosensory triggers) cause a modified motor pattern.

2.3 Intrinsic Laryngeal Call Generation

In contrast to the expiratory and inspiratory callers described above, pipids such as *Xenopus* generate calls completely underwater. This obviously necessitates uncoupling of the vocal mechanisms from the constraints of actual respiration (Yager 1982, 1992, 1996; Tobias and Kelley 1987; Kelley 2004; Tobias et al. 2004). The calls of *Xenopus laevis* consist of pulse trains (click trains) which are generated in the larynx, that is, by contraction of the laryngeal muscles and separation of a pair of cartilaginous discs. As in expiratory calling species, the laryngeal elements seem to contribute to the structure of the calls. *Xenopus laevis* is one of the few anuran species in which females also produce calls to advertise their fertility or pending oviposition (Tobias et al. 1998; Emerson and Boyd 1999). Male-specific vocalizations include advertisement calls, growling, and amplectant calls. Female-specific calls include release calls ("ticking") given by nongravid females in response to male clasping (Kelley and Tobias 1999; Yamaguchi and Kelley 2000). The larynges of both sexes show a striking dimorphism (Sassoon et al. 1986, 1987), not only in laryngeal size but also in muscle fiber type and synaptic strength of neuromuscular junctions. The calls of the two genders differ fundamentally in sex-specific temporal structure and amplitude profile. The temporal structure (i.e., the click rate), is directly correlated with the contraction rate of the laryngeal muscles: this suggests the importance of the CNS in production of the different call patterns.

3. The Nervous System

The kind of call generation mechanism determines which motor nerves are involved in calling. The trigeminal, facial, glossopharyngeal, vagal, and hypoglossal nerves (N. V, VII, IX–X, XII) innervate the buccal force pump and the larynx (Fig. 4.6). In inspiratory calling species, these cranial nerves are the only nerves involved in calling. In addition, in expiratory call generation, spinal motor nerves innervate the muscles that produce the expiratory air stream. As an exception, in Pipidae, only N. IX–X are involved in calling due to the intrinsic laryngeal call production.

The neural circuitry underlying vocalization has been studied extensively. Schmidt first identified two areas that were involved in generating call activity. One is located in the dorsal tegmentum, rostral to the motor nucleus of the trigeminal nerve (Schmidt 1973, 1974); he named this area pretrigeminal nucleus (PrV). The PrV forms a dense neuropil that extends from the ventricle to the anterior border of the trigeminal nerve, and neurons therein send their dendrites deeply into the white matter (Fig. 4.8). Electrical stimulation of the PrV in *R. pipiens* elicited similar neural activity patterns in the laryngeal nerve to those patterns detected by recording electroneurograms (or electromyograms) during calling. When the PrV was lesioned, frogs could no longer utter release calls, either by tactile stimulation of the animals or by electrical stimulation of the mesencephalic tegmental areas (in lesioned animals, electrostimulation of the preoptic area also fails to evoke advertisement calls; see below). Results of 2-deoxy-glucose-experiments further demonstrated the importance of PrV in release calling (Schmidt 1981); in particular, prolonged release calling (evoked by clasping behind the forelimbs) increased the metabolic rate of the PrV. Finally, Schmidt (1980) also found that the PrV stains strongly for succinic dehydrogenase (SDH; an indicator for high metabolic activity) in seven species of anurans from four families and two suborders of frogs and toads; this is the same area in *X. laevis* (which was named "dorsal tegmental area of the medulla") that was reported to concentrate testosterone (Kelley et al. 1975). Taken together, the PrV is a prominent component of the central pattern generator for the release call and the advertisement call. According to Schmidt (1992), a second semi-independent pattern generator, located in the reticular formation at the level of the IX–X complex (cranial nerves IX, X, and XI), may be involved in call triggering.

Further studies showed that the medullary pattern generators are not likely responsible for advertisement call triggering. Lesion and electrical stimulation experiments in *Rana pipiens*, *R. clamitans*, and *Bufo americanus* by Schmidt (1966a, 1968b, 1971, 1973) demonstrated that the preoptic area (in the rostroventral diencephalon) fulfills the prerequisites for initializing call readiness and call triggering. Electrical stimulation of the ventral magnocellular preoptic nucleus in restrained acute preparations (or in unrestrained, chronic preparations) can evoke calls that resemble a frog's natural advertisement calls. Removal of the dorsal and medial thalamus, hypothalamus, optic tectum, and cerebellum did not abolish advertisement calling as long as the preoptic area was intact.

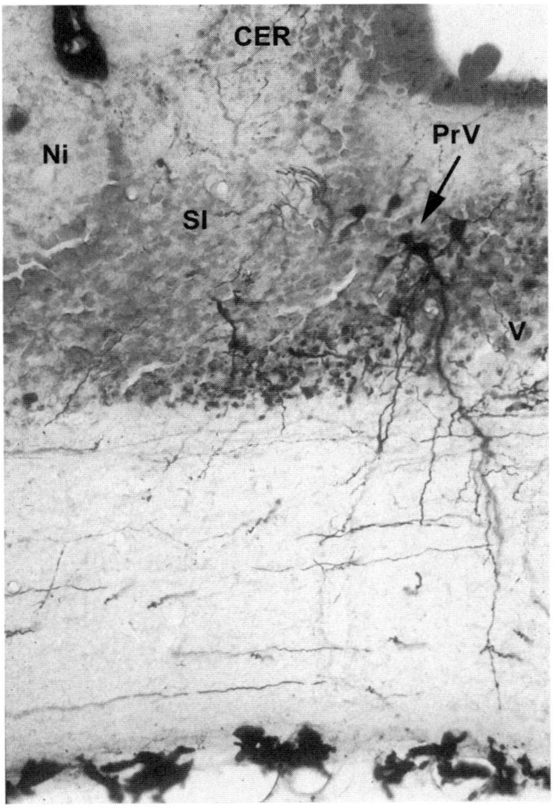

FIGURE 4.8. Retrogradely labeled neurons in the PrV after an injection of HRP into the contralateral motor nuclei N.V/N.VII. Sagittal section. From Walkowiak (1992). With kind permission of Ethology, Ecology and Evolution.

Knorr (1976) later re-examined Schmidt's findings in *Hyla arborea savignyi.* He stimulated different regions of the preoptic area using a brain atlas for this species (Knorr and Schneider 1975) as a guide and found that stimulation of some loci in the preoptic nucleus and the hypothalamus elicited motor outputs (such as acceleration of breathing, laryngeal movements, partial filling of the vocal sac with air, and extension of the forelegs): these served to prepare calling. In this state, male frogs readily produced advertisement calls when they were acoustically stimulated with conspecific calls. Stimulation of other loci (e.g., the dorsal part of the posterior preoptic nucleus and in the anterior preoptic nucleus) induced calling itself with a latency of several seconds up to minutes (Fig. 4.9). On the other hand, stimulation of the magnocellular nucleus of the preoptic area did not induce calling activity. Wada and Gorbman (1977a,b) corroborated some of Knorr's findings; specifically, using lower current amplitudes they could elicit

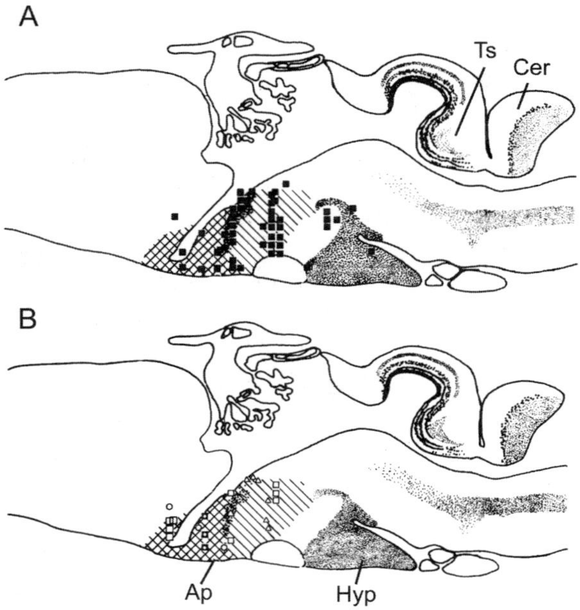

FIGURE 4.9. Diagrammatic representation of a sagittal section of a *Hyla arborea* brain taken from ~100 μm lateral to the midline. At the indicated sites, electrical stimulation elicited a state of readiness to call (■), advertisement call production (□), adoption of the spawning posture (○), and mucus secretion (△). Reprinted from Behav Processes, 1, Knorr A, Central control of mating call production and spawning in the tree frog *Hyla arborea savignyi* (Audouin): Results of electrical stimulation of the brain 295–317, 1976, with permission from Elsevier.

advertisement call activity in normal and castrated *R. pipiens* most effectively when the electrodes were placed in the anterior preoptic area.

Stimulation of the preoptic area not only elicited calling activity but also some other behaviors (Knorr 1976), for example, an egg-laying posture in males that does not occur under normal conditions (Fig. 4.9). Moreover, calling activity could be induced in females, which, in this species, normally do not utter calls. These are to be expected because different motor patterns are represented in the preoptic/hypothalamic nuclei; which particular behavior is elicited is dependent on the sex and hormonal state, however (Schmidt 1966b; Urano and Gorbman 1981).

Anatomical (i.e., hodological) studies in *X. laevis* (Wetzel et al. 1985) and *B. orientalis* (Walkowiak 1992) using standard tract tracers support and supplement the physiological data described above. Injections of HRP into motor nuclei V, VII, IX–X, and XII retrogradely labeled cells in the corresponding contralateral motor nuclei, bilaterally in reticular formation nuclei, the PrV and the secondary isthmal nucleus, and ipsilaterally in the ventral tegmentum and the torus semicircularis. Injections of HRP into the area of the secondary isthmal nucleus in *B.*

orientalis resulted in retrograde labeling of cells in the ipsilateral cerebellum, the contralateral reciprocal nuclei, and different nuclei in the reticular formation, the ipsilateral tegmentum, the ipsilateral central and ventral thalamus, and the ipsilateral preoptic area. Iontophoretic injection into the laminar nucleus of the torus semicircularis labeled ipsilaterally descending neurons that gave off collaterals projecting dorsally towards the PrV. Additionally, Wetzel et al. (1985) described retrogradely labeled neurons in the striatum after injection of HRP–WGA into the dorsal tegmental area of the medulla (DTAM).

In a more recent study, Brahic and Kelley (2003) used fluorescent dextran amines to re-examine the anterograde and retrograde connectivity patterns of the vocal pathway in *X. laevis* in an isolated brain preparation (Luksch et al. 1996). They confirmed some of the connections described above, for example, the reciprocal connection between motor neurons and the PrV of the DTAM and the ventral striatum. The dorsal raphé also projects to vocalization-related motor units. In this latter study, neither sex differences nor the direct connection between the preoptic area and DTAM were observed. The connections between the preoptic area and the medullary vocal circuits appear to be indirect, via the nucleus raphé and the ventral striatum. According to Endepols et al. (2004), the origin of the striatal projection is the globus pallidus (the output structure of the basal ganglia), instead of the striatum *sensu strictu*.

4. Hormonal Control of Vocalization

Calling strongly depends on the hormonal state of the animal (for review see Emerson and Boyd 1999). The classical experiments in ethoendocrinology involved castration and/or treatment with androgens; such procedures abolish and restore advertisement calling, respectively (Schmidt 1966b; Palka and Gorbman 1973; Kelley and Pfaff 1976; Wetzel and Kelley 1983). Besides androgens, corticosteroids have been shown to also play a role in male calling. For example, calling males of *Bufo woodhousii* and *B. cognatus* have a higher circulating corticosterone concentration than silent satellite males (Leary et al. 2004). Also, arginine vasotocin has a stimulating effect on advertisement calling in several species (Schmidt and Kemnitz 1989; Penna et al. 1992; Boyd 1994; Marler et al. 1995; Propper and Dixon 1997; Chu et al. 1998), and at the same time it abolishes release calling in *R. pipiens* and *R. catesbeiana* (Diakow 1978; Boyd 1992). Finally, injections of human chorionic gonadotropin also bring anurans [*Bomina bombina* (Walkowiak 1988b), *B. orientalis* (Mohr and Schneider 1993), and *Hyla* cinerea and *H. versicolor* (Schmidt 1966)] into calling condition. In contrast, administration of prostaglandin abolishes advertisement calling in *Bufo americanus* and *Rana pipiens* (Schmidt and Kemnitz 1989).

Neurons in the vocal control centers of the brain concentrate gonadal steroids. Dihydrotestosterone-concentrating cells are found in the PrV of *X. laevis* (Kelley 1980, 1981), and the reticular formation, motor nuclei IX–X, and some spinal nuclei. The thalamus, the striatum, and the amygdala contain androgen receptors (Kelley 1980), and so does the preoptic area (Boyd and Ebersole 1997). According to Emerson and Boyd (1999), it is the interaction of arginine vasotocin with steroid hormones that acts on higher brain centers involved in vocalization.

5. Audio–Motor Integration

Conspecific calls can influence frogs' vocalization in several different ways. Silent males may be stimulated to utter advertisement or aggressive calls. This behavior, known as the evoked vocal response, has been described for many species in different families (see Wells and Schwartz, Chapter 3; Gerhardt and Huber 2002). To avoid acoustic overlap, call-timing of individual males may lead to antiphonal calling; that is, males try to place their calls at silent intervals of neighbors' calls. Two mechanisms—reset and inhibition—are responsible for timing the call pacemaker (Loftus-Hills 1973, 1974; Zelick and Narins 1982). Detailed analysis of the vocal response and antiphonal calling of *Bombina* showed that the interaction between the auditory and the vocal pathways is influenced by two distinguishable filter systems with different reaction times (Walkowiak 1988b). The latency of inhibition in antiphonal calling can still be effective 60 to 80 ms after the onset of an acoustic signal, but the latency of the vocal response may be several seconds due to the fact that calling is initiated by an increase of air volume in the buccal cavity and the lungs which takes place more slowly (Walkowiak 1992). A shorter reaction time of ca. 40 ms has been reported for *Smilisca sila* (Ryan 1986), a synchronized-calling species, however.

These findings suggest that audio–motor interaction is organized at several different levels of the brain. Three identified shortcuts at the brainstem level may account for a fast auditory input into the vocalization pathway (Aitken and Capranica 1984; Walkowiak 1992; Strake et al. 1994; Walkowiak and Luksch 1994): (1) direct input from the superior olive into the secondary isthmal nucleus and PrV, (2) descending fibers from the torus semicircularis may contact dendritic arborizations of pretrigeminal neurons, and (3) descending toral fibers could activate reticular neurons which in turn activate the secondary isthmal nucleus and the PrV.

Preliminary intracellular recordings in an isolated brain preparation from different branchial motorneurons during fictive respiration or calling have unraveled a short-latency effect of auditory nerve or auditory midbrain stimulation (Walkowiak unpublished). Most neurons were inhibited by brainstem auditory stimulation (Fig. 4.10). This is noteworthy because spontaneous fictive lung inflations also start with inhibition of the levator motorneurons (Kogo and Remmers 1994; Walkowiak unpublished).

Detailed data concerning audio–motor interactions are available for the midbrain (see also Wilczynski and Endepols, Chapter 8), with the most concrete data coming from intracellular recording and filling experiments in the torus semicircularis and tegmentum of *Bombina orientalis* and *Discoglossus pictus* (Walkowiak and Luksch 1994; Luksch and Walkowiak 1998; Endepols and

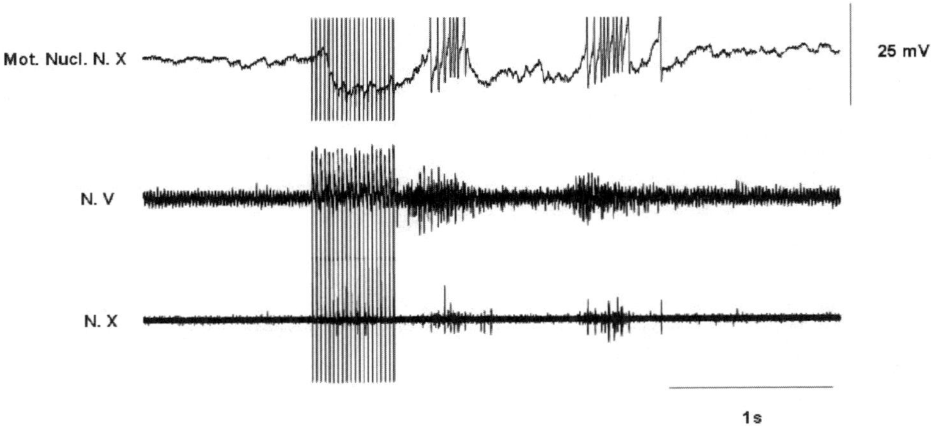

FIGURE 4.10. Upper trace: Intracellular recording of the membrane potential of a vagal motor neuron in an isolated brain of *Bombina orientalis* in response to electrical stimulation of the contralateral auditory nerve. Note the strong hyperpolarization ca. 75 ms after the onset of the stimulation and the subsequent bursts of action potentials. Middle and lower trace: Extracellular recordings of the electrical activities of N.V and N.X respectively. From Walkowiak (unpublished).

Walkowiak 1999, 2001). Auditory afferents primarily terminate in the principal nucleus of the torus semicircularis, a nucleus with mostly intrinsic projections. Cells of the laminar nucleus and the magnocellular nucleus extend their dendrites into the neuropil thus getting auditory input; these are the main projection neurons and their axonal collaterals reach auditory centers in the thalamus, tegmentum, periaqueductal gray (i.e., the secondary isthmal nucleus and PrV), reticular formation, and—in the case of neurons in the laminar nucleus—the spinal cord (Fig. 4.11). Dorsal tegmental cells receive direct auditory projections as well, and their axons project to different targets including the laminar nucleus. These data imply that individual toral and tegmental neurons serve both auditory and premotor functions.

Some axons of the laminar nucleus neurons project to the telencephalon and terminate in the ipsilateral striatum and contralateral limbic structures, for example, the nucleus accumbens, the lateral septum, and the diagonal band of Broca (Endepols and Walkowiak 2001; Roden et al. 2005; Endepols et al. 2005). Interestingly, lesions in these forebrain areas in *Hyla versicolor* alter the calling behavior of males. Specifically, the probability of vocal response is increased and the antiphonal call-timing is affected; that is, calls are often uttered during other vocalizations and not in the silent intervals thus producing acoustic overlap. If the lesions include the ventral striatum, calling activity is abolished altogether (Walkowiak et al. 1999).

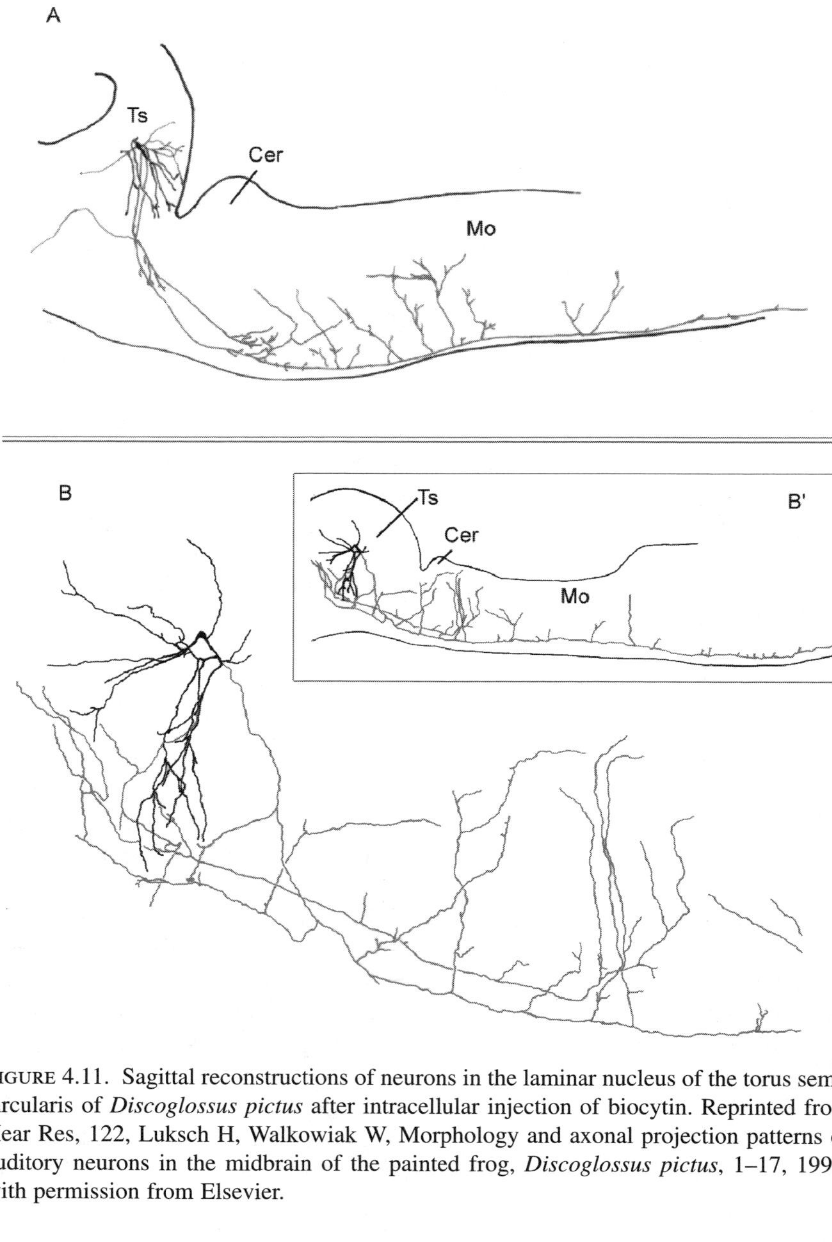

FIGURE 4.11. Sagittal reconstructions of neurons in the laminar nucleus of the torus semicircularis of *Discoglossus pictus* after intracellular injection of biocytin. Reprinted from Hear Res, 122, Luksch H, Walkowiak W, Morphology and axonal projection patterns of auditory neurons in the midbrain of the painted frog, *Discoglossus pictus*, 1–17, 1998, with permission from Elsevier.

The preoptic area and the ventral hypothalamus also receive auditory projections (Urano and Gorbman 1981; Allison and Wilczynski 1991; Allison 1992). Conspecific calls elicit stronger firing activities in neurons therein compared to synthetic simple signals. In contrast to the brainstem auditory nuclei, their neural responses do not reflect the temporal structure of the acoustic stimuli. Activation of the preoptic area gives rise to long latency and long-lasting vocal responses (Schmidt 1968b; Knorr 1976; Wada and Gorbman 1977b).

6. Summary and Conclusions

Anuran amphibians have evolved different calling mechanisms. Archaeobatrachian species lack vocal sacs and some of the species employ inspiratory call mechanisms, with lung inflation as the origin of airborne sound production. Expiratory call mechanisms are found in neobatrachian species that have evolved vocal sacs into which the expired air from the lungs is forced. In pipids, a reduced laryngeal mechanism (which appears to be highly derived) is involved in call production.

The acoustical characteristics of anurans' calls are determined by the mechanical properties of the larynx and the sound-emitting apparatus, as well as the central nervous system. The nervous system can control the temporal parameters of calls, that is, the frequency modulation, the pulse frequency, and the amplitude modulation. The vocal control centers are located at multiple levels of the brain. The central pattern generator resides in the brainstem, comprising the pretrigeminal nucleus in the DTAM and cell populations in the reticular formation (close to the motor nuclei IX–X). The call trigger that regulates the readiness to call resides in the preoptic area. Several forebrain areas, especially the striatum and the nuclei of the limbic system, serve to control calling, regulation of call strategy, and motivation.

Audio–motor interaction is not exclusively organized hierarchically, but is rather parallel and highly distributive (Fig. 4.12). Already at the level of the medulla, short-latency auditory input can be found in premotor and motor units. All call-related higher brain centers mentioned above also receive auditory input. Interestingly, neurons in the auditory as well as vocal control centers concentrate sex hormone receptors and neuromodulators; these may be responsible for seasonal (Goense and Feng 2005) and endogenous control of vocal activity.

To better understand the neural mechanisms of call generation and audio–motor interactions, further research is necessary. Physiological studies are necessary to elucidate the underlying mechanisms and validate the findings from the classical behavioral and endocrinological studies. A promising approach would be using a combination of intracellular studies in isolated brain preparations and in vivo analysis.

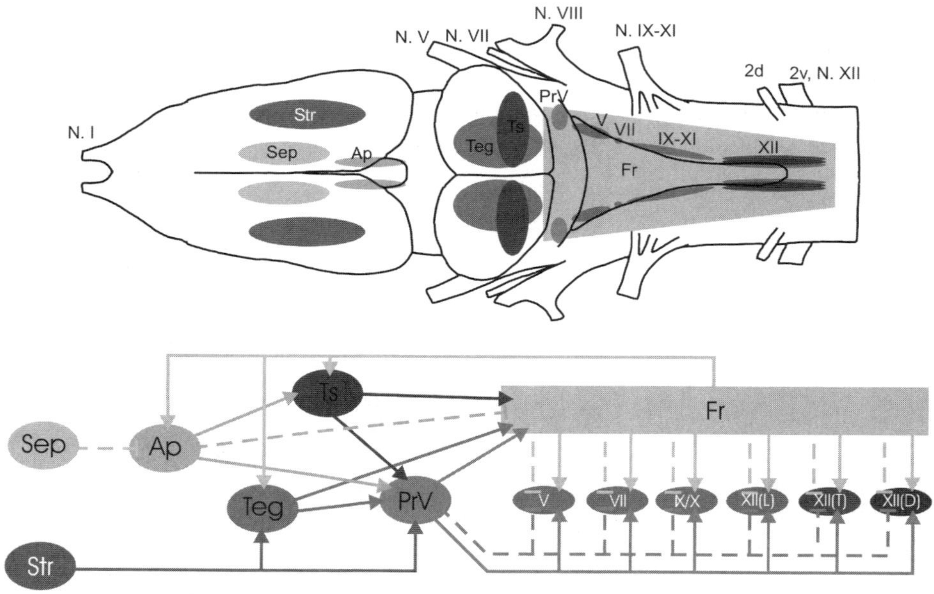

FIGURE 4.12. Upper part: dorsal schematic view of a *Bombina* brain with the nuclei and brain areas involved in vocal control and audio–motor integration. Lower part: proposed circuitry of audio–motor integration and triggering of vocalizations. ⟶ excitatory connection, ----- inhibitory connection.

Abbreviations:

V	N. trigeminus
VII	N. facialis
IX	N. glossopharyngeus
V	N. vagus
XII	N. hypoglossus
Ad lar	Aditus larynges
AP	Area praeoptica
C ary	Cartilago arytaenoidea
CER	Cerebellum
cor brach brev	Musculus coraco-brachialis brevis
cor brach long	Musculus coraco-brachialis longus
cor rad	Musculus coraco-radialis
delt	Musculus deltoideus
DTAM	Pretrigeminal nucleus of the dorsal tegmental area of the medulla
fl c rad	Musculus flexor carpi radialis
Fr	Formatio reticularis

hy	Hyoid
lar	Larynx
M dil lar, mdl	Musculus dilatator laryngis
M hyo lar	Musculus hyo-laryngeus
Mo	Medulla oblongata
M sphi ant, msphia	Musculus spincter anterior
M sphi post	Musculus spincter posterior
mggl	Musculus genioglossus
mhgl	Musculus hyoglossus
mgh	Musculus geniohyoideus
mia	Musculus intermandibularis anterior
mih	Musculus interhyoideus
mip	Musculus intermandibularis posterior
mom	Musculus omohyoideus
mpa	Musculus petrohyoidei anteriores
mpp	Musculi petrohyoideus posterior
mst	Musculus sternohyoideus
Ni	Mucleus isthmi
obl ext	Musculus obliqus externus
pal	Processus alare of the hyoid
ppl	Processus postero-lateralis of the hyoid
ppm	Processus postero-medialis
P post-med	Processus postero-medialis
Pa cr	Pars cricoidea of the Cartilago cico-trachealis
Pa tr	Pars trachealis of the Cartilago cico-trachealis
pect	Musculus pectoralis
PrV	Nucleus praetrigeminalis
R lar brev, rb X	Ramus laryngis brevis N. X
R lar long, rl X	Ramus laryngis longus N. X
rec	Musculus rectus abdominis
rih VII	Ramus interhyoideus of N. VII
rmi V	Ramus mandibularis internus N. V
rmi VII	Ramus mandibularis internus N. VII
Sep	Septum
Si	Nucleus secundarius isthmi
Str	Striatum
trans	Musculus transverses
Teg	Tegmentum
Ts	Torus semicircularis

References

Aitken PG, Capranica RR (1984) Auditory input to a vocal nucleus in the frog *Rana pipiens*: Hormonal and seasonal effects. Exp Brain Res 57:33–39.

Akef MSA, Schneider H (1985) Vocalization, courtship and territoriality in the Chinese fire-bellied toad *Bombina orientalis* (Anura, Discoglossidae). Zool Jb Physiol 89: 119–136.

Allison JD (1992) Acoustic modulation of neural activity in the preoptic area and ventral hypothalamus of the green treefrog (*Hyla cinerea*). J Comp Physiol A 171:387–395.

Allison JD, Wilczysnki W (1991) Thalamic and midbrain auditory projections to the preoptic area and ventral hypothalamus in the green treefrog (*Hyla cinerea*). Brain Behav Evol 38:322–331.

Boyd SK (1992) Sexual differences in hormonal control of release calls in bullfrogs. Horm Behav 26:522–535.

Boyd SK (1994) Arginine vasotocin facilitation of advertisement calling and call phonotaxis in bullfrogs. Horm Behav 28:232–240.

Boyd SK, Ebersole TJ (1997) Distribution of androgen receptors in the brain of two anuran amphibians. Soc Neurosci Abstr 23:2381.

Brahic CJ, Kelley DB (2003) Vocal circuitry in Xenopus laevis: Telencephalon to laryngeal motor neurons. J Comp Neurol 464:115–130.

Chu J, Marler CA, Wilczynski W (1998) The effects of arginine vasotocin on the calling behavior of male cricket frogs in changing social contexts. Horm Behav 34: 248–261.

De Jongh HJ, Gans C (1969) On the mechanism of respiration in the bullfrog, *Rana catesbeiana*: A reassessment. J Morph 127:259–290.

Diakow C (1978) A hormonal basis for breeding behaviour in female frogs: Vasotocin inhibits the release call of *Rana pipiens*. Science 199:1456–1457.

Eichelberg H, Schneider H (1973) Die Feinstruktur der Kehlkopfmuskeln des Laubfrosches Hyla arborea arborea (L.) im Vegleich zu einem Skelettmuskel. Z Zellforsch 141:223–233.

Emerson SB, Boyd SK (1999) Mating vocalizations of female frogs: Control and evolutionary mechanisms. Brain Behav Evol 53:187–197.

Endepols H, Walkowiak W (1999) Influence of descending forebrain projections on processing of acoustic signals and audiomotor integration in the anuran midbrain. Eur J Morph 37:182–184.

Endepols H, Walkowiak W (2001) Integration of ascending and descending inputs in the auditory midbrain of anurans. J Comp Physiol A 186:1119–1133.

Endepols H, Roden K, Luksch H, Dicke U, Walkowiak W (2004) Dorsal striatopallidal system in anurans. J Comp Neurol. 468:299–310.

Endepols H, Roden K, Walkowiak W (2005) Hodological characterization of the septum in anuran amphibians: II. Efferent connections. J Comp Neurol 483:437–457.

Endepols H, Walkowiak W, Luksch H (2000) Chemoarchitecture of the anuran auditory midbrain. Brain Res Rev 33:179–198.

Gaupp E (1896) Ecker's A, Wiederheim's R Anatomie des Frosches. 1. Abteilung: Lehre vom Skelett und vom Muskelsystem. 3rd ed. Braunschweig Vieweg.

Gerhardt HC, Huber F (2002) Acoustic Communication in Insects and Anurans. Common Problems and Diverse Solutions. Chicago and London: University of Chicago Press.

Girgenrath M, Marsh RL (1997) In vivo performance of trunk muscles in tree frogs during calling. J Exp Biol 200:3101–3108.

Goense J, Feng AS (2005) Seasonal changes in frequency tuning and temporal processing in single neurons in the frog auditory midbrain. J Neurobiol 65:22–36.

Gridi-Papp M (2003) Mechanism, behavior and evolution of calling in four North American treefrogs. PhD thesis, University of Texas at Austin.

Kelley DB (1980) Auditory and vocal nuclei in the frog brain concentrate sex hormones. Science 207:553–555.

Kelley DB (1981) Location of androgen-concentrating cells in the brain of *Xenopus laevis*: Autoradiography with ^{3}H-dihydrotestosterone. J Comp Neurol 199:221–231.

Kelley DB (2004) Vocal communication in frogs. Curr Opin Neurobiol 14:751–757.

Kelley DB, Pfaff DW (1976) Hormone effects on male sex behavior in adult South African clawed frogs, *Xenopus laevis*. Horm Behav 7:159–182.

Kelley DB, Tobias ML (1999) The vocal repertoire of *Xenopus laevis*. In: Hauser M, Konishi M (eds.) Neural Mechanisms of Communication. Cambridge, MA: MIT Press, pp. 9–35.

Kelley DB, Morrell JI, Pfaff DW (1975) Autoradiographic localization of hormone-concentrating cells in the brain of an amphibian, *Xenopus laevis*. I, Testosterone. J Comp Physiol 164:47–62.

Knorr A (1976) Central control of mating call production and spawning in the tree frog *Hyla arborea savignyi* (Audouin): Results of electrical stimulation of the brain. Behav Processes 1:295–317.

Knorr A, Schneider H (1975) Beitrag zur Morphologie des Gehirns von *Hyla arborea arborea* (L.) (Hylidae, Anura). Bonn Zool Beitr 26:370–382.

Kogo N, Remmers JE (1994) Neural organization of the ventilatory activity in the frog, *Rana catesbeiana*. II. J Neurobiol 25:1080–1094.

Leary CJ, Jessop TS, Garcia AM, Knapp R (2004) Steroid hormone profiles and relative body condition of calling and satellite toads: Implications for proximate regulation of behavior in anurans. Behav Ecol 15:313–320.

Loftus-Hills JJ (1973) Neural mechanisms underlying acoustic behaviour of the frog *Pseudophryne semimarmorata* (Anura: Leptodactylidae). Anim Behav 21:781–787.

Loftus-Hills JJ (1974) Analysis of an acoustic pacemaker in Strecker's chorus frog, Pseudacris streckeri (Anura: Hylidae). J Comp Physiol 90:75–87.

Lörcher (1969) Vergleichende bio-akustische Untersuchungen an der Rot- und Gelbbauchunke *Bombina bombina* (L.) und *Bombina v. variegata* (L.) Oecologia 3:84–124.

Luksch H, Walkowiak W (1998) Morphology and axonal projection patterns of auditory neurons in the midbrain of the painted frog, *Discoglossus pictus*. Hear Res 122: 1–17.

Luksch H, Walkowiak W, Munoz A, ten Donkelaar HJ (1996) The use of in vitro preparations of the isolated amphibian central nervous system in neuroanatomy and electrophysiology. J Neurosci Methods 70:91–102.

Marler CA, Chu J, Wilczynski W (1995) Arginine vasotocin injection increases probability of calling in cricket frogs, but causes call changes characteristic of less aggressive males. Horm Behav 29:554–570.

Martin WF (1971) Mechanics of sound production in toads of genus *Bufo*: Passive elements. J Exp Zool 176:274–294.

Martin WF (1972) Evolution of vocalization of the genus *Bufo*. In: Blair WF (ed) Evolution in the Genus Bufo. Austin: University of Texas Press, pp. 279–309.

Martin WF, Gans C (1972) Muscular control of the vocal tract during release signaling in the toad *Bufo valliceps*. J Morph 137:1–28.

Mohr S, Schneider H (1993) Analysis of the call pacemaker if the Chinese fire-bellied toad, *Bombina orientalis* (Boulenger, 1890) (Amphibia, Anura), and its operation during auditory stimulation. Zool Jb Physiol 97:215–231.

Palka YS, Gorbman A (1973) Pituitary and testicular influenced sexual behaviour in male frogs, *Rana pipiens*. Gen Comp Endocrinol 21:148–151.

Paulsen K (1965) Beiträge zur Anatomie und Physiologie des Froschkehlkopfes (mit besonderer Berücksichtigung der Stimmbildung). Z Wiss Zool 172:1–16.

Paulsen K (1967) Das Prinzip der Stimmbildung in der Wirbeltierreihe und beim Menschen. Akad Verl Ges, Frankfurt, pp. 1–143.

Penna M, Capranica RR, Somers J (1992) Hormone-induced vocal behavior and midbrain auditory sensitivity in the green treefrog, *Hyla cinerea*. J Comp Physiol A 170:73–82.

Propper CR, Dixon TB (1997) Differential effects of arginine vasotocin and gonadotropin-releasing hormone on sexual behaviors in an anuran amphibian. Horm Behav 32: 99–104.

Roden K, Endepols H, Walkowiak W (2005) Hodological characterization of the septum in anuran amphibians: I. Afferent connections. J Comp Neurol 483:415–436.

Ryan MJ (1986) Synchronized calling in a treefrog (*Smilisca sila*). Brain Behav Evol 29: 196–206.

Ryan MJ, Drewes RC (1990) Vocal morphology of the *Physalaemus pustulosus* species group (Family: Leptodactylidae): Morphological response to sexual selection for complex calls. Biol J Linn Soc 40:37–52.

Sassoon D, Gray G, Kelley DB (1987) Androgen regulation of muscle fiber type in the sexually dimorphic larynx of *Xenopus laevis*. J Neurosci 7:3198–3206.

Sassoon D, Segil N, Kelley DB (1986) Androgen-induced myogenesis and chondrogenesis in the larynx of *Xenopus laevis*. Dev Biol 113:135–140.

Schmid E (1977) The vocal apparatus of *Bombina bombina* (L.), *Bombina v. variegata* (L.) and *Alytes o. obstetricans* (Laur.). Zool Jb Anat 98:171–180.

Schmid E (1978) Contribution to the morphology and histology of the vocal cords of Central European anurans (Amphibia). Zool Jb Anat 99:133–150.

Schmidt RS (1965) Larynx control and call production in frogs. Copeia 1965:143–147.

Schmidt RS (1966a) Central mechanisms of frog calling. Behaviour 26:252–285.

Schmidt RS (1966b) Hormonal mechanisms of frog calling. Copeia 1966:637–644.

Schmidt RS (1968a) Chuckle calls of the Leopard frog (*Rana pipiens*). Copeia 1968: 561–569.

Schmidt RS (1968b) Preoptic activation of frog mating behaviour. Behviour 30:239–257.

Schmidt RS (1971) A model of the central mechanism of male anuran acoustic behaviour. Behaviour 39:288–317.

Schmidt RS (1972) Action of intrinsic laryngeal muscles during release calling in leopard frog. J Exp Zool 181:233–244.

Schmidt RS (1973) Central mechanisms of frog calling. Amer Zool 13:1169–1177.

Schmidt RS (1974) Neural correlates of frog calling—trigeminal tegmentum. J Comp Physiol A 92:229–254.

Schmidt RS (1980) Succinic dehydrogenase staining of anuran pretrigeminal nucleus. Brain Behav Evol 17:411–418.

Schmidt RS (1981) Pretrigeminal nucleus of leopard frog concentrates 2-deoxy-D-glucose during release calling. Exp Brain Res 42:212–214.

Schmidt RS (1992) Neural correlates of frog calling: production of two semi-independent generators. Behav Brain Res 50:17–30.

Schmidt RS, Kemnitz CP (1989) Anuran mating calling circuits: Inhibition by prostaglandin. Horm Behav 23:361–367.

Schneider H (1970) Morphologie des Larynx von *Hyla a. arborea* (L.) und *Hyla meridionalis* Boetgger (Amphibia, Anura). Z Morph Ökol Tiere 66:299–309.

Schneider H, Hussein F, Akef MSA (1986) Comparative bioacoustic studies in the yellow-bellied toad, *Bombina variegata* (L.), and relationships of European and Asian species and subspecies of the genus *Bombina* (Anura, Amphibia). Bonn Zool Beitr 37:49–67.

Strake J (1995) Atmung und Rufgenerierung bei *Bombina orientalis* und *Discoglossus pictus* (Discoglossoidea, Anura): Morphologische und physiologische Untersuchungen. Thesis: University of Cologne.

Strake J, Luksch H, Walkowiak W (1994) Audio-motor interface in anurans. Eur J Morph 32:122–126.

Tobias ML, Kelley DB (1987) Vocalizations of a sexually dimorphic isolated larynx: Peripheral constraints on behavioral expression. J Neurosci 7:3191–3197.

Tobias ML, Barnard C, O'Hagan R, Horng SH, Rand M, Kelley DB (2004) Vocal communication between male *Xenopus laevis*. Anim Behav 67:353–365.

Tobias ML, Viswanathan S, Kelley DB (1998) Rapping, a female receptive call, initiates male/female duets in the South African clawed frog. Proc Natl Acad Sci USA 95: 1870–1875.

Trueb L, Cannatella DC (1982) The cranial osteology and hyolaryngeal apparatus of *Rhinophrynus dorsalis* (Anura: Rhinophrynidae) with comparison to recent pipid frogs. J Morph 171:11–40.

Urano A, Gorbman A (1981) Effects of pituitary hormonal treatment of responsiveness of anterior preoptic neurons in male leopard frogs, *Rana pipiens*. J Comp Physiol 141:163–171.

Wada M, Gorbman A (1977a) Relation of mode of administration of testosterone to evocation of male sex behaviour in frogs. Horm Behav 8:310–319.

Wada M, Gorbman A (1977b) Mate calling induced by electrical stimulation in freely moving leopard frogs, *Rana pipiens*. Horm Behav 9:141–149.

Wahl (1969) Untersuchungen zur Bio-Akustik der Wasserfrosches *Rana esculaenta* (L.). Oecologia (Berl.) 3:14–55.

Walkowiak W (1980) The coding of auditory signals in the Torus semicircularis of the fire-bellied toad and the grass frog: Responses to simple stimuli and to conspecific calls. J Comp Physiol A 138:131–148.

Walkowiak W (1988a) Neuroethology of anuran call recognition. In: Fritzsch B, Ryan MJ, Wilczynski W, Hetherington TE, Walkowiak W (eds) The Evolution of the Amphibian Auditory System. New York: Wiley, pp. 485–509.

Walkowiak W (1988b) Two auditory filter systems determine the calling behavior of the fire-bellied toad. A behavioural and neurophysiological characterization. J Comp Physiol A 164:31–41.

Walkowiak W (1992) Acoustic communication in the fire-bellied toad: An integrative neurobiological approach. Ethol Ecol Evol 4:63–74.

Walkowiak W, Luksch H (1994) Sensory motor interfacing in acoustic behaviour of anurans. Amer Zool 34:685–695.

Walkowiak W, Berlinger M, Schul J, Gerhardt HC (1999) Significance of forebrain structures in acoustically guided behavior in anurans. Eur J Morph 37:177–181.

Weber E (1974) Vergleichende Untersuchungen zur Bioakustik von *Discoglossus pictus* OTTH 1837 und *Discoglossus sardus* TSCHUDI, 1837 (Discoglossidae, Anura). Zool Jb Physiol 78:40–84.

Weirich D, Strake J, Walkowiak W (1989) Motor patterns of respiration and vocalization in the fire-bellied toad. In: Elsner N, Singer W (eds) Dynamics and Plasticity in Neural Systems. Proceedings of the 17th Göttingen Neurobiology Conference. Stuttgart, New York: Thieme Verlag, p. 230.

Wells KD (2001) The energetics of calling in frogs. In: Ryan MJ (ed) Anuran Communication. Washington Smithsonian Institution Press, pp. 45–60.

Wetzel DM, Kelley DB (1983) Androgen and gonadotropin effects on male mate calls in South African clawed frogs, *Xenopus laevis*. Horm Behav 17:388–404.

Wetzel DM, Haerter UL, Kelley DB (1985) A proposed neural pathway for vocalization in South African clawed frogs, *Xenopus laevis*. J Comp Physiol A 157:749–761.

Yager D (1982) A novel mechanism for underwater sound production in *Xenopus borealis*. Am Zool 22:887.

Yager DD (1992) A unique sound production mechanism in the pipid anuran *Xenopus borealis*. Zool J Linn Soc 104:351–375.

Yager DD (1996) Sound production and acoustic communication in fishes and frogs. In: Tinsley RC, Kobel HR (eds) The Biology of *Xenopus*. New York: Oxford University Press.

Yamaguchi A, Kelley DB (2000) Generating sexually differentiated vocal patterns: Laryngeal nerve and EMG recordings from vocalizing male and female African clawed frogs (*Xenopus laevis*). J Neurosci 20:1559–1567.

Zelick RD, Narins PM (1982) Analysis of acoustically evoked call suppression behaviour in a neotropical treefrog. Anim Behav 30:728–733.

5
Recognition and Localization of Acoustic Signals

H. Carl Gerhardt and Mark A. Bee

1. Introduction

Frogs and toads (Order Anura) are among the most vocal of vertebrates. Yet in most species only the male produces long-range signals. Such advertisement calls are produced predominately in the context of reproduction, serving to keep rivals at bay while attracting gravid females (see Wells and Schwartz, Chapter 3). Although they may use other sensory modalities at close range (e.g., Narins et al. 2003; Rosenthal et al. 2004), frogs and toads display highly specific vocal and phonotactic responses to playbacks of acoustic signals alone. Here we focus on the recognition and localization of acoustic signals and some of the fitness consequences of vocal communication. We also consider some hypotheses about broad-scale patterns of evolution of communication systems.

Although some species of frogs and toads produce simple tones or trills, the acoustic signals of other species have relatively complex spectral and temporal properties that are equivalent in complexity to the acoustic elements of signals produced by higher vertebrates, including the phonetic segments of human speech (Gerhardt 1978a; Gerhardt and Huber 2002). There is no evidence that young frogs and toads learn to produce calls or to selectively respond to them by listening to them during their development (Doherty and Gerhardt 1984a). Adults of some territorial species, however, learn to recognize the calls of their neighbors and adjust their aggressive responses accordingly (Section 4.2). Frogs and toads of many species are common and easily approached for observation in the field. Individuals of both sexes also respond reliably to playbacks of prerecorded and computer-generated signals in both the field and laboratory. These attributes of frogs and toads and their relatively simple communication systems have made them favorite subjects for a wide range of research: the biophysics of sound production, peripheral and central mechanisms of sound reception and sound pattern recognition, mechanisms of sound localization, environmental effects on acoustic signals, male–male competition for calling sites and territories, signal timing interactions, female mate choice, interspecific interactions, and speciation (Gerhardt and Huber 2002).

One disadvantage of studying anurans is that no widely used, standard methods exist for conditioning the animals to respond to sounds [see Feng and Schul, Chapter 11 and Elephandt et al. (2000) for a method that has been used in a completely aquatic frog]. As a consequence, if an animal fails to respond to playbacks, it may well have detected the stimulus but is unmotivated to respond. For example, nongravid females or males and females tested outside of the breeding season are unlikely to respond to playbacks of conspecific calls. More significantly, a frog or toad in breeding condition may detect a stimulus and fail to respond because the stimulus is not recognized as a biologically meaningful signal. This explanation is supported if the same animal responds readily to stimuli of known biological relevance. A positive response indicates that the animal has both detected and recognized the stimulus. Similarly, in simultaneous playbacks of two or more stimuli that each elicit responses in isolation, selective responses by individuals indicate that some difference between the sounds has been resolved [i.e., exceeds the just-noticeable difference (JND)] and that the difference is meaningful to the animals [i.e., exceeds what Nelson and Marler (1990) have termed the just-meaningful difference (JMD)]. When animals fail to respond selectively, the difference in the stimuli is below the JND, the JMD, or both. In practice, reproductively active frogs and toads often respond selectively in tests of synthetic signals modeled after conspecific calls in which differences in the value of a single acoustic property probably just exceed the resolving power of their auditory system.

In this chapter, we highlight behavioral studies of communication in the field and laboratory, emphasizing sexual selection and its implications for the evolution of signals and receiver responses in contemporary populations. We also consider interactions between species and other factors that have contributed to geographic variation in communication systems and speciation. Throughout the chapter we indicate ways in which acoustic selectivity in anurans reflects the underlying mechanisms or suggests hypotheses about such mechanisms. This approach is particularly well established in studies of sound localization, which we cover in the last section of the chapter.

2. Biological Contexts of Acoustic Communication

Communication is shaped in part by the external environment, that is, the communication channel (see Feng and Schul, Chapter 11), and by the fitness consequences of sending and responding to signals. For males of most anuran species, vocal communication is the principal means by which individuals compete to attract females or to acquire resources required by females. Calling is also energetically costly and may attract predators as well as potential mates. For females, important fitness consequences arise from mate choice. Females that respond to signals of inferior conspecific males or males of other species will usually waste time, gametes, or produce less viable or less fit offspring. Moving around to find and assess different males is costly in terms of time and predation risks. A primary environmental pressure on mate acquisition in both genders arises from the background noise of the large aggregations of calling males in which communication usually takes place.

2.1 Chorusing

During the height of the breeding season, aggregations of calling males, which are termed "choruses," contain dozens to tens of thousands of individuals (Duellman and Trueb 1986). Choruses are focal points for sexual selection (Section 2.2), and the background noise generated by the chorus represents a significant challenge for the detection and recognition of signals by individual males and females (Schwartz 1993; Sections 2.2 and 2.3; Wells and Schwartz, Chapter 3; Feng and Schul, Chapter 11). Directional hearing and the spatial separation of calling individuals are two factors that can ameliorate this problem to some extent (Section 8).

Choruses potentially serve as beacons for other males and for females (reviewed in Gerhardt and Huber 2002). Signaling males would presumably benefit by attracting more females by joining a chorus than by calling alone. Other males and females would benefit because the location of a suitable breeding site may be facilitated, and females could be sure that reproductively active males were present. Although sound intensity does not increase linearly as more males call, the continuity of sound output from a chorus would nearly always be greater than that of single males because individuals do not call continuously but rather in bouts. That is, a male produces a series of calls (the bout) over a period of several to many minutes, which is followed by a variable period of silence before the next series.

Surprisingly little empirical support is available for the hypothesis that other males and females use chorus sounds to find the breeding site (see Feng and Schul, Chapter 11) or to make a decision about when to breed. Indeed, in the barking treefrog (*Hyla gratiosa*), the usual number of females arrived at the pond during the usual time period on nights when the chorus was experimentally nullified as when the chorus was allowed to form (Murphy 2003). Because there was a positive correlation between the numbers of males and females arriving on nights when no chorusing occurred, frogs of both genders probably use the same or a similar set of environmental conditions to decide when to go to the pond for breeding (Murphy 2003). Similar experimental studies of other species are needed, however, before the chorus-attraction hypothesis can be confidently rejected.

2.2 Sexual Selection

Sexual selection operates on variation within one gender—usually males—in mating success. The two commonly recognized forms of sexual selection are male–male competition and female choice (Andersson 1994). Male–male competition in anurans causes variation in mating success because not all males are equally capable of defending a calling space or territory or for calling for prolonged periods of time in one or more choruses. Many studies have shown that chorus attendance is the most reliable predictor of mating success (reviewed in Gerhardt 1994a; Halliday and Tejedo 1995), a predictable consequence even if females arrive at the pond and mate randomly with the first conspecific male whose calls they detect (Friedl and Klump 2005). Female choice also results in variation in mating success and is expected when females can freely choose their mate based on male qualities, the resources he defends, or both. In species that form resource-based leks (Alexander 1975), females are likely to rely almost exclusively on differences in the properties of the conspecific advertisement call.

As shown below (Section 5), playback experiments have identified the particular properties that females assess and the most effective range of values of such relevant properties. An important question remains, however. To what extent are females able to assess these call properties in nature? Choruses are noisy places, and females are unlikely to be able to detect the calls of more than a few males from any one area (Gerhardt and Klump 1988; see Wells and Schwartz, Chapter 3; Feng and Schul, Chapter 11). Even in relatively simple experimental conditions, females were somewhat less selective in choosing among four different synthetic calls than when choosing between two such calls (Gerhardt 1987, 1991). Choruses are also dangerous places, and moving from area to area subjects female frogs and toads to a variety of sit-and-wait predators, such as snakes, turtles, fish, giant water bugs, and even large frogs of other species (reviewed in Gerhardt and Huber 2002). These factors probably limit the degree of choice exercised by females in nature and may account for the paucity of field studies that have robustly demonstrated female choice based on male calls or other, correlated, attributes such as size (Wells and Schwartz, Chapter 3).

2.3 Interspecific Interactions

The calls of other species can have two significant effects on communication. First, if heterospecific males are much more common than conspecific males, then their calls can constitute a significant source of background noise. This is especially true if the spectral properties of heterospecific calls are similar to those of conspecific calls, but the potential for masking interference is high even when carrier frequencies differ by as much as an octave because the calls of many species of frogs and toads are so intense (Gerhardt 1975). Second, if the properties of the calls of heterospecific males are similar to those of conspecific calls, females might be attracted to these signals and risk mismating, which almost always involves a loss of fitness (Section 7.2). In principle, however, the discrimination against heterospecific signals is not a process that is distinctly different from (female) mate choice vis-à-vis different conspecific signals (Ryan and Rand 1993a). Indeed, females would be expected to prefer conspecific calls even if there were no history of selection against heterospecific signals simply because of coevolutionary forces (e.g., Gerhardt 1994a; further discussion in Section 7.2).

3. Statistical Analysis of Calls

Because advertisement calls are so essential for reproductive success and potentially convey different kinds of information about the signaler, statistical descriptions of patterns of variation in their physical properties are a logical starting point for analyzing vocal communication in anurans. These analyses delimit the potential for these properties to reliably encode attributes such as species, geographical origin, sex, size, individual identity, motivational state, and physical and even genetic fitness. A complete analysis considers patterns of variation at the individual, local population, and geographical levels (Gerhardt 1991). At the individual level some kinds of properties typically show very little variation (static properties) and others are highly variable (dynamic properties). Between these two ends of the continuum are some properties that can be either stereotyped or highly variable and hence do not readily fall into either of these arbitrary categories (Gerhardt and Huber 2002). These analyses are germane to this chapter because patterns of variation in signal properties are almost certainly shaped over evolutionary time by the preferences and recognition mechanisms of receivers.

3.1 Static Properties

Static properties are highly invariant from call to call within and between bouts of calling by an individual. Typically these properties include spectral features such as the fundamental frequency or carrier frequency or frequencies (frequency components or bands with the greatest relative amplitude) and fine-scale temporal properties such as the repetition rate, duration, and rise–fall characteristics of the short sounds (pulses) that make up the calls of many anuran species. Within-individual variability in these properties of advertisement calls is thus constrained, even though the values of a static property such as pulse rate may be significantly different in other kinds of calls, such as in aggressive or encounter calls (see Figs. 5.1 and 5.2). This last observation suggests that in addition to constraints arising from physics or call-production mechanisms, receivers (especially females) may impose limits on within-male and among-male variation in static properties (Gerhardt 1991; Section 5.2.1).

3.2 Dynamic Properties

Individual frogs and toads readily alter other properties of advertisement calls within and between calling bouts. These are usually gross-temporal properties such as the rate of calling, the rate of call-note production (if calls have such subunits), and the duration of calls or call-notes (if present; Gerhardt and Huber 2002). Although neuromuscular mechanisms are likely to set an upper limit on the values of dynamic properties, comparative studies indicate that for some species, there are energetic constraints as well (Wells 2001; see Wells and Schwartz, Chapter 3). Much of the variation in dynamic properties can be traced

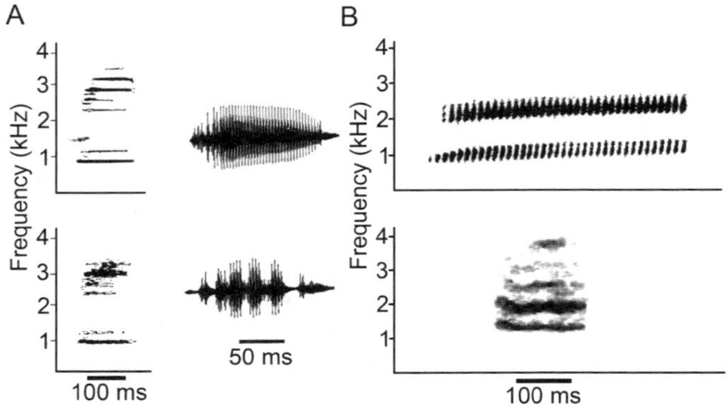

FIGURE 5.1. Examples of advertisement and aggressive calls in treefrogs. (A) Sonograms of an advertisement call (top) and aggressive call (bottom) from a male of the green treefrog (*Hyla cinerea*). Oscillograms to the right of the sonograms show that the aggressive call (bottom) has a strongly pulsed structure throughout the duration of the signal compared to the advertisement call (top), which has a pulsatile beginning (with a higher rate of amplitude modulation). (B) Sonograms of a typical advertisement call (top) of a gray treefrog (*Hyla chrysoscelis*) and a typical aggressive call (bottom). Advertisement calls are long trills with two main frequency bands and regularly repeated short pulses, whereas aggressive calls are much shorter and have more prominent harmonics.

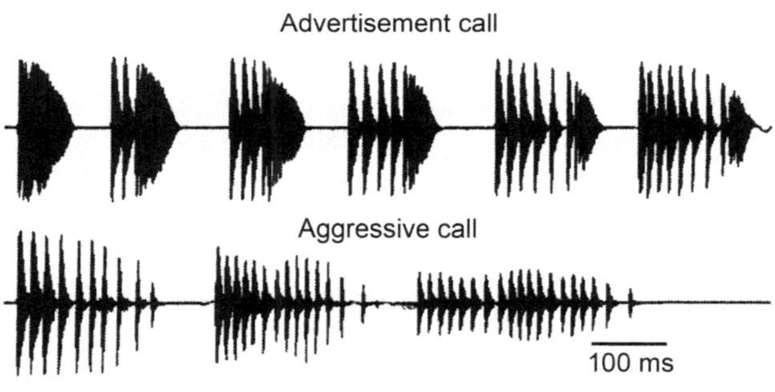

FIGURE 5.2. Example of a graded series of calls. Oscillograms in the top trace show an advertisement call (first call to left) of the reed frog (*Hyperolius marmoratus*) and a series of intermediate calls that differ in the number of cycles of pulsing. Bottom trace shows examples of aggressive calls, which are pulsed throughout. From Grafe (1995).

to interactions among males that are competing for females. Males in dense choruses are more likely to produce energetically costly long-duration or high-rate calls than males calling alone or in sparse choruses (e.g., Wells and Taigen 1986). Such signals may be more easily detected against the chorus background, and even in quiet conditions, females usually prefer longer, high-rate alternatives (see Sections 5.2.2 and 7.3).

4. Male–Male Communication

Male–male competition is an important form of sexual selection (Andersson 1994). In anurans, competitive interactions among males for access to females, calling sites, and breeding territories involve acoustic communication. Compared to previous work with female frogs and toads (Sections 5 to 8), however, far fewer studies have investigated signal recognition and localization by males. Much of the work on vocal communication by male frogs has focused on signal-timing interactions (see Wells and Schwartz, Chapter 3), discrimination among different call types (e.g., Brenowitz et al. 2001), and information transfer during aggressive contests (e.g., Burmeister et al. 2002). As noted earlier, males use distinct aggressive vocalizations in defense of calling sites and territories, and these calls function to deter encroachment or intrusion by rivals (Whitney 1980). Hence male receivers clearly discriminate among advertisement and aggressive calls. In addition, vocal signals can provide important information about the signaler—body size, fighting ability, physiological condition, individual identity—that can be used by male receivers to determine the most appropriate behavioral response in agonistic encounters.

4.1 Recognition of Different Call Types

Usually the two most common vocalizations produced by males are advertisement and aggressive calls, and these call types often differ in the presence/absence or rate of amplitude modulation (AM) (Fig. 5.1). Field playback tests with males and females of the Pacific treefrog, *Hyla regilla,* indicate that both sexes discriminate between a "diphasic" advertisement call and an aggressive call (reviewed in Brenowitz et al. 2001). These two call types have nearly identical frequency spectra, but differ in pulse repetition rate.

Aggressive responses rapidly habituated when males of *H. regilla* were stimulated for several minutes with playbacks of either advertisement calls or aggressive calls at sound pressure levels that exceeded the subject's threshold for responding aggressively to the playback. Importantly, however, habituation was associated with a temporary elevation in a male's "aggressive threshold" that was specific to the type of call used as a stimulus. For example, thresholds for eliciting aggressive responses to advertisement calls from males that had previously habituated to playbacks of this call type were 7 dB higher than in a preplayback period, whereas aggressive thresholds for responding to the aggressive call increased less than 2 dB. The reverse asymmetry in elevated thresholds was found when aggressive calls were used as the habituating stimulus. These results support the hypothesis that call type recognition is mediated by the processing of temporal properties of advertisement and aggressive calls in separate channels in the central auditory system. Electrophysiological recordings in the auditory midbrain (torus semicircularis) suggest that call recognition may depend on "counting" neurons that selectively respond to certain pulse repetition rates and temporally integrate over several interpulse intervals of the correct duration before responding (Adler and Rose 1998; Edwards et al. 2002).

4.2 Recognition of Different Callers

Across a wide range of taxa, territorial animals often respond less aggressively toward their adjacent territorial neighbors compared to unfamiliar conspecifics or "strangers" based on learning to recognize their neighbors by voice (reviewed in Temeles 1994). This ability is thought to represent an adaptation that allows territory residents to avoid wasting time and energy in unnecessary agonistic interactions with neighbors that possess their own territory and respect a mutual territory boundary. This form of social recognition appears to be more common among species that defend long-term breeding or multipurpose territories (Temeles 1994). Frogs probably represent one of the best taxa for future studies of the mechanisms and evolution of vocally mediated neighbor recognition systems because of the overwhelming importance of acoustic signals in male–male interactions and because the defense of long-term breeding or multipurpose territories has almost certainly had multiple evolutionary origins in anurans.

4.2.1 Territorial Neighbor Recognition in Frogs

In a field playback experiment, Davis (1987) demonstrated that male bullfrogs (*Rana catesbeiana*) respond less aggressively to the advertisement calls of an adjacent territorial neighbor compared to those of an unfamiliar conspecific broadcast from the boundary of the neighbor's territory (Fig. 5.3A). Davis (1987)

FIGURE 5.3. Territorial neighbor recognition in bullfrogs, *Rana catesbeiana*. (A) Male bullfrogs responded with more aggressive calls and closer approaches toward the speaker in response to broadcasts of the advertisement calls of a stranger compared to those of a neighbor (*left*) and in response to broadcasts of the neighbor's advertisement calls from a novel location compared to broadcasts from the direction of the neighbor's territory (*right*). (B) Results from an habituation–discrimination experiment showing changes in an aggression index during an initial habituation phase (stimulus periods 1–30) and a subsequent discrimination phase (DP). The aggression index was determined from a principal components analysis of aggressive calling, movements, and approaches toward the speaker and is expressed on a log-scale as a percentage of the mean response during the first three stimulus periods (block 1) of the habituation phase. Over the 30 repeated stimulus periods of the habituation phase (approx 4 h), aggressive responses decreased significantly. With additional presentations of the same stimulus from the same location during the DP of the control condition (*upper left*), responses remained at low levels. Habituated responses recovered during the DP in responses to the same stimulus broadcast from a different location (*lower left*), a stimulus with a different fundamental frequency broadcast from the same location (*upper right*), or a stimulus with a different fundamental frequency broadcast from a different location (*lower right*). Data in (A) redrawn from Davis (1987) and data in (B) redrawn from Bee and Gerhardt (2002).

also demonstrated that bullfrogs, like many territorial songbirds, associate their neighbor's advertisement calls with the direction from which the neighbor normally called (Fig. 5.3A). The ability of male frogs to discriminate among the calls of neighbors and strangers, or presumed familiar and unfamiliar calls, has been tested subsequently in three additional species. In the dendrobatid *Colostethus beebii*, males approached the playback speaker and emitted aggressive calls during playbacks of the calls of nonneighbors, but remained stationary, called antiphonally, or ignored playbacks of the calls of neighbors (Bourne et al. 2001). Males of the agile frog, *Rana dalmatina*, increased the duration of their calls by adding pulses in responses to playbacks of presumed unfamiliar calls compared to the presumed familiar calls of a male that was resident in the same pond (Lesbarrères and Lodé 2002). Neighbor–stranger discrimination was also recently

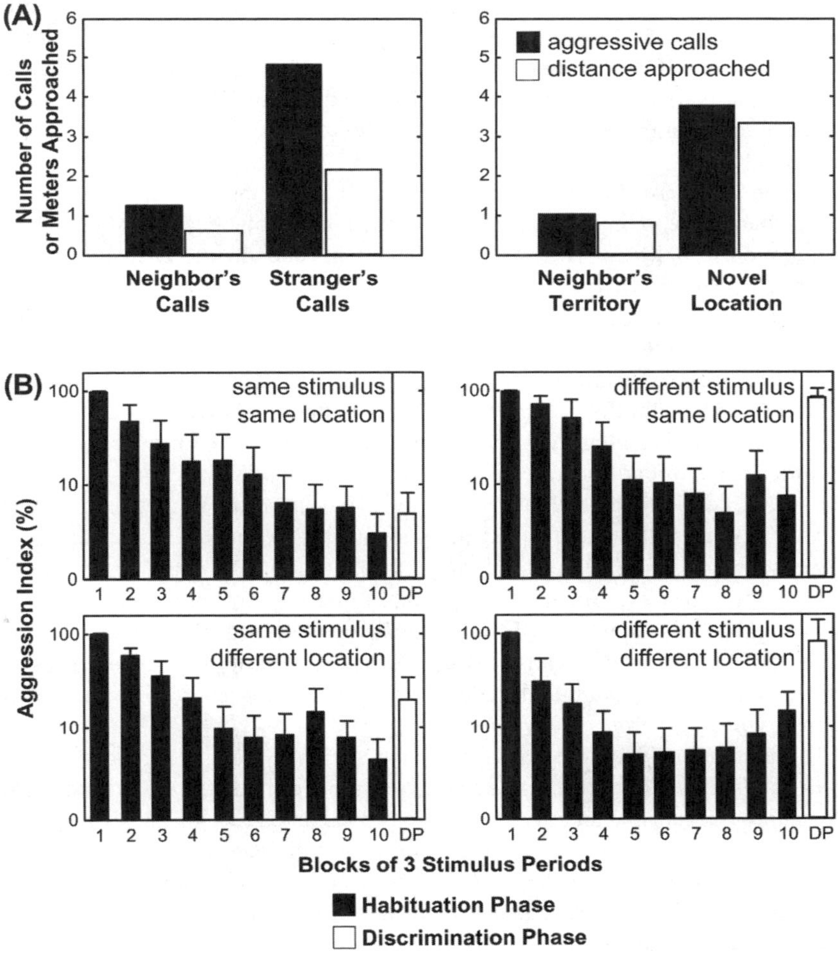

tested in a second dendrobatid, the strawberry dart-poison frog (*Dendrobates pumilio*). Despite evidence that males of this species defend long-term breeding territories and that advertisement calls are individually distinct, territorial males did not behaviorally discriminate between the calls of a nearby neighbor and a stranger (Bee 2003a).

4.2.2 Acoustic and Perceptual Bases of Neighbor Recognition in Bullfrogs

In acoustical and statistical analyses of bullfrog advertisement calls, fundamental frequency and correlated spectral properties were the most individually distinct call properties and contributed most toward assigning calls to the correct individuals (Bee and Gerhardt 2001a; Bee 2004). Other call properties also varied significantly more among individuals than within individuals, but not to the same degree as fundamental frequency. By comparing the distributions of among-individual and within-individual differences in fundamental frequency, Bee (2004) predicted that the JMD in this call property should be about 4%. The rationale was that, even if males could discriminate smaller differences in frequency (i.e., have a smaller JND), they should not act on these smaller differences because fundamental frequency could vary up to 4% (coefficient of variation) within individuals.

Bee and Gerhardt (2001b,c, 2002; Bee 2003b) used the habituation–discrimination paradigm to investigate the perceptual basis of neighbor–stranger discrimination in bullfrogs (Fig. 5.3B). Changes in fundamental frequency on the order of 5 to 10% elicited a recovery of habituated aggression, but 2% changes in frequency failed to do the same. These results accord well with predicted JMDs for this call property based on patterns of individual variation (Bee 2004). Additional playback experiments revealed a number of important related findings. With repeated playbacks, males formed long-lasting memories of a call that were specific to a fundamental frequency (Bee and Gerhardt 2001c); males were able to associate a familiar fundamental frequency with a particular sound source location (Bee and Gerhardt 2001b); and males could discriminate between familiar and unfamiliar calls that differed in fundamental frequency independently of sound source location (Bee and Gerhardt 2002; Fig. 5.3B). Together, the results of these playback experiments suggest that repeated exposure to the calls of a new neighbor allows male bullfrogs to encode memories of the fundamental frequency of the neighbor's calls that could, in turn, allow them to discriminate among familiar and unfamiliar callers based on differences in this individually distinct voice property.

The patterns of response habituation and stimulus-specific recovery demonstrated in field playback tests can exclude a number of potential hypotheses about the neural mechanisms underlying the vocally mediated recognition of neighbors in bullfrogs. For example, the observation that habituated responses recover to changes in fundamental frequency, location, or both, excludes the hypothesis that reduced aggression toward a neighbor results from fatigue or from changes in overall arousal or aggressive motivation. Likewise, the stimulus-specificity and location-specificity of habituation rule out peripheral explanations (e.g., sensory adaptation) and hypotheses based on a simple form of neural habituation in the ascending auditory system. Findings from other behavioral and neurophysiological studies of bullfrogs are consistent with this view (Megela and Capranica 1983; Boatright-Horowitz et al. 2000). Additional physiological studies are needed to characterize the form of neural plasticity in the central nervous system that mediates voice recognition in bullfrogs.

4.3 Assessments of Resource Holding Power

In escalated physical fights between male frogs, larger body size confers advantages and usually determines contest outcome (e.g., Davies and Halliday 1978; Howard 1978; Arak 1983; Wagner 1989b). Hence, large males are said to have higher "resource holding power (or potential)" (RHP; reviewed in Maynard Smith 1982). The fundamental frequency and correlated spectral properties of frog calls depend on the size of the vocal folds in the larynx, which is related to overall body size (Martin 1972), and these properties are often the best acoustic predictors of male body size (Robertson 1986; Wagner 1989a; Bee and Gerhardt 2001a). Davies and Halliday (1978) first showed that the perception of size-related differences in fundamental frequency can influence the course of aggressive contests in a laboratory study of common toads (*Bufo bufo*). Several field playback studies have since demonstrated that male frogs in some species persistently attack or direct aggressive vocalizations toward the high-frequency calls of a simulated small frog and stop calling, retreat, or adopt a satellite posture in response to the low-frequency calls of a large frog (e.g., Robertson 1986; Given 1987; Arak 1983; Wagner 1989b). Such results are consistent with the hypothesis that male frogs can acoustically assess the RHP of their competitive rivals. The extent to which this applies to frogs in general, however, requires additional study.

Exceptions to the general pattern of size assessment in frogs have been provided in recent playback studies of cricket frogs (*Acris crepitans*; Burmeister et al. 2002) and North American bullfrogs (*Rana catesbeiana*; Bee 2002). In bullfrogs, for example, males responded similarly to synthetic calls that acoustically simulated a large male, a small male, or a male of about the same size. Moreover, there is now evidence from several frog species that the fundamental frequency or spectral properties of calls can be altered, especially during actual or simulated social interactions among competing males (reviewed in Bee and Bowling 2002; see also Wells and Schwartz, Chapter 3). Although a few species increased or decreased the fundamental frequency of their advertisement calls (e.g., Lopez et al. 1988), males of most other species either lowered the fundamental frequency of their calls (e.g., Wagner 1989b; Grafe 1995; Bee et al. 2000; Bee and Bowling 2002; Burmeister et al. 2002) or shifted the acoustic energy within their broadband signals toward lower frequencies (e.g., Given 1999). In general, we still do not know how receivers respond to males that alter the spectrum or temporal fine-structure of their calls; thus, the significance of these alterations in relation to either assessments of RHP or female choice remains unclear and deserves further study.

5. Selective Phonotaxis in Females

Females of many species of frogs and toads reliably show phonotaxis to playbacks of prerecorded advertisement calls or synthetic sounds modeled after these natural signals (Gerhardt 1988, 1994a). These experiments have identified the relevant acoustic properties of signals necessary to attract gravid females for mating and the most effective values of these relevant properties. These data bear upon or generate hypotheses about the underlying mechanisms of call recognition, sexual selection, and broad-scale patterns of evolution.

5.1 Relevant and Irrelevant Properties

Even though some acoustic properties of advertisement calls are highly stereotyped and invariant within and between males of a given species, they do not appear to be required for effective attraction of gravid females. Examples include the frequency modulation of calls and pulses (Gerhardt 1978a, 2005b; Doherty and Gerhardt 1984b; Hödl et al. 2004) and the pulsatile beginning of the call of the green treefrog (*H. cinerea*; Fig. 5.4). Conclusions from these studies are based on playback experiments in which females were offered a choice of two synthetic calls, one with and one without these features. Previous experiments had showed that the synthetic call with the property in question was as attractive as typical prerecorded calls. Because females did not show preferences, the difference in the acoustic stimuli must not be critical or even relevant for mate choice, at least under the low-noise conditions of the laboratory. As emphasized above (Section 1), however, these experiments do not prove that the animals cannot discriminate between the test signals.

Closely related species often differ in which aspects of advertisement calls are relevant. For example, whereas females of *H. cinerea* do not require a pulsatile beginning, females of their sister taxon *H. gratiosa* do require this feature (Gerhardt 1981). In the diploid–tetraploid sibling species of gray treefrog, females of *H. chrysoscelis* (diploid) evaluate calls on the basis of pulse rate and females of *H. versicolor* (tetraploid), on the basis of pulse duration and intervals (Schul and Bush 2002). Females of *H. versicolor* are selective for the species difference in pulse shape and make choices based on subtle differences in pulse rise-time (Gerhardt and Doherty 1988; Gerhardt and Schul 1999) when the pulse duration of alternatives is at species-typical values. Females of *H. chrysoscelis* were selective for species differences in pulse shape only when the pulse duration of alternatives was increased so that absolute differences in rise-time were comparable to those discriminated by *H. versicolor* (Gerhardt 2005b). These results suggest that preferences can be hidden or revealed depending on evolu-

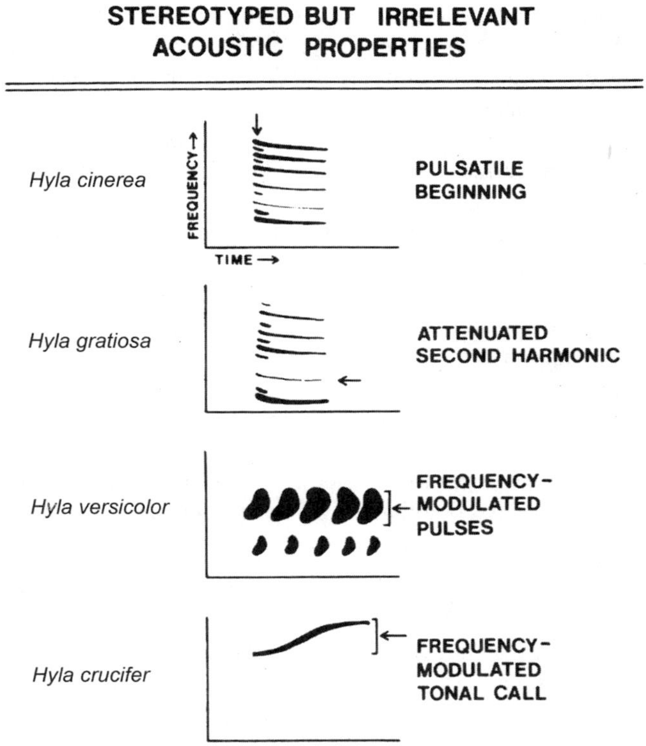

FIGURE 5.4. Examples of properties that are typical of advertisement calls in four species of frogs but which are apparently irrelevant for recognition or preferences in females of the same species. The arrows (from the top figure to the bottom) indicate the irrelevant properties in the cartoons of sonograms: (1) a pulsatile beginning; (2) an attenuated second harmonic; (3) frequency modulated pulses; (4) a frequency modulated tonal call. Modified from Gerhardt (1988).

tionary changes in the quantitative values of a property relative to the resolving power of the auditory system or the magnitude of a JMD.

5.2 Preference Functions

Preference functions quantify how variation in the values of relevant properties affects female preferences. Typically this is accomplished by playback experiments in which the value of one property at a time is varied in a systematic fashion (Wagner 1995; Bush et al. 2001; Gerhardt and Huber 2002). In single-speaker playbacks, responses are scored relative to responses to some "standard" synthetic call that has properties typical of the advertisement call of the species. In multiple-speaker tests, the standard call is one of the alternatives. It is important to test enough values of the property in question to establish the shape and peak (if any) of the preference function. Even if females prefer values from one end of the range of variation to values at the other end, the selection on the trait might still be weak. In studies of frequency preferences in *Physalaemus pustulosus*, for example, females preferred a low-frequency synthetic call that was about 1.5 standard-deviation units below the mean in the population to a high-frequency alternative that was about 2.5 standard-deviation units above the mean (Bosch et al. 2000). But no preference was found when the frequencies of the two alternatives were 1.5 standard-deviation units above and below the mean, respectively. Thus, the mating success of relatively few individuals is likely to be reduced, and individuals with calls having values at or close to the mean (the majority of males) would also be at least as likely to attract mates as any other male. Estimates based on playbacks in natural and semi-natural conditions also suggested that, even though females of the gray treefrog (*H. versicolor*) prefer long- to short-duration calls in simple laboratory conditions (Gerhardt et al. 1996), only males with very short-duration calls are likely to have significantly reduced chances of attracting females in nature (Schwartz et al. 2001).

5.2.1 Unimodal Preference Functions

Many anuran preference functions are unimodal (Appendix 4 in Gerhardt and Huber 2002). That is, there is an optimal range of values, and signals with lower or higher values are less attractive. Typically the most effective range matches or is close to average values of advertisement calls of conspecific males in the same population. In terms of sexual selection, this form of preference function represents stabilizing selection if the match with the mean is very close or weakly directional selection if the preferred range is somewhat higher or lower than the mean. Unimodal preference functions are typical for static properties such as carrier frequency (e.g., Doherty and Gerhardt 1984b; Gerhardt 1991, 2005a; Grafe 1997; Márquez and Bosch 1997; Castellano and Giacoma 1998) and pulse rate (e.g., Arak 1988; Castellano and Giacoma 1998; Bush et al. 2001; Fig. 5.5). Furthermore, values of these properties that are close to the mean in conspecific advertisement calls are usually preferred by females and correspond with sensory tuning in the frequency and time domains, an observation that inspired the matched-filter concept (Capranica and Moffat 1983). Gerhardt and Schwartz (2001) provide a critical discussion, with an emphasis on nonlinear effects.

5.2.2 Strongly Directional Preference Functions

Preference functions for the gross temporal properties (duration of long calls; call rate) of anuran advertisement calls are often highly directional with respect to the range of variation in conspecific calls: values at one end of the range or even beyond are more attractive than average values and values at the other end of the range (Appendix 4 in Gerhardt and Huber 2002). There are many examples in other animals as well, and these kinds of functions led Ryan and Keddy-Hector

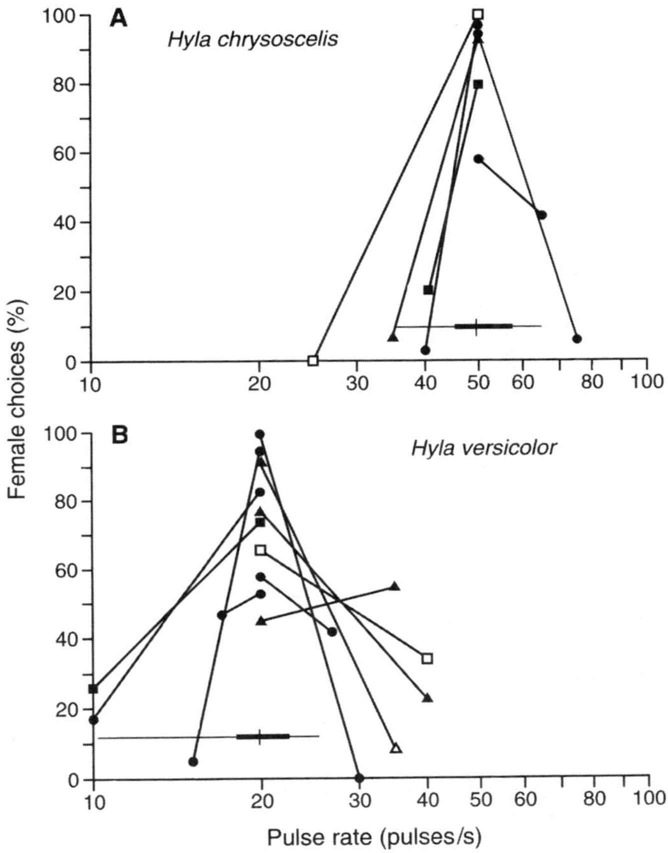

FIGURE 5.5. Examples of unimodal preference functions in gray treefrogs based on the natural pattern of variation in pulse rate, in which pulse duration and interpulse intervals maintain about the same 1 : 1 ratio over the range of natural variation. (A) Preference in the diploid gray treefrog, *Hyla chrysoscelis.* Each line connects two points showing the percentage of females choosing each alternative. For example 100% of the females chose the "standard" call of 50 pulses/s and 0% chose the alternative of 25 pulses/s. (B) Preference function for the tetraploid gray treefrog *Hyla versicolor.* The symbols indicate the relative intensities of the two alternatives. □: the SPL of the standard call was 18 dB less than that the alternative; ▲: the SPL of the standard call was 12 dB less than that of the alternative; ■: the SPL of the standard call was 6 dB less than that of the alternative. •: the SPL of the standard call and the alternative was the same (85 dB). The diagrams between (A) and (B) show the ranges of variation in pulse rate uncorrected for temperature (thin lines), the range of variation corrected to 20°C (thick horizontal lines), and the mean pulse rate at 20°C (the vertical line). Females of *H. chrysoscelis* are highly selective in rejecting calls with lower pulse rates in the range of *H. versicolor*, whereas many females of *H. versicolor* chose alternatives with pulse rates within the range of variation of *H. chrysoscelis* calls. However, in these experiments, the alternatives all had pulses with the conspecific shape. Females of *H. versicolor* were as selective as females of *H. chrysoscelis* when alternatives with higher pulse rates also had pulses with the shape typical of *H. chrysoscelis* pulses (Diekamp and Gerhardt 1995).

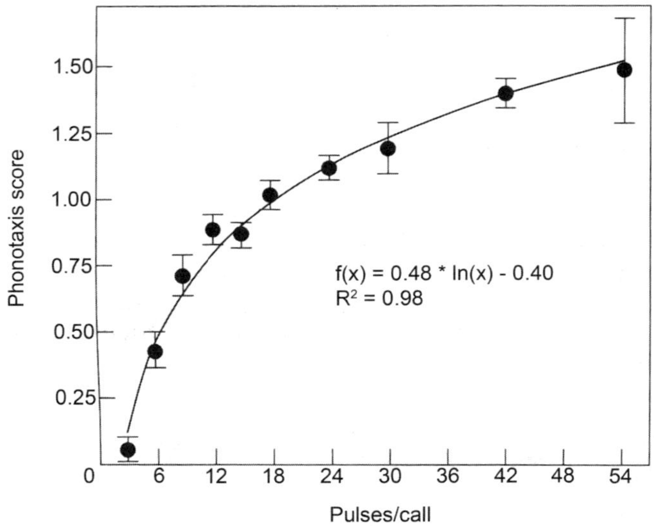

FIGURE 5.6. Preference function based on the relative speed of phonotaxis by females of *H. versicolor* to single-stimulus presentation of synthetic calls that differed in pulse number (= duration of the call). The average pulse number in advertisement calls in the population was about 18 per call. Even though females responded more rapidly to calls with more than the average number of pulses, the increase in speed of the response was not linearly related to the increase in pulse number. Similar results were obtained in two-stimulus choice tests (Gerhardt et al. 1996, 2000); preference strength was greater for the longer of a pair of alternatives with fewer than the average number of pulses in comparison with the preference strength for the longer of a pair of alternatives with more than the average number of pulses. From Bush et al. (2001).

(1992) to propose a strong correlation between signal attractiveness and the quantity of stimulation of an animal's sensory system (see also Section 7.3). However, these functions are usually not linear and open-ended: as values are increased even more, the effectiveness of the stimulus levels off or decreases (e.g., Gerhardt et al. 1996, 2000; Beckers and Schul 2004; Fig. 5.6).

5.3 Interaction of Preferences Based on Multiple Properties

Several different properties of male signals vary independently within calling bouts and among males. The overall attractiveness of a signal may thus depend on how females weigh the values of these different properties. There have been two approaches to this problem. Some studies using synthetic calls have simultaneously varied the values of two relevant properties. In gray treefrogs, systematic variation in pulse duration and the interpulse interval yielded very different preference spaces for the two cryptic species (Schul and Bush 2002). In another study of one of these species (*H. chrysoscelis*), females weighed pulse rate (stabilizing preference function) more heavily than call duration (directional preference function) in tests of individuals from populations sympatric with *H. versicolor* (Gerhardt 1994b; Section 7.2). Finally, experiments using synthetic calls showed that simultaneous variation in pulse rate and shape affected the strength of preferences in females of *H. versicolor* (Diekamp and Gerhardt 1995).

Another, recent approach has been to test signals that differ in multiple acoustic properties, regardless of whether variation in a given property is known to affect female preferences in isolation. It could well be that variation in a property that seemingly has no effect on signal attractiveness in isolation could have an effect in combination with variation in other properties. Ryan and Rand (2003) used multidimensional scaling of 15 different call variables from recordings of 300 calls of 50 males of the Túngara frog (*Physalaemus pustulosus*) to define an acoustic space. They then chose as test signals the calls of individual frogs that were representative of this space and assayed the relative attractiveness to females of these test calls to estimate a perceptual space. Not surprisingly, the perceptual space did not coincide with the signal space because that would have required that all the properties that defined the acoustic space be equally weighed by females. The perceptual space was less variable than the signal space (hence potentially reducing rather than enhancing acoustic variation), and signal similarity sometimes but not always predicted relative attractiveness.

5.4 Temperature Effects

Frogs are ectotherms. Hence their physiology is strongly affected by ambient temperature. Because some species breed over a relatively wide range of temperature, some call properties can vary significantly within and between individuals. The most strongly affected properties are temporal features such as pulse and call rate and the duration of pulses and calls: rates increase and durations decrease as temperature increases. Temperature can also affect receivers. One of the two inner ear organs of anurans—the amphibian papilla—has hair cells whose frequency response is significantly influenced by temperature (reviewed in Lewis and Narins 1999).

In two species of anurans (gray treefrogs: *Hyla chrysoscelis* and *H. versicolor*), female preferences for fine-scale temporal properties (pulse rate, duration or both) are affected by temperature and change in a fashion that parallels temperature-dependent changes in the values of these properties in male calls (Gerhardt 1978b, 2005b). This phenomenon has been termed "temperature coupling" (Gerhardt 1978b). Thus, females at some particular temperature are most likely to be attracted to males of their own species calling at about the same temperature.

Temperature effects on frequency tuning and preferences can sometimes have a very different result. Whereas temperature has a direct effect on the preferences of females of the green treefrog (*H. cinerea*) for the frequency of the low-frequency peak in the advertisement call of the male (the lower the temperature, the lower the preferred frequency), the low-frequency peak in the call is hardly affected at all (Gerhardt and Mudry 1980). At low temperatures, females even preferred calls with a low-frequency peak in the range of variation of another species to calls with a peak typical of conspecific males. This kind of mismatch is likely to be rare in nature, however, because female green treefrogs are seldom reproductively active at the lower end of the temperature range at which some males call. The temperature-dependence of peripheral tuning could, however, be viewed as a "preadaptation" for evolutionary change in the frequency of vocalizations (see Section 7.3) should there be a shift in the breeding season of populations or species to different temperature regimes.

6. Biological Consequences of Female Selectivity

In this section we consider the main fitness consequences of selective phonotaxis on males and females at the level of populations. Although it is obvious that female preferences exert selection on male signals, the fitness consequences of mate choice on females is just as important for predicting the evolution of the communication system. Females are thus expected to use acoustic criteria that provide reliable information about the suitability of prospective mates.

6.1 Selection on Male Call Properties

As discussed above (Section 5.2), numerous studies have demonstrated that female preferences can potentially exert selection on male calls: preferred values of particular properties are often higher or lower than the mean in the population (reviewed in Gerhardt and Huber 2002, their Appendix 4). The extent to which such selection can cause evolutionary change will depend on the existence of heritable variation underlying phenotypic variation in these acoustic properties. Unfortunately, there have been no formal estimates of the heritability of call-properties.

6.2 Benefits of Female Choice

Sexual selection theory distinguishes between direct and indirect benefits of female choice (Andersson 1994). Direct benefits include factors that increase the number of offspring and the health of the offspring and female. For example, in an Australian species (*Uperoleia laevigata*), increased fertilization success derives from size-assortative mating, which presumably results in better apposition of the male and female cloacae (Robertson 1990). Females showed size-dependent frequency preferences that were consistent with the choice of males of the correct relative size. Direct benefits in the form of increased fertilization success also occurred in female spadefoot frogs (*Spea multiplicata*) that chose males producing calls at a higher-than-average rate (Pfennig 2000). There is, however, little evidence that female choice of calls is correlated with parasite- or disease-free males (Hausfater et al. 1990).

Indirect benefits of female choice include enhanced viability, genetic quality, and sexual attractiveness of male offspring. Welch et al. (1998) used a half-sib design to show that the offspring of male gray treefrogs (*H. versicolor*) that produced preferred long calls had greater viability (survival, growth rate, size at metamorphosis) than did the offspring of males that produced relatively unattractive short calls. Because the clutches of females were divided and fertilized externally (artificially and randomly) by sperm collected from males of the two classes, there was no possibility for females to influence differences in the offspring. Thus, the differences in offspring quality had to be caused by genetic differences in the males. Such "good genes" effects might also be augmented by increased attractiveness of the sons of the long-caller fathers depending on the heritability of this call property. More recent field experiments demonstrated genotype-by-environment effects, in which the expression of increased viability depended on the rearing environment (Welch 2003).

7. Broad-Scale Patterns of Evolution

Geographic variation in acoustic signals and female preferences has been documented in numerous studies of frogs and toads (reviewed in Gerhardt and Huber 2002). Given the lack of evidence for learning of calls or preferences during development, these geographic differences are probably heritable differences in these traits, and reflect recent evolutionary history. A common pattern for acoustic signals is the differentiation of one or more properties at two ends of the geographical distribution of a wide-ranging species, with clinal variation (gradual change) in geographically intermediate populations. Clinal variation is usually indicative of gene flow. Nonclinal patterns of geographic variation are more interesting and may reflect selection arising from habitat acoustics (Section 7.1) or interactions between species whose ranges partially overlap (Section 7.2).

The comparative approach, whereby the phylogenetic relationships of groups of related species and appropriate outgroup taxa are analyzed using a variety of algorithms, generates hypotheses about the evolution of signals and preferences over longer periods of time (reviewed in Harvey and Pagel 1991). At the very least, these methods can often distinguish between similarities that reflect genetic relatedness and similarities that are caused by convergence, implicating similar evolutionary responses to the same environmental factors in relatively distantly related taxa. By the same logic, divergence is more interesting when it occurs in closely related taxa rather than distantly related ones, again because of the possibility of identifying environmental causes. These methods can also be used to generate hypotheses about patterns of coevolution of signals and receiver biases or preferences (Section 7.3).

7.1 Geographical Variation and Habitat Acoustics

Both spectral and temporal properties of acoustic signals are subject to degradation as they propagate through natural environments (Feng and Schul, Chapter 11). For anurans, the best evidence for environmental effects on signal structure concerns two subspecies of cricket frog, *Acris c. crepitans* and *Acris c. blanchardii*. The first subspecies is mainly found in still bodies of water in pine forests and produces calls of higher frequency carriers, longer duration, and slower rates than the other subspecies, whose breeding sites are located in more open habitats (Ryan and Wilczynski 1991). Males of *A. c. blanchardii* in an isolated area of pine forest tended to produce calls that were similar to those of the other subspecies, suggesting a habitat effect. Nevertheless, the calls of this subspecies also showed enormous and unpredictable variation in structure in various parts of the rest of its distribution, which were generally open grasslands of some form.

Little evidence for habitat effects has been found in studies in which the calls of various species of anurans have been played back in different habitats (Penna and Solís 1998; Kime et al. 2000; Feng and Schul, Chapter 11). This state of affairs is perhaps not so surprising if frogs and toads do not use long-range acoustic signals to locate breeding areas (Section 2.1). Rather, these signals are probably intense (and hence potentially long-range) because individuals are selected to make their signals detectable against the loud and complex chorus environment in which most calling occurs (e.g., Gerhardt and Klump 1988). Research exploring the effects of environmental degradation of signals in preferred and nonpreferred habitats on male and female receivers is badly needed.

7.2 Reproductive Character Displacement

When the communication systems of two species do not diverge sufficiently in areas of allopatry to prevent misidentification of calls when contact is reestablished, then selection arising from the negative consequences of mismatings is expected to occur. The predicted result is that the communication system of one or both of the taxa will change so that differences between the species in areas of sympatry will be greater than in areas of allopatry (Dobzhansky 1940). This geographical pattern has been termed reproductive character displacement. A full discussion of the terminology, theoretical background, and empirical data is provided by Noor (1999) and Gerhardt and Huber (2002).

Hybridization in frogs and toads is commonly detected because male hybrids produce distinctive advertisement calls (Gerhardt 1974). Although these individuals reach sexual maturity, hybrid matings usually result in at least some reduction in the viability and fertility of the offspring, and in some combinations, no viable offspring are produced (reviewed in Blair 1964). Male hybrids that reach maturity produce calls that are less attractive to females of the parental species, and hybrids of both genders are probably at a disadvantage in terms of their ecological adaptations (e.g., Mecham 1960).

Frogs and toads have provided several robust examples of divergence in communication systems in sympatric areas that are consistent with reproductive character displacement (Gerhardt and Huber 2002). Two classic examples involve shifts in the putative species-identifying properties (pulse rate) of advertisement calls (e.g., *Litoria verreauxii* and *L. ewingii*; *Pseudacris feriarum* and *P. nigrita*; Littlejohn 1965; Fouquette 1975). Alternatively, selection on females is likely to be more severe than on males because a female may lose its only chance of successful reproduction if it mismates whereas males potentially have more opportunities. Hence selection on the female side of communication systems might be more common than currently recognized by evolutionary biologists. Two studies are available that have sampled females from multiple populations. In the gray treefrog (*H. chrysoscelis*), females from populations where its genetically incompatible and cryptic sister species (*H. versicolor*) was also present weighed differences in pulse rate (which distinguish the calls of the two species) much more heavily than they did call duration (which overlaps broadly in the calls of the two species; Gerhardt 1994b). Females from allopatric areas much more frequently chose an alternative with a long call duration that had a pulse rate within or close to the range of variation in *H. versicolor*. The intensity-independence of preferences for synthetic conspecific calls was also greater in sympatric areas than in allopatric areas (Gerhardt 1994b). A similar pattern was found in female green treefrogs (*H. cinerea*) vis-à-vis synthetic calls that simulated those of a closely related congener, the barking treefrog (*H. gratiosa*; Fig. 5.7; Höbel and Gerhardt 2003).

7.3 Preexisting Sensory Biases

Sensory systems are seldom tuned precisely to species-typical values of the pertinent properties of conspecific calls, and some significant mismatches have been found (reviewed in Gerhardt and Huber 2002). Conspecific signals are thus not necessarily more effective in stimulating sensory systems than signals with other acoustic properties. Indeed, the early ethologists identified many examples of supernormal stimuli, which were often characterized by their greater size, loudness, duration, or repetition rate in comparison with typical conspecific signals. Ryan and his colleagues, studying the Túngara frog (*Physalaemus pustulosus*) species group, proposed the "sensory exploitation" hypothesis, which holds that a preexisting bias for extra stimulation by acoustic appendages arose early in the evolutionary history of this group of frogs (reviewed in Ryan and Rand 1993b). Mutations giving rise to the production of appendages by males of some species were then immediately favored by females and the acoustic appendages ultimately became established as part of the vocal repertoire.

The evidence for sensory exploitation in anurans comes from playback experiments with two species (Ryan et al. 1990). Males of *P. pustulosus* add "chucks" to their whine, which can attract females in the absence of chucks; chucks alone are ineffective. Whines alone are less attractive than combinations of whines plus chucks or whines plus other acoustic appendages. Males of *P. coloradorum* do

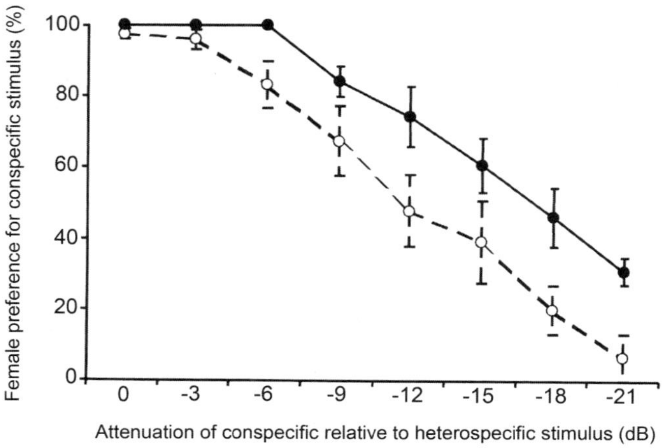

FIGURE 5.7. Reproductive character displacement in preference strength in females *of Hyla cinerea*. Females were given a choice between synthetic calls modeled after conspecific calls and those of *H. gratiosa*, and the relative intensity of the conspecific model was reduced until there was no longer a preference. Preference strength was defined as the difference at which the preference persisted. The solid points and line show the results for females from sympatric populations and the open circles and dotted line, the results from allopatric populations. Notice that when the SPL of the conspecific model was reduced by 6 dB or more relative to the heterospecific model, higher proportions of females from sympatric populations chose the conspecific model than did females from allopatric populations. From Höbel and Gerhardt (2003).

not produce chucks, but females prefer whines plus chucks. Together with a phylogenetic hypothesis indicating that chucks and other appendages evolved in only some species rather than being lost in *P. coloradorum* and its close relatives, this is the main evidence for preexisting biases in this group of Neotropical frogs. Almost any kind of extra acoustic stimulation within the hearing range, including noise bursts and "double-whine" calls sometimes produced by males of *P. coloradorum*, resulted in preferences in *P. pustulosus*. Surprisingly, there was only a weak, nonsignificant trend for a preference for double-whines to a single whine in *P. coloradorum*, suggesting that the form of the bias has not been completely conserved.

Ryan and his colleagues suggest that preferences observed in other species of frogs, insects, and other animals that are apparently based on increased sensory stimulation are consistent with predictions of the preexisting bias hypothesis (e.g., Ryan and Keddy-Hector 1992). The fact that anurans have two auditory organs with different ranges of frequency sensitivity offers an opportunity for testing more than just the relative attractiveness of signals of longer duration, higher rate, and with additional elements (appendages). That is, under the preexisting bias hypothesis we might predict a trend for the evolution of calls with bimodal spectra, with one peak exciting the amphibian papilla, and the other simultaneously exciting the basilar papilla. Consistent with this idea, synthetic signals with two peaks were more effective in eliciting behavioral responses than were signals with one peak that primarily stimulated only one organ in five species of anurans in which conspecific calls have bimodal spectra (*Hyla cinerea, H. gratiosa, H. chrysoscelis, H. versicolor*, and *Rana catesbeiana*; Lombard and Straughan 1974; Gerhardt 1981, 2005a; Capranica and Moffat 1983). Perhaps the ancestors of some or all of these species had a single peak that stimulated only one organ effectively. This idea has been tested in female cricket frogs (*Acris crepitans*; Witte et al. 2001) and midwife toads (*Alytes cisternasii*; Bosch and Boyero 2003), in which males produce signals with a single spectral peak that matches the tuning of the basilar papilla, by adding a second low-frequency component with a frequency in the range to which the amphibian papilla is tuned. Rather than enhancing signal attractiveness, however, females preferred calls having only the high-frequency peak. Witte et al. (2001) suggest that the "general rule" that the most attractive signals better stimulate the sensory system be restricted to encompass only the frequency range of the sensory system that normally detects conspecific signals. But this argument seemingly contradicts the proposed scenario in *P. pustulosus* because the "new" signal that presumably enhanced attractiveness when it first appeared—the chuck—primarily stimulates the basilar papilla, whereas the whine primarily stimulates the amphibian papilla. More studies of groups of species whose phylogenetic relationships are well known, are needed to assess the generality of the preexisting bias hypothesis.

A sensory phenomenon that can constrain the effectiveness of adding frequency components not found in conspecific calls arises from a common, if not universal, property of the peripheral auditory system of vertebrates: tone-on-tone inhibition (reviewed in Feng and Schellart 1999). In green treefrogs and bullfrogs (*H. cinerea* and *R. catesbeiana*) addition of components with frequencies intermediate between the two spectral peaks typical of conspecific calls reduced signal attractiveness to females (Gerhardt 1974; Gerhardt and Höbel 2005) and depressed evoked calling in males (Capranica and Moffat 1983), respectively. In both species, the addition of a second, higher-frequency tone suppressed the activity of low-frequency-tuned neurons innervating the amphibian papilla in response to stimulation at their characteristic frequency (Capranica and Moffat 1983). In green treefrogs, these extra components probably suppress the activity of neurons that are actually tuned to frequencies substantially below the frequency of the low-frequency spectral peak in conspecific calls (Gerhardt and Höbel 2005).

8. Sound Localization

Frogs and toads must localize potential mates and rivals as well as recognize their acoustic signals. Moreover, sound localization, in conjunction with spatial separation of calling males, can help to ameliorate the problems of signal detection and recognition in dense choruses. For example, males that maintain some minimum distance, or spacing, between themselves and their neighbors can benefit from doing so through decreased masking interference by the signals of nearby individuals (Schwartz and Gerhardt 1995) and improvements in a female's ability to locate signaling males (Telford 1985) and to exercise preferences for individuals with attractive signals (Dyson and Passmore 1992). Call amplitude, which decreases as a function of increasing distance, appears to function as a proximate cue for estimating intermale distances in frogs (Wilczynski and Brenowitz 1988; Brenowitz 1989).

The traditional vertebrate model of azimuthal sound localization involves binaural comparisons of time and intensity cues that arise when sound reaches the external surfaces of the two tympanic membranes at different times and causes differences in their displacement. These differences exist whenever the sound does not arrive from directly in front, in back, or above the animal, and intensity differences will exist under some conditions provided that the wavelength of the sound is sufficiently short to allow diffraction by the head and other structures (pinnae) to create sound shadows.

Sound localization in anurans provides a significant challenge to this model for three main reasons. First, the head dimensions and hence ear separation in very small species are so small that the external time cues are minute and external intensity cues may be nonexistent (Rheinlaender et al. 1979). Anurans also lack pinnae. Second, unlike higher vertebrates, there are prominent secondary pathways whereby sound can reach the inner surface of the tympanic membrane. The situation is complicated by the fact that the frog can probably control to some extent how sound travels through these pathways (Rheinlaender et al. 1981). Third, there is evidence that sound can also reach the inner ear organs of the frog via extratympanal pathways (e.g., Jørgensen and Christensen-Dalsgaard 1997). Such a multiinput system remains a significant challenge to researchers interested in the biophysical bases for sound localization in these animals.

8.1 Estimates of Localization Accuracy

The first experiments estimating the horizontal accuracy of sound localization quantified the head and body orientation of females (*Hyla cinerea*) and males (*Colostethus nubicola*) under closed-loop conditions (Gerhardt and Rheinlaender 1980; Rheinlaender et al. 1979). That is, because the animals were free to move from a starting point to a loudspeaker, they had the possibility of updating directional information obtained during previous movements. Most individuals showed a zigzag pattern of hopping or crawling in which the leading ear (closest to the direction of the sound source) alternated between movements. On average, females of *H. cinerea* did not turn more than about 35° to the right or left of the speaker axis before turning or hopping back across the axis. Given the usual separation distance of the ears, the maximum time cue during these approaches would have been of the order of 10 to 20 microseconds, and the time difference would have been even less in some individual approaches that were especially accurate. Females also showed the same degree of accuracy in approaches to syn-

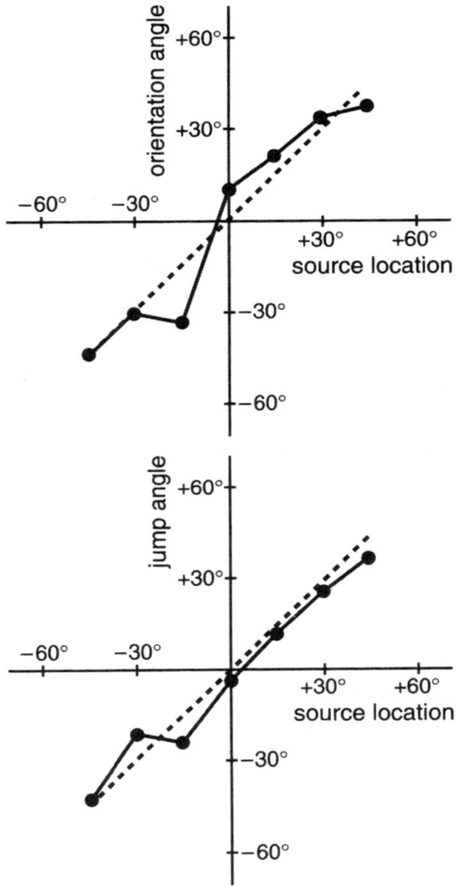

FIGURE 5.8. Open-loop accuracy of sound localization in barking treefrogs (*Hyla gratiosa*). The dotted lines represent the angles expected if the frogs' orientations were perfectly aligned toward the sound source. Orientation and jump angles (only the first turn or jump was considered) increased with the initial disparity between the frogs' body axes and the speaker position. If the frogs merely lateralized the sound source, then the size of these angles would be independent of the angle of incidence once a critical angle that allows lateralization is exceeded. From Klump and Gerhardt (1989).

thetic calls consisting of a toneburst of 900 Hz, whose wavelength greatly exceeds the head dimensions. Hence females suffered no decrement in localization accuracy when *external differences* in intensity were nonexistent. In barking treefrogs (*H. gratiosa*) tested under open-loop conditions, the magnitude of orientation movements was positively correlated with the initial deviation of the frog's position (long axis) with respect to the loudspeaker location (Fig. 5.8; Klump and Gerhardt 1989).

Arboreal frogs also showed the ability to distinguish between sound sources that were elevated from sources at the same level as the frog's starting point. In two species (*Hyla cinerea* and *Hyperolius marmoratus*), the accuracy of locating elevated sound sources was estimated to be significantly less than accuracy in the horizontal plane (Gerhardt and Rheinlaender 1982; Passmore et al. 1984), but in another species (*Hyla versicolor*), the three-dimensional errors were not significantly higher than two-dimensional ones (Jørgensen and Gerhardt 1991).

Whereas females of *H. cinerea* often elevated their heads and then scanned when the speaker was elevated relative to their position, this behavior was rarely observed in *H. versicolor.*

8.2 Localization Mechanisms: Hypotheses and Data

The behavioral results suggest that the anuran auditory system must be able to use exceedingly small time cues, uses some form of pressure-difference mechanism and extratympanic input, or both (reviewed by Gerhardt and Huber 2002). Laser vibrometry and physiological studies of green treefrogs (*H. cinerea*) indicate the existence of binaural intensity cues of the order of 2 to 3 dB in the +30° frontal area where the frogs gain sufficient directional information to make correctional orientation movements (Rheinlaender et al. 1981; Michelsen et al. 1986). These intensity cues are available in the low-frequency range (around 900 Hz) in which external cues are probably nonexistent, thus supporting the hypothesis that these differences are generated by some form of pressure-difference mechanism. Moreover, such intensity differences would also create usable time cues generated by latency shifts. For example, in the leopard frog (*Rana pipiens*) a change of just 1 dB can result in a latency shift of 100 to 600 microseconds (Feng 1982). Klump et al. (2004) showed that auditory nerve fibers can reliably phase-lock to the 300-Hz periodicity typical of the amplitude envelope of the advertisement call of *H. cinerea*. Moreover, they estimated that neural delays, presumably generated by binaural intensity differences, could magnify the physical time differences by an order of magnitude. Such differences could allow accurate sound localization and would reliably affect neurons in the central auditory system that have been hypothesized to mediate directional hearing (Feng and Capranica 1978).

9. Summary

Vocal communication in anurans has long served as an important model system for understanding the evolution of sensory systems and the underlying mechanisms. Detection and recognition of signals is usually handicapped by the complex background noise generated by the chorusing in which most competition and mate choice takes place. Male calling tactics, such as spacing and alternation of calls, improve the detectability of their signals in such an environment, and directional hearing may enhance the detection and recognition of temporally overlapping signals and noise. The physical properties of advertisement calls and their variation within and between individuals potentially provide important biological information to receivers. Playback experiments using synthetic calls confirm that females use some of this potential information in mate choice, and other experiments have shown that these choices can also influence a female's reproductive success. Field playback studies of male–male communication have elucidated some of the mechanisms mediating call type recognition and have shown that males can acquire information about a competitor's resource holding power and individual identity based on patterns of individual variation in acoustic signals. Studies of anurans have been especially influential in showing that the reciprocal selection pressures exerted by senders and receivers, the external environment (including other species), and sensory mechanisms can all influence the evolution of communication systems.

Additional research is badly needed to address several open questions. First, is there any learning by juvenile frogs of either gender in species where the young normally hear adults of the same species? If so, what mechanisms ensure that the young learn the signals of conspecifics rather than those of other species calling at the same time and place?

Second, how widespread is learning to recognize the individually distinct calls of a nearby calling male in the context of territorial neighbor recognition? Studies that integrate behavioral and physiological approaches with the comparative method should shed considerable light on both the mechanisms and evolution of vocally mediated social recognition in frogs.

Third, do anurans use the full resolving power of their auditory system in call recognition? This question cannot be answered without the development of reliable methods for conditioning frogs and toads to respond to sound playbacks.

Fourth, how does stimulation of the basilar or amphibian papilla affect signal recognition in species whose calls primarily stimulate only one of these inner ear organs? So far experiments have addressed this question only in two species in which conspecific calls stimulate the basilar papilla, and the additional stimulation reduced signal attractiveness. Would the same be true for species in which conspecific calls primarily stimulate the amphibian papilla? A related question is: is the simultaneous stimulation of both papillae always optimal in species with calls having bimodal spectra? Comparative studies of closely related species and appropriate outgroup taxa could address these questions in a phylogenetic context and hence provide tests of the preexisting bias hypothesis.

Fifth, are either of the auditory organs specialized for resolving temporal differences in calls? For example, experiments in gray treefrogs suggest that subtle temporal differences are resolved only when they are primarily processed by the amphibian papilla (Gerhardt and Schul 1999). Is this true for other species and the discrimination of more salient temporal differences?

Sixth, what role does tone-on-tone inhibition play in call recognition and the evolution of acoustic communication? The reduction in midfrequency energy in the calls of bullfrogs and green treefrogs reflects the adverse effects of such inhibition on signal effectiveness, but other species have broadband spectra without such attenuation (Gerhardt and Höbel 2005). Finally, what is the relationship between processes and mechanisms underlying signal recognition and localization. Do frogs process these two kinds of information by serial processing as described for crickets (see Gerhardt and Huber 2002; Feng and Schul, Chapter 11) or might signals from different locations be perceived as separate acoustic images as in higher vertebrates?

Acknowledgments. We thank Sarah Humfeld, Johannes Schul, and Albert Feng for comments on the manuscript and Sarah Humfeld for helping to prepare the figures. HCG's research has been generously supported by the National Science Foundation and National Institutes of Health; MAB's research has been supported by the National Science Foundation in the form of a Graduate Research Fellowship, a Doctoral Dissertation Improvement Grant, and an International Research Fellowship Award (INT 0107304).

References

Adler TB, Rose GJ (1998) Long-term temporal integration in the anuran auditory system. Nat Neurosci 1:519–523.

Alexander RD (1975) Natural selection and specialized chorusing behavior in acoustical insects. In: Pimentel D (ed) Insects, Science and Society. New York: Academic, pp. 35–77.

Andersson M (1994) Sexual Selection. Princeton, NJ: Princeton University Press.

Arak A (1983) Sexual selection by male-male competition in natterjack toad choruses. Nature 306:261–262.

Arak A (1988) Female mate selection in the natterjack toad: Active choice or passive attraction? Behav Ecol Sociobiol 22:317–327.

Beckers OM, Schul J (2004) Phonotaxis in *Hyla versicolor* (Anura: Hylidae): The effect of absolute call amplitude. J Comp Physiol A 190:869–876.

Bee MA (2002) Territorial male bullfrogs (*Rana catesbeiana*) do not assess fighting ability based on size-related variation in acoustic signals. Behav Ecol 13:109–124.

Bee MA (2003a) A test of the "dear enemy effect" in the strawberry dart-poison frog (*Dendrobates pumilio*). Behav Ecol Sociobio 154:601–610.

Bee MA (2003b) Experience-based plasticity of acoustically evoked aggression in a territorial frog. J Comp Physiol A 189:485–496.

Bee MA (2004) Within-individual variation in bullfrog vocalizations: Implications for a vocally mediated social recognition system. J Acoust Soc Amer 116:3770–3781.

Bee MA, Bowling AC (2002) Socially mediated pitch alteration by territorial male bullfrogs, *Rana catesbeiana*. J Herp 36:140–143.

Bee MA, Gerhardt HC (2001a) Neighbour-stranger discrimination by territorial male bullfrogs (*Rana catesbeiana*): I. Acoustic basis. Anim Behav 62:1129–1140.

Bee MA, Gerhardt HC (2001b) Neighbour-stranger discrimination by territorial male bullfrogs (*Rana catesbeiana*): II. Perceptual basis. Anim Behav 62:1141–1150.

Bee MA, Gerhardt HC (2001c) Habituation as a mechanism of reduced aggression between neighboring territorial male bullfrogs (*Rana catesbeiana*). J Comp Psychol 115:68–82.

Bee MA, Gerhardt HC (2002) Individual voice recognition in a territorial frog (*Rana catesbeiana*). Proc Roy Soc Lond B 269:1443–1448.

Bee MA, Perrill SA, Owen PC (2000) Male green frogs lower the pitch of acoustic signals in defense of territories: A possible dishonest signal of size? Behav Ecol 11: 169–177.

Blair WF (1964) Evolution at populational and interpopulational levels: Isolating mechanisms and interspecies interactions in anuran amphibians. Quart Rev Biol 39: 333–344.

Boatright-Horowitz SL, Horowitz SS, Simmons AM (2000) Patterns of vocal interactions in a bullfrog (*Rana catesbeiana*) chorus: Preferential responding to far neighbors. Ethology 106:701–712.

Bosch J, Boyero L (2003) Double stimulation of the inner ear organs of an anuran species (*Alytes cisernasii*) with simple tonal advertisement calls. J Zoology 260:347–351.

Bosch J, Rand AS, Ryan MJ (2000) Signal variation and call preferences in the túngara frog, *Physalaemus pustulosus*. Behav Ecol Sociobiol 49:62–66.

Bourne GR, Collins AC, Holder AM, McCarthy CL (2001) Vocal communication and reproductive behavior of the frog *Colostethus beebei* in Guyana. J Herp 35:272–281.

Brenowitz EA (1989) Neighbor call amplitude influences aggressive behavior and intermale spacing in choruses of the Pacific treefrog (*Hyla regilla*). Ethology 83:69–79.

Brenowitz EA, Rose GJ, Adler T (2001) The neuroethology of acoustic communication in pacific treefrogs. In: Ryan MJ (ed) Anuran Communication. Washington DC: Smithsonian Institution Press, pp. 145–155.

Burmeister SS, Ophir AG, Ryan MJ, Wilczynski W (2002) Information transfer during cricket frog contests. Anim Behav 64:715–725.

Bush SL, Gerhardt HC, Schul J (2001) Pattern recognition and call preferences in treefrogs (Anura: Hylidae): A quantitative analysis using a no-choice paradigm. Anim Behav 63:7–14.

Capranica RR, Moffat AJM (1983) Neurobehavioral correlates of sound communication in anurans. In: Ewert JP, Capranica RR, Ingle DJ (eds) Advances in Vertebrate Neuroethology. New York: Plenum, pp. 701–730.

Castellano S, Giacoma C (1998) Stabilizing and directional female choice for male calls in the European green toad. Anim Behav 56:275–287.

Davies NB, Halliday TR (1978) Deep croaks and fighting assessment in toads *Bufo bufo*. Nature 274:683–685.

Davis MS (1987) Acoustically mediated neighbor recognition in the North American bullfrog, *Rana catesbeiana*. Behav Ecol Sociobiol 21:185–190.

Diekamp BM, Gerhardt HC (1995) Behavioral and neurophysiological studies of acoustic communication in the gray treefrog *Hyla versicolor*. J Comp Physiol A 177:173–190.

Dobzhansky T (1940) Speciation as a stage in evolutionary divergence. Am Natur 74: 312–321.

Doherty JA, Gerhardt HC (1984a) Acoustic communication in hybrid treefrogs: Sound production by males and selective phonotaxis of females. J Comp Physiol A 154: 319–330.

Doherty JA, Gerhardt HC (1984b) Evolutionary and neurobiological implications of selective phonotaxis in the spring peeper (*Hyla crucifer*). Anim Behav 32:875–871.

Duellman WE, Trueb L (1986) Biology of Amphibians. New York: McGraw-Hill.

Dyson ML, Passmore NI (1992) Effect of intermale spacing on female frequency preferences in the painted reed frog. Copeia 1992:1111–1114.

Edwards CJ, Adler TB, Rose GJ (2002) Auditory midbrain neurons that count. Nat Neurosci 5:934–936.

Elephandt A, Eistettler I, Fleig A, Gunther E, Hainich S, Hepperle S, Traub B (2000). Hearing threshold and frequency discrimination in the purely aquatic frog Xenopus laevis (Pipidae): Measurement by means of conditioning. J Exp Biol 203:3621–3629.

Feng AS (1982) Quantitative analysis of intensity-rate and intensity-latency functions in peripheral auditory nerve fibers of northern leopard frogs (*Rana p. pipiens*) Hear Res 6:242–246.

Feng AS, Capranica RR (1978) Sound localization in anurans. II. Binaural interaction in superior olivary nucleus of the green tree frog (Hyla cinerea). J Neurophysiol 41:43–54.

Feng AS, Schellart NAM (1999) Central auditory processing in fish and amphibians. In: Fay RR, Popper AN (eds) Comparative Hearing: Fish and Amphibians. New York: Springer, pp. 218–268.

Fouquette MJ (1975) Speciation in chorus frogs. I. Reproductive character displacement in the *Pseudacris nigrita* complex. Syst Zool 24:16–23.

Friedl TWP, Klump GM (2005) Sexual selection in the lek-breeding European treefrog (*Hyla arborea*): Body size, chorus attendance, random mating, and good genes. Anim Behav 70:1141–1154.

Gerhardt HC (1974) Vocalizations of some hybrid treefrogs: Acoustic and behavioral analyses. Behaviour 49:130–151.

Gerhardt HC (1975) Sound pressure levels and radiation patterns of the vocalizations of some North American frogs and toads. J Comp Physiol A 102:1–12.

Gerhardt HC (1978a) Discrimination of intermediate sounds in a synthetic call continuum by female green tree frogs. Science 199:1089–1091.

Gerhardt HC (1978b) Temperature coupling in the vocal communication system of the gray treefrog *Hyla versicolor*. Science 199:992–994.

Gerhardt HC (1981) Mating call recognition in the barking treefrog (*Hyla gratiosa*): Responses to synthetic calls and comparisons with the green treefrog (*Hyla cinerea*). J Comp Physiol A 144:17–25.

Gerhardt HC (1987) Evolutionary and neurobiological implications of selective phonotaxis in the green treefrog (*Hyla cinerea*). Anim Behav 35:1479–1489.

Gerhardt HC (1988) Acoustic properties used in call recognition by frogs and toads. In: Fritzsch B, Ryan MJ, Wilczynski W, Hetherington TE, Walkowiak W (eds) The Evolution of the Anuran Auditory System. New York: Wiley, pp. 455–483.

Gerhardt HC (1991) Female mate choice in treefrogs: Static and dynamic acoustic criteria. Anim Behav 42:615–635.

Gerhardt HC (1994a) The evolution of vocalization in frogs and toads. Annu Rev Ecol Syst 25:293–324.

Gerhardt HC (1994b) Reproductive character displacement of female mate choice in the grey treefrog *H. chrysoscelis*. Anim Behav 47:959–969.

Gerhardt HC (2005a) Acoustic spectral preferences in two cryptic species of gray treefrogs: Implications for mate choice and sensory mechanisms. Anim Behav 70: 39–48.

Gerhardt HC (2005b) Advertisement-call preferences in diploid-tetraploid treefrogs (*Hyla chrysoscelis* and *Hyla versicolor*): Implications for mate choice and the evolution of communication systems. Evolution 59:395–408.

Gerhardt HC, Doherty JA (1988) Acoustic communication in the gray treefrog, *Hyla versicolor*: Evolutionary and neurobiological implications. J Comp Physiol A 162: 261–278.

Gerhardt HC, Höbel G (2005) Mid-frequency suppression in the green treefrog (*Hyla cinerea*): Mechanisms and implications for the evolution of acoustic communication. J Comp Physiol A 191:707–714.

Gerhardt HC, Huber F (2002) Acoustic Communication in Insects and Anurans: Common Problems and Diverse Solutions. Chicago: University of Chicago Press.

Gerhardt HC, Klump GM (1988) Masking of acoustic signals by the chorus background noise in the green treefrog: A limitation on mate choice. Anim Behav 36:1247–1249.

Gerhardt HC, Mudry KM (1980) Temperature effects on frequency preferences and mating call frequencies in the green treefrog (*Hyla cinerea*) (Anura: Hylidae). J Comp Physiol A 137:1–6.

Gerhardt HC, Rheinlaender J (1980) Accuracy of sound localization in a miniature dendrobatid frog. Naturwissenschaften 67:362–363.

Gerhardt HC, Rheinlaender (1982) Localization of an elevated sound source by the green treefrog. Science 217:663–664.

Gerhardt HC, Schul J (1999) A quantitative analysis of behavioral selectivity for pulse-rise time in the gray treefrog, *Hyla versicolor*. J Comp Physiol A 185:33–40.

Gerhardt HC, Schwartz JJ (2001) Auditory tuning and frequency preferences in anurans. In: Ryan MJ (ed) Anuran Communication. Washington: Smithsonian Institution Press, pp. 73–85.

Gerhardt HC, Dyson ML, Tanner SD (1996) Dynamic acoustic properties of the advertisement calls of gray treefrogs: Patterns of variability and female choice. Behav Ecol 7:7–18.

Gerhardt HC, Tanner SD, Corrigan CM, Walton HC (2000) Female preference functions based on call duration in the gray treefrog (*Hyla versicolor*). Behav Ecol 11:663–669.

Given MF (1987) Vocalizations and acoustic interactions of the carpenter frog, *Rana virgatipes*. Herpetologica 43:467–481.

Given MF (1999) Frequency alteration of the advertisement call in the carpenter frog, *Rana virgatipes*. Herpetologica 55:304–317.

Grafe TU (1995) Graded aggressive calls in the African reed frog *Hyperolius marmoratus* (Hyperoliidae). Ethology 101:67–81.

Grafe TU (1997) Costs and benefits of mate choice in the lek-breeding reed frog, *Hyperolius marmoratus*. Anim Behav 53:1103–1117.

Halliday TR, Tejedo M (1995) Intrasexual selection and alternative mating behaviour. In: Heatwole H, Sullivan BK (eds) Amphibian Biology: Vol. 2: Social Behaviour. Chipping Norton, UK: Surrey Beatty, pp. 419–468.

Harvey PH, Pagel MD (1991) The Comparative Method in Evolutionary Biology. Oxford: Oxford University Press.

Hausfater G, Gerhardt HC, Klump G (1990) Parasites and mate choice in gray treefrogs, *Hyla versicolor*. Am Zool 30:299–311.

Höbel G, Gerhardt HC (2003) Reproductive character displacement in the communication system of green treefrogs (*Hyla cinerea*). Evolution 57:894–904.

Hödl W, Amézquita A, Narins PM (2004). The role of call frequency and the auditory papillae in phonotactic behavior in male dart-poison frogs *Epipedobates femoralis* (Dendrobatidae). J Comp Physiol A 190:823–829.

Howard RD (1978) Evolution of mating strategies in bullfrogs, *Rana catesbeiana*. Evolution 32:850–871.

Jørgensen MB, Christensen-Dalsgaard (1997) Directionality of auditory nerve fiber responses to pure tone stimuli in the grassfrog, *Rana temporaria*. I. spike rate responses. J Comp Physiol A 180:493–502.

Jørgensen MB, Gerhardt HC (1991) Directional hearing in the gray treefrog *Hyla versicolor*: Eardrum vibrations and phonotaxis. J Comp Physiol A 169:177–183.

Kime NM, Turner WR, Ryan MJ (2000) The transmission of advertisement calls in central American frogs. Behav Ecol 11:71–83.

Klump GM, Gerhardt HC (1989) Sound localization in the barking treefrog. Naturwissenschaften 76:35–37.
Klump GM, Benedix JH, Gerhardt HC, Narins PM (2004) AM representation in green treefrog auditory nerve fibers: Neuroethological implications for pattern recognition and sound localization. J Comp Physiol A 190:1011–1021.
Lesbarrères D, Lodé T (2002) Variations in male calls and responses to an unfamiliar advertisement call in a territorial breeding anuran, *Rana dalmatina*: Evidence for a "dear enemy" effect. Ethol Ecol Evol 14:287–295.
Lewis ER, Narins PM (1999) The acoustic periphery of amphibians: Anatomy and physiology. In: Fay RR, Popper AN (eds) Comparative Hearing: Fishes and Amphibians. New York: Springer, pp. 101–154.
Littlejohn MJ (1965) Premating isolation in the *Hyla ewingi* complex (Anura: Hylidae). Evolution 19:234–243.
Lombard ER, Straughan IR (1974) Functional aspects of anuran middle ear structures. J Exp Biol 61:57–71.
Lopez PT, Narins PM, Lewis ER, Moore SW (1988) Acoustically-induced call modification in the white-lipped frog, *Leptodactylus albilabris*. Anim Behav 36:1295–1308.
Márquez R, Bosch J (1997) Female preference in complex acoustical environments in the midwife toads *Alytes obstetricans* and *Alytes cisternasii*. Behav Ecol 8:588–594.
Martin WF (1972) Evolution of vocalization in the genus *Bufo*. In: Blair WF (ed) Evolution in the Genus *Bufo*. Austin: University of Texas Press, pp. 279–309.
Maynard Smith J (1982) Evolution and the Theory of Games. Cambridge UK: Cambridge University Press.
Mecham JS (1960) Introgressive hybridization between two southeastern tree frogs. Evolution 14:447–457.
Megela AL, Capranica RR (1983) A neural and behavioral study of auditory habituation in the bullfrog, *Rana catesbeiana*. J Comp Physiol A 151:423–434.
Michelsen A, Jørgensen M, Christensen-Dalsgaard J, Capranica RR (1986) Directional hearing of awake, unrestrained treefrogs. Naturwissenschaften 73:682–683.
Murphy CG (2003) The cause of correlations between nightly numbers of male and female barking treefrogs (*Hyla gratiosa*) attending choruses. Behav Ecol 14:274–281.
Narins PM, Hödl W, Grabul DS (2003) Bimodal signal requisite for agonistic behavior in a dart-poison frog, *Epipedobates femoralis*. Proc Natl Acad Sci USA 100:577–580.
Nelson DA, Marler P (1990) The perception of birdsong and an ecological concept of signal space. In: Stebbins WC, Berkley MA (eds) Comparative Perception, Vol II: Complex Signals. New York: Wiley, pp. 443–478.
Noor MAF (1999) Reinforcement and other consequences of sympatry. Heredity 83: 503–508.
Passmore NI, Capranica RR, Telford SR, Bishop PJ (1984) Phonotaxis in the painted reed frog (*Hyperolius marmoratus*). The localization of elevated sound sources. J Comp Physiol A 154:189–197.
Penna M, Solís R (1998) Frog call intensities and sound propagation in the South American temperate forest region. Behav Ecol Sociobiol 42:371–381.
Pfennig KS (2000) Female spadefoot toads compromise on mate quality to ensure conspecific matings. Behav Ecol 11:220–227.
Rheinlaender J, Gerhardt HC, Yager D, Capranica RR (1979) Accuracy of phonotaxis in the green treefrog (*Hyla cinerea*). J Comp Physiol A 133:247–255.

Rheinlaender J, Walkowiak W, Gerhardt HC (1981) Directional hearing in the green treefrog: A variable mechanism? Naturwissenschaften 67:430–431.

Robertson JGM (1986) Male territoriality, fighting and assessment of fighting ability in the Australian frog *Uperoleia rugosa*. Anim Behav 34:763–772.

Robertson JGM (1990) Female choice increases fertilisation success in the Australian frog, *Uperoleia laevigata*. Anim Behav 39:639–645.

Rosenthal GG, Rand AS, Ryan MJ (2004) The vocal sac as a visual cue in anuran communication: An experimental analysis using video playback. Anim Behav 68:55–58.

Ryan MJ, Keddy–Hector A (1992) Directional patterns of female mate choice and the role of sensory biases. Am Natur 139:S4–S35.

Ryan MJ, Rand AS (1993a) Species recognition and sexual selection as a unitary problem in animal communication. Evolution 47:647–657.

Ryan MJ, Rand AS (1993b) Sexual selection and signal evolution: The ghost of biases past. Phil Trans R Soc Lond B 340:187–195.

Ryan MJ, Rand AS (2003) Sexual selection in female perceptual space: How female túngara frogs perceive and respond to complex population variation in acoustic mating signals. Evolution 57:2608–268.

Ryan MJ, Wilczynski W (1991) Evolution of intraspecific variation in the advertisement call of a cricket frog (*Acris crepitans*, Hylidae). Biol J Linnean Soc 44:249–271.

Ryan MJ, Fox JH, Wilczynski W, Rand AS (1990) Sexual selection for sensory exploitation in the frog *Physalaemus pustulosus*. Nature 343:66–67.

Schul J, Bush SL (2002) Non-parallel coevolution of sender and receiver in the acoustic communication system of treefrogs. Proc R Soc Lond B 269:847–852.

Schwartz JJ (1993) Male calling behavior, female discrimination and acoustic interference in the Neotropical treefrog *Hyla microcephala* under realistic acoustic conditions. Behav Ecol Sociobiol 32:401–414.

Schwartz JJ, Gerhardt HC (1995) Directionality of the auditory system and call pattern recognition during acoustic interference in the gray treefrog, *Hyla versicolor*. Aud Neurosci 1:195–206.

Schwartz JJ, Buchanan B, Gerhardt HC (2001) Female mate choice in the gray treefrog (*Hyla versicolor*) in three experimental environments. Behav Ecol Sociobiol 49: 443–455.

Telford SR (1985) Mechanisms and evolution of inter-male spacing in the painted reed frog (*Hyperolius marmoratus*). Anim Behav 33:1353–1361.

Temeles EJ (1994) The role of neighbors in territorial systems: When are they dear enemies? Anim Behav 47:339–350.

Wagner Jr WE (1989a) Social correlates of variation in male calling behavior in Blanchard's cricket frog, *Acris crepitans blanchardi*. Ethology 82:27–45.

Wagner Jr WE (1989b) Fighting, assessment, and frequency alteration in Blanchard's cricket frog. Behav Ecol Sociobiol 25:429–436.

Wagner Jr WE (1995) Measuring female mating preferences. Anim Behav 55:1029–1042.

Welch AM (2003) Genetic benefits of a female mating preference in gray tree frogs are context-dependent. Evolution 57:883–893.

Welch AM, Semlitsch RD, Gerhardt HC (1998) Call duration as an indicator of genetic quality in male gray tree frogs. Science 280:1928–1930.

Wells KD (2001) The energetics of calling in frogs. In: Ryan MJ (ed) Recent Advances in Anuran Communication. Washington DC: Smithsonian Institution Press, pp. 45–60.

Wells KD, Taigen TL (1986) The effects of social interactions on calling energetics in the gray treefrog (*Hyla versicolor*). Behav Ecol Sociobiol 19:9–18.

Whitney CL (1980) Role of the encounter call in spacing of Pacific tree frogs, *Hyla regilla*. Can J Zool 58:75–78.

Wilczynski W, Brenowitz EA (1988) Acoustic cues mediate inter-male spacing in a Neotropical frog. Anim Behav 36:1054–1063.

Witte K, Ryan MJ, Wilczynki W (2001) Changes in frequency structure of a mating call decrease its attractiveness to females in the cricket frog *Acris crepitans blanchardi*. Ethology 107:685–699.

6
Pathways for Sound Transmission to the Inner Ear in Amphibians

MATTHEW J. MASON

1. Introduction

The tympanic middle ear, featuring a tympanic membrane and air-filled middle ear cavity, is used to transfer sound vibrations from air to the fluid-filled inner ear in many tetrapods. Because the tympanic membrane is light and the air behind it in the middle ear cavity is compressible, airborne sound pressure fluctuations cause the membrane to vibrate. The hair cells that will ultimately transduce mechanical vibrations into electrical signals are located within the inner ear, the entrance to which is the oval window. In mammals, vibrations of the tympanic membrane are transferred to the inner ear by means of three auditory ossicles, the malleus, incus, and stapes. The footplate of the stapes lies within the oval window. In birds, reptiles, and frogs the malleus and incus are represented, respectively, as the articular and quadrate bones of the jaw-joint, and the tympanic membrane is coupled to the stapes by means of a cartilaginous extrastapes instead. The term "stapes" rather than "columella" is used here with reference to non-mammals because this ossicle is believed to be homologous in all tetrapods.

The impedance-matching properties of the tympanic middle ear improve the efficiency of sound energy transfer from air to inner ear fluids, so the tympanic ear is seen as an adaptation for detecting airborne sound. Although this is the primary acoustic sense in birds and most mammals, the detection of sound traveling in water or as seismic vibrations through the ground does not require a tympanic middle ear, because impedance matching in these cases is less of a problem. Many species of amphibians (and reptiles) lack both tympanic membrane and middle ear cavity and must therefore use alternative pathways for sound transmission, pathways which may or may not involve residual elements of the middle ear apparatus.

This review considers how acoustic vibrations traveling in air, water, or solid substrate might be transmitted to the inner ear of amphibians. The anatomy of the middle ear of frogs, urodeles, and caecilians is described and several areas of apparent confusion addressed. The physiology of both tympanic and extratympanic hearing in amphibians is then discussed, including a consideration of the role of the opercularis system, unique to frogs and urodeles. Finally, underwater audition in larval and adult amphibians is examined.

2. The Anatomy of the Middle Ear in Amphibians

We first consider the anatomy of the "complete" tympanic ear as found in many adult frogs, for example, *Rana catesbeiana* (American bullfrog), before looking at the reduced middle ear apparatus of some other anuran species, and finally the middle ear structures in urodeles and caecilians. The ear apparatus of amphibian larvae is discussed briefly in Section 6.4 (Underwater Hearing in Tadpoles).

2.1.1 The Middle Ear Cavity and Tympanic Membrane

The middle ear cavity (tympanic cavity) in frogs (Fig. 6.1) is an air-filled diverticulum of the pharynx, derived from the hyoid pouch (Lombard and Bolt 1979). It is roughly conical in shape, the base of the cone being the tympanic membrane, and it is open to the mouth cavity ventrally via a short, wide, and permanently open Eustachian tube. The tympanic membrane is often visible on the side of the head of a frog, just behind the eye (Figs. 6.2A,B,C). It is supported around its perimeter by an annular cartilage which is itself supported rostrally and dorsally by the squamosal (Ecker 1889; Bolt and Lombard 1985; Wever 1985). The membrane is made of thin nonglandular skin and is histologically composed of endodermal, fibrous mesodermal, and epidermal layers (Witschi 1949; Hetherington 1987b). Connective tissue fibers radiate outwards from the insertion of the extrastapes, and smooth muscle is found in the periphery. The internal (endodermal) surface is made of columnar epithelium, continuous with the mucosa of the middle ear cavity (Ecker 1889).

In two species of ranid frogs, *Huia cavitympanum* (Sabah huia frog) from southeast Asia (Noble 1931; Inger 1966; Yang 1991) and *Amolops tormotus* (concave-eared torrent frog) from China (Feng et al. 2002), the tympanic membrane has sunk beneath the surface of the head, resulting in the presence of what might be described as a short external auditory meatus (Fig. 6.2D); a meatus is lacking in other amphibians.

2.1.2 The Stapes

A stapes (Figs. 6.1, 6.3) is found in at least some members of all three amphibian orders. In frogs, it typically consists of a pars interna and a pars media (Wever 1985). The pars interna is a thick cartilaginous disk that occupies the anterior half of the oval window (*fenestra ovalis*). In some frogs, a flange of the pars interna extends caudally to lie underneath the operculum (Lombard and Straughan 1974; Wever 1979; Hetherington et al. 1986; Mason and Narins 2002b). The ossified pars media has a wide base that is firmly attached to the pars interna, ventrolateral to which is a ridge that articulates with the rim of the oval window (Bolt and Lombard 1985). The base of the pars media and the pars interna together form

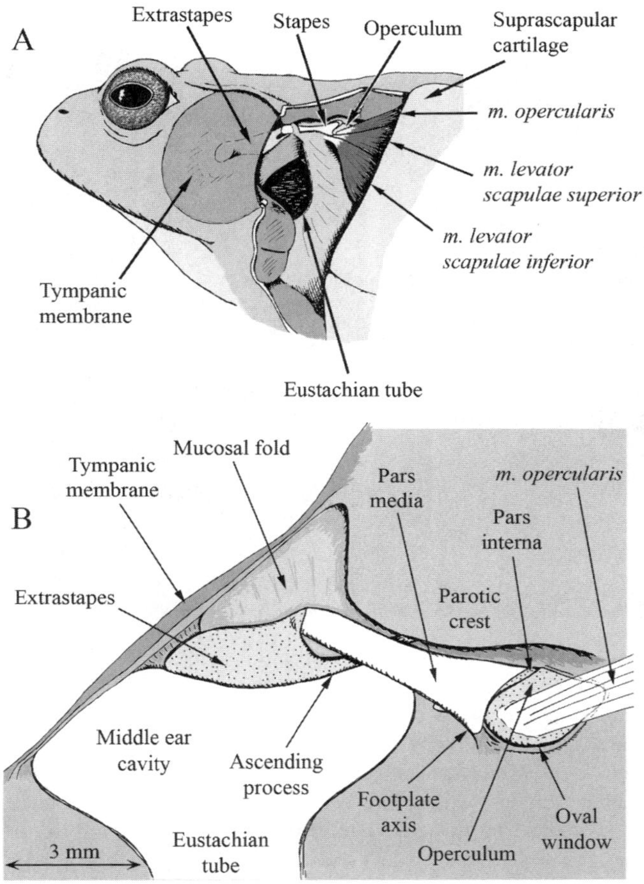

FIGURE 6.1. Middle ear structures in *Rana catesbeiana*, a frog with tympanic ears. A: Semi-diagrammatic representation of the exposed left middle ear structures, seen from caudolaterally. B: Semi-diagrammatic transverse section of the left middle ear of a female frog, seen from caudally. Cartilage is stippled. The pars media and pars interna are collectively referred to as the stapes. The hinge-like articulation between pars media and otic capsule, along the ventral rim of the oval window, forms the rotatory axis of the stapes footplate. Figure 6.1B from Mason and Narins (2002a), with permission of The Company of Biologists Ltd.

the footplate of the stapes. The pars media tapers to form the shaft of the stapes, projecting dorsolaterally.

The extrastapes, also known as the pars externa, is a cartilaginous rod that projects ventrolaterally from the lateral tip of the pars media to insert by means of loose connective tissue onto the center of the tympanic membrane (Fig. 6.1B). From the ventral surface of the extrastapes, a slim flattened strip of cartilage

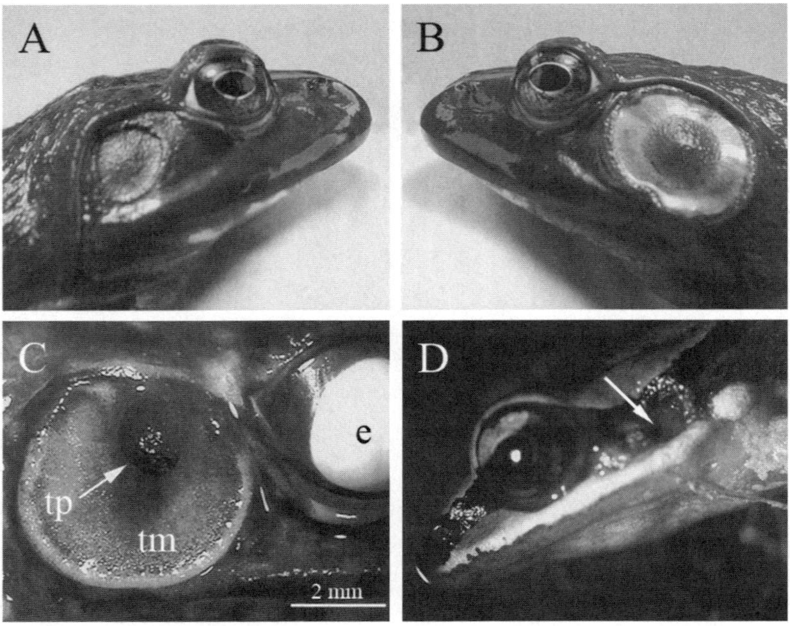

FIGURE 6.2. External ear structures in frogs (not to scale). A, B: Heads of *Rana catesbeiana*, female and male, respectively, showing dimorphism in the size of their tympanic membranes. From Mason et al. 2003, with permission of S. Karger AG, Basel. C: Lateral view of the head region of a breeding male *Petropedetes parkeri* (Parker's water frog), showing the eye, e, the tympanic membrane, tm, and the tympanic papilla, tp. From Narins et al. 2001, with permission of The Company of Biologists Ltd. D: Head of *Amolops tormotus*; the arrow points to the external auditory meatus. Courtesy of Prof. A.S. Feng.

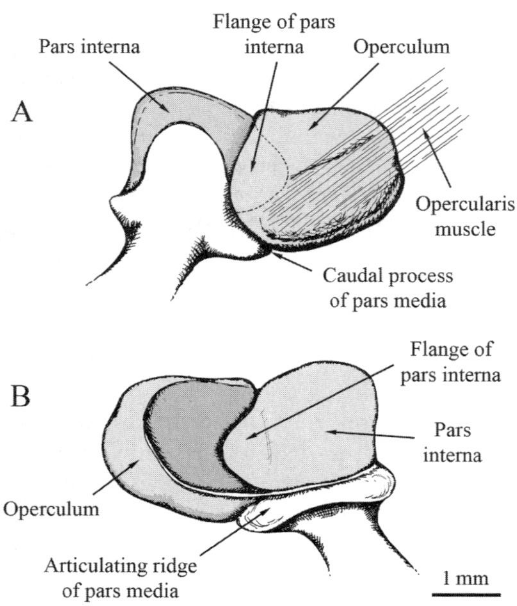

FIGURE 6.3. Left stapes footplate and operculum of *Rana catesbeiana*, seen from (A) dorsolateral and (B) ventromedial views. Cartilage is shaded. The operculum in A is drawn as if semi-transparent, to reveal the flange of the pars interna. From Mason and Narins (2002b), with permission of The Company of Biologists Ltd.

called the ascending process (*processus ascendens plectri*) extends rostrally and dorsally to insert on the underside of the parotic crest (*crista parotica*), the lateral part of the paroccipital process of the otic capsule (Bolt and Lombard 1985; Wever 1985; Hetherington 1987b). The ascending process is also referred to as the "suprastapedial process" (Goodrich 1930) and the "plectral ligament" (Moffat and Capranica 1978).

The extrastapes and lateral part of the pars media are suspended from the roof of the middle ear cavity within a thin fold of mucosa; the rest of the stapes lies medial to the cavity, buried within connective tissue.

2.1.3 The Operculum

An otic operculum (*operculum fenestrae ovalis*; Figs. 6.1, 6.3) is found in some frogs and salamanders; it is unrelated to the operculum in fish. In frogs with tympanic ears, the operculum occupies only the caudal half of the oval window, the rostral half containing the pars interna of the stapes. The operculum is typically cartilaginous but may be partially or completely ossified in some species (de Villiers 1934; Hetherington et al. 1986). It is anchored to the rim of the oval window by means of a synchondrosis along its dorsocaudal margin, this acting as a hinge joint (Hetherington et al. 1986; Mason and Narins 2002b). The operculum usually possesses a thickened projecting ridge referred to as the muscular process, for the insertion of the opercularis muscle (Hetherington et al. 1986).

The operculum in frogs is described as fitting into a notch in the caudal part of the stapes footplate (Wever 1979, 1985; Bolt and Lombard 1985; Hetherington et al. 1986; Hetherington 1987b). This impression arises from the fact that the ventral rim of the operculum is attached to a thickened band of connective tissue, which passes rostrally between the pars interna and pars media of the stapes (Wever 1979, 1985; Mason and Narins 2002b), together with the presence of a flange of the pars interna, which extends medial to the operculum in some species (Fig. 6.3).

Some frogs possess a lateral chamber to the inner ear, separated from the otic cavity proper by a bony shelf (Wever 1985; Hetherington et al. 1986). The passageway between the lateral chamber and the rest of the inner ear is a small hole in the caudal part of this shelf, usually located just medial to the operculum (Hetherington et al. 1986; Mason and Narins 2002b). Bolt and Lombard (1985) interpret this hole as the true oval window and refer to the lateral chamber as the "pericapsular space," filled with a diverticulum of the periotic cistern. A similar morphology exists in most plethodontid and some salamandrid urodeles (Lombard 1977).

2.1.4 The Middle Ear Muscles

Most frogs possess a *m. opercularis* derived from the *m. levator scapulae superior* and innervated by spinal nerves 2 and 3 (Duellman and Trueb 1986; Hetherington 1987b). This long narrow muscle originates on the suprascapular cartilage and passes rostroventrally to insert on the operculum (Figs. 6.1, 6.3). Although Wever (1979, 1985) asserts that the *m. opercularis* possesses smaller and more translucent fibers than the *m. levator scapulae superior*, other authors have commented that the two muscles are often difficult to distinguish, especially near their origin (Hetherington et al. 1986; Hetherington 1987b). In *Ascaphus truei* (tailed frog), Hetherington et al. (1986) found a portion of the *m. levator scapulae inferior* inserting on the operculum in addition to the *m. opercularis*. This has been observed as an anomaly in one specimen of *Rana catesbeiana* (personal observation).

Wever (1979, 1985) described a "columellar muscle" in the frog ear, originating on the suprascapular cartilage immediately adjacent to the *m. opercularis* and inserting on a caudal process of the stapes footplate. However, subsequent studies have failed to identify a discrete "columellar muscle" (Hetherington et al. 1986; Hetherington 1987b; Hetherington and Tugaoen 1990). In *Rana catesbeiana*, the *m. levator scapulae superior* inserts along the ventral rim of the oval window by means of a broad band of connective tissue, and it is the rostral part of this aponeurosis that is attached to the caudal process of the pars media of the stapes (Hetherington and Lombard 1983a; Mason and Narins 2002b). Muscle fibers inserting onto the stapes footplate via a tendon were identified in only 4 out of 53 anuran species studied by Hetherington et al. (1986), and these authors feel that the "columellar muscle" is best considered a slip of the *m. levator scapulae superior*, present only occasionally in frogs.

2.1.5 Reduction of the Ear Apparatus in Frogs

The tympanic membrane in frogs is often poorly differentiated from the surrounding skin (Fox 1995), although other components of the tympanic ear may be present. Some frogs, however, have lost tympanic membrane, middle ear cavity, and stapes and are termed "earless," although the inner ear can be well developed. Most of these species are small, of snout-vent length below 20 to 30 mm (Hetherington 1992b). Jaslow et al. (1988) found that 11 of 21 anuran families include at least one "earless" species. Examples include the discoglossid toads in the genus *Bombina*, which have lost their tympanic membranes and middle ear cavities: *Bombina bombina* (European fire-bellied toad) has lost all trace of a stapes (Fig. 6.4A), whereas *B. orientalis* (Eastern fire-bellied toad) and *B. variegata* (yellowbelly toad) each possess a vestigial stapes with a ligamentar attachment to the hyoid apparatus (Wever 1985; Jaslow et al. 1988; Smirnov 1991). Other well-studied "earless" frogs include the leiopelmatids *Ascaphus* and *Leiopelma* (de Villiers 1934; Wagner 1934; Stephenson 1951; Wever 1985) and most species of the bufonid genus *Atelopus* (McDiarmid 1971; Wever 1985; Jaslow et al. 1988).

Despite assertions to the contrary (Jaslow et al. 1988), the opercular apparatus is not always present in frogs. Loss of the operculum and opercularis muscle, at least in adults, has been documented in *Xenopus* (Fig. 6.4B), *Pipa*,

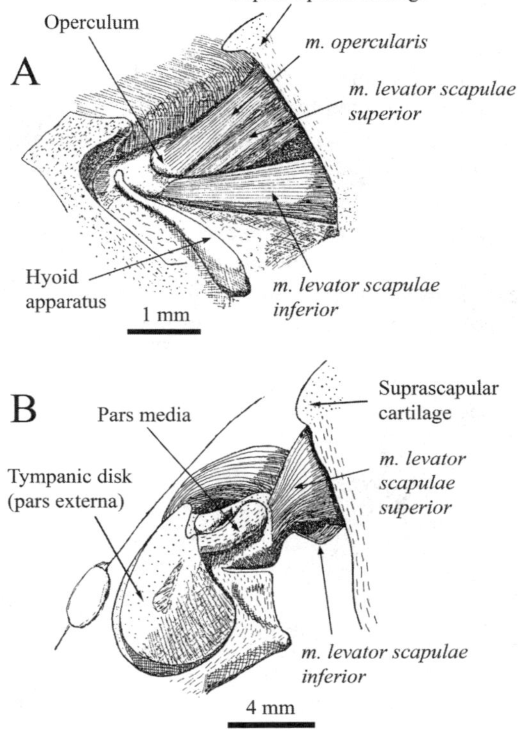

FIGURE 6.4. Middle ear structures in frogs that lack elements found in typical tympanic ears, lateral views. A: *Bombina bombina*, which lacks both stapes and middle ear cavity. B: *Xenopus laevis*, which has a "tympanic disc" derived from the extrastapes instead of a tympanic membrane, and which lacks the opercularis system. From Eiselt (1941); scales are approximate.

Hymenochirus, and *Pseudhymenochirus*, all strictly aquatic frogs in the family Pipidae (van Seters 1922; de Villiers 1932; Eiselt 1941; Spannhof 1954; Cannatella and Trueb 1988; Smirnov and Vorobyeva 1988). Although a small operculum is identifiable in the *Xenopus* embryo (de Villiers 1932; Spannhof 1954; Sedra and Michael 1957), it fuses with the edge of the otic capsule and is not believed to have any sensory function in the adult (Wever 1985; Elepfandt 1996b). There is no tympanic membrane in pipids, but the extrastapes expands to form a cartilaginous "tympanic disk," covered by fatty tissue and skin (Spannhof 1954; Sedra and Michael 1957; Wever 1985). An ascending process of the extrastapes is lacking (de Villiers 1932).

It has been claimed that leiopelmatids also lack an opercular apparatus (Duellman and Trueb 1986; Stebbins and Cohen 1995), but in fact these frogs possess both operculum and opercular muscle (de Villiers 1934; Wagner 1934; Stephenson 1951; Wever 1985). Reports suggesting that some frog species possessing an operculum lack an opercularis muscle have been strongly challenged (see Eiselt 1941).

2.2 The Middle Ear of Urodeles

The middle ear apparatus of urodeles is very variable in structure, both between and within families (Fig. 6.5). The most "complete" urodele middle ear, as found in primitive hynobiids and ambystomatids, includes two movable elements within the oval window, the operculum and stapes (Monath 1965). The stapes is typically ossified, whereas the operculum may be either cartilaginous or bony (Monath 1965). The stapes generally possesses a rod-like process called the stylus, which may be connected by ligaments to the squamosal or palatoquadrate, or, as in *Siren*, to the ceratohyal (Kingsbury and Reed 1909). Urodeles lack extrastapes, tympanic membrane, and middle ear cavity. Although Schmalhausen (1968) claimed to have identified an incipient middle ear cavity developing

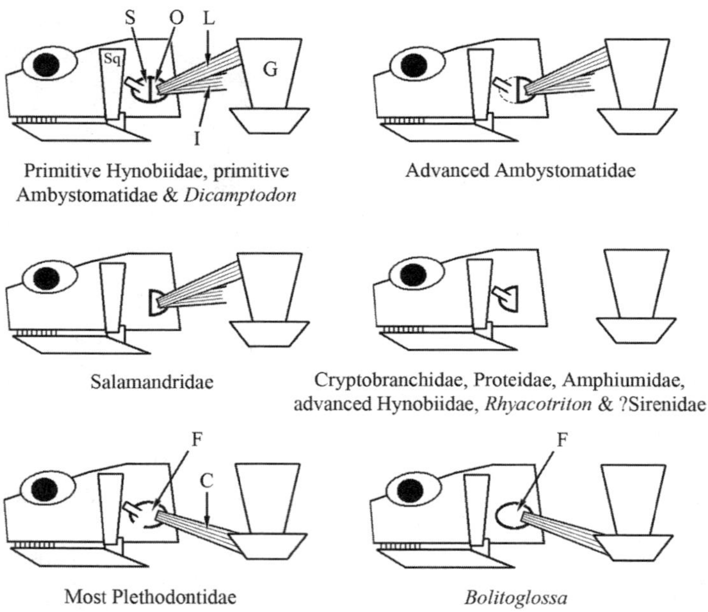

FIGURE 6.5. Diagrammatic representations of the ear structures in urodeles, according to the interpretations of Monath (1965) and Kingsbury and Reed (1909); some other interpretations in the literature differ in certain respects from what is shown. Only "representative" types are portrayed. Muscular attachments (only represented when inserting on the operculum or fenestral plate) in particular have been greatly simplified: the *m. levator scapulae* and *m. cucullaris* are actually subdivided in many species, whereas the *m. intertransversarius capitis inferior* does not insert on the operculum in all species within the groups in which this muscle is represented. Key: C = *m. cucullaris*; F = fenestral plate (homologies of which are unclear); G = shoulder girdle (of which suprascapula is dorsal); I = *m. intertransversarius capitis inferior*; L = *m. levator scapulae*; O = operculum; S = stapes; Sq = squamosal.

in hynobiid embryos, this interpretation has been questioned (Hetherington 1988b).

Many species of urodeles have only one movable skeletal element in their oval window. There has been much debate in the literature about whether, in each group examined, this single element is derived from a fusion of the stapes footplate with the operculum, whether one or the other element has fused with the rim of the oval window, or whether one has simply failed to develop. The term "fenestral plate" has been used to avoid reference to the possible homologies of the single element found in plethodontid salamanders (Monath 1965). Despite bearing a muscular connection to the pectoral girdle like an operculum, the plethodontid fenestral plate might actually be derived from the stapes (Hetherington 1988b). Not only is the "opercularis" muscle of plethodontids of a different embryological origin to that of other salamanders (see below), but in most species the fenestral plate appears to be hinged to the otic capsule at its anteroventral margin, rather than caudally or dorsally as in other urodeles (Smith 1968; Hetherington et al. 1986).

The urodele operculum (or fenestral plate) is connected to the pectoral girdle by means of an "opercularis" muscle. In most urodele families, this muscle represents the (entire) *m. levator scapulae*, originating as either one or two heads on the pectoral girdle, whereas in plethodontids it is derived from part of the *m. cucullaris major*, originating on the scapulocoracoid (Dunn 1941; Monath 1965; Hetherington et al. 1986). Certain species from several families have a second muscle also inserting on the operculum or fenestral plate, this representing part of the *m. intertransversarius capitis inferior*, which originates from the transverse processes of the first few presacral vertebrae (Monath 1965; Hetherington et al. 1986). All three muscles were found to insert on the fenestral plate of one specimen of the plethodontid *Pseudotriton montanus* (eastern mud salamander), but this was not the case in the other two specimens of this species examined (Monath 1965).

Members of the Sirenidae, Cryptobranchidae, Proteidae, and Amphiumidae, all neotenic and aquatic, are said to lack both operculum and opercularis muscle, as are certain members of several other urodele families (Kingsbury and Reed 1909; Dunn 1941; Monath 1965). Some authors, however, have identified these structures in members of the Sirenidae (Reed 1920; Wever 1985).

Urodeles, unlike frogs, lack a round window. It is believed that inner ear fluid, displaced when the fenestral plate is pushed inwards, is transmitted through the perilymphatic duct into the cranial cavity, passing from there through the cerebrospinal fluid to the contralateral ear (Wever 1978, 1985). Coupling between the ears thereby provides a pressure relief route.

2.3 The Middle Ear of Caecilians

Like urodeles, caecilians lack both tympanic membrane and middle ear cavity (Wever 1975; Wever and Gans 1976). Despite early reports indicating the presence of an operculum fused to the stapes, it is now believed that all caecilians

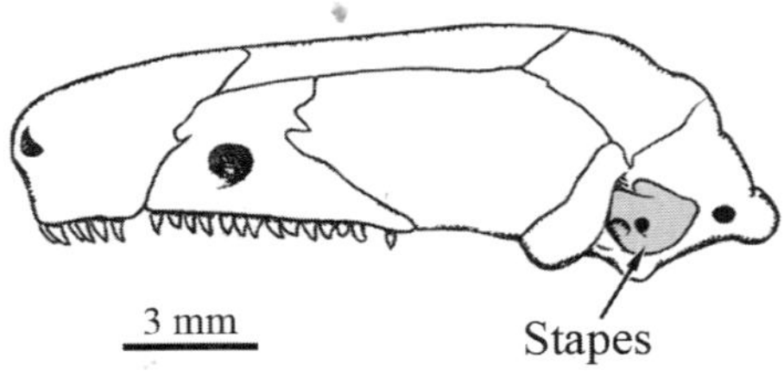

FIGURE 6.6. Skull of a caecilian (*Gymnopis multiplicata*, = *G. proxima*), showing the relatively large stapes (shaded). Drawn from a photograph in Taylor (1969), with permission of The University of Kansas Science Bulletin.

lack an operculum and opercularis muscle (Lombard and Bolt 1979; Hetherington 1988b). The oval window is occupied solely by the very large, ossified footplate of the stapes (Fig. 6.6), from the anterior end of which a short shaft projects rostrolaterally to articulate with the quadrate (de Jager 1939b, 1947; Wever 1975; Wever and Gans 1976). The stapes is overlaid laterally by skin and a thin layer of muscle (Wever 1975).

The anteroventral part of the stapes footplate is more firmly articulated with the rim of the oval window than is the rest of its perimeter, by means of a synchondrosis (de Jager 1939b; Wever 1975). Although de Jager (1939b) believed that the stapes of *Dermophis mexicanus* (Mexican caecilian) is incapable of movement within the oval window, Wever (1975) found that the stapes was very mobile in this same species, the cartilaginous region representing a hinge-point. Several reports suggest that the articulation between stapes and quadrate in caecilians is basically synovial (de Jager 1939a,b, 1947). However, this may be subject to ontogenetic change: de Villiers (1938) found a synovial articulation in one specimen of *Boulengerula boulengeri* (Boulenger's caecilian) but a synostosis in a second, and de Jager (1939b) found the articulation to be synchondrotic in two *Dermophis* species, with only a remnant of the synovial cavity. To add to the confusion, Wever (1985) describes the articulation as being ligamentar in caecilians.

Scolecomorphus species, in the monogeneric family Scolecomorphidae, lack both stapes and oval window (de Villiers 1938; Brand 1956; Taylor 1969). They are the only living amphibians known to lack all components of the middle ear apparatus.

Like urodeles, caecilians lack a round window for pressure relief as the stapes vibrates within the oval window. Instead, caecilians are believed to use a re-entrant fluid circuit similar to that seen in certain reptiles. The fluid displaced by the medial side of the stapedial footplate flows through the inner ear, into the cranial cavity via the perilymphatic duct, and from the cranial cavity back to the lateral surface of the stapes footplate (Wever 1975; Wever and Gans 1976). Fluid vibrations may also be communicated through the cranial cavity to the contralateral side, as in salamanders, this representing a second relief pathway (Wever 1985).

2.4 Evolution of the Middle Ear in Amphibians

It has been argued that the tympanic ear in frogs evolved independently of the tympanic ear in other vertebrate classes, and that the lack of a tympanic ear in urodeles and caecilians represents the retention of a primitive tetrapod condition (Lombard and Bolt 1979; Bolt and Lombard 1985). If so, the possession of a tympanic ear might either have been plesiomorphic for frogs (de Villiers 1934; Bolt and Lombard 1985), or the tympanic ear might have evolved within the crown-group Anura, the lack of tympanic ears in the "primitive" leiopelmatid frogs representing a primary condition (Stephenson 1951). However, Schmalhausen (1968) argues on embryological grounds that urodeles have lost a tympanic ear rather than being primitively earless. Unfortunately, the fossil history of the urodeles and caecilians is currently too poorly known to shed much light on this debate (Carroll 2001).

Reduction of the tympanic middle ear in frogs has been associated with factors including absence of vocalization, fossoriality, or living in noisy environments (McDiarmid 1971; Jaslow et al. 1988; Hetherington 1992b). Although any of these explanations might be plausible for a given group of frogs, Jaslow et al. (1988) conclude that there are no systematic, environmental, or behavioral factors consistently associated with the reduction of the tympanic ear. Indeed, Smirnov and Vorobyeva (1988) and Smirnov (1991) argue that the loss of the tympanic middle ear in many species of frogs might not be adaptive at all. Because air-borne sound can be transmitted to the inner ear by extratympanic routes (see Section 4), loss of the tympanic system might not seriously compromise airborne hearing and might occur simply as the by-product of a general paedomorphic trend. Middle ear reduction in frogs is typically restricted to smaller species, in which extratympanic detection of air-borne sound is likely to be more important (Hetherington 1992b). It has been proposed that the lack of the tympanic ear in urodeles is also a corollary of neoteny (de Villiers 1934).

The loss of the opercularis system in both frogs and urodeles is strongly associated with strictly aquatic habits. This is discussed further in Section 5 (The Opercularis System).

3. Physiology of the Tympanic Middle Ear in Frogs

The following represents a very brief overview of tympanic ear function in tetrapods: for more detailed reviews, the reader is referred to Dallos (1973) and Relkin (1988).

Impedance is the ratio of sound pressure to velocity, and is a characteristic property of a given medium. The proportion of sound energy that enters one medium from another is determined by the relative impedances of the two media, the remainder being reflected at the boundary. The impedance of air is much smaller than the impedance of the fluid in the inner ear, and calculations suggest that only around 3% of incident sound energy in air would enter the mammalian cochlea if sound impinged onto it directly (Relkin 1988). The impedance-matching function of the middle ear is to ensure that the impedance of the tympanic membrane at the frequencies of biological interest is close to that of the air, so that as much sound energy as possible is absorbed rather than reflected.

The middle ear apparatus will contribute to the impedance of the tympanic membrane, and will affect its frequency response. The impedance at the membrane may be separated into a reactive component, contributed by the mass and stiffness of the ear, and a resistive component. At low frequencies, the stiffness of the membrane and of the middle ear air cushion behind it is expected to dominate the overall impedance at the membrane, whereas at high frequencies the mass of the membrane and auditory ossicle(s) attached to it is expected to be dominant (but see Overstreet and Ruggero, 2001, for a criticism of this concept as applied to mammals). At intermediate frequencies, the contributions of mass and stiffness tend to cancel each other out and the reactive component of the impedance at the tympanic membrane is thus minimized. The low overall impedance that results is, it is hoped, close to that of air, so that the efficiency of sound transfer from the air is maximized.

The impedance seen at the vertebrate tympanic membrane when the reactive components are minimized is largely resistive, and the main source of resistance is usually the inner ear. The effective resistance of the inner ear, as seen at the tympanic membrane, may be reduced by decreasing the pressure or by increasing the velocity at the membrane, compared to values at the stapes footplate. This is achieved by means of the larger surface area of the tympanic membrane compared to the stapes footplate (the area ratio), and by the mechanical leverage provided by the middle ear ossicles (the lever ratio).

3.1 Response of the Tympanic Membrane

When sound is presented to the frog ear through a closed-field acoustic coupler, the displacement response of the membrane shows a low-pass characteristic, whereas when presented free-field with the mouth closed, the response is band-pass in nature (Pinder and Palmer 1983). The difference is due to the fact that sound in the free-field can reach the inside of the tympanic membrane, via the mouth cavity and wide Eustachian tubes. The head of *Rana temporaria* (grass frog) was found to be essentially transparent to sound at frequencies below 2 kHz (Aertsen et al. 1986). Sound reaching the middle ears via the lungs has also been documented in several species (see Section 4.1). At low frequencies, sound pressure on each side reduces the pressure difference across the tympanic membrane, diminishing its overall response, whereas high-frequency sound cannot reach the inside of the membrane. At intermediate frequencies, the pressure on either side of the membrane adds constructively to increase the pressure difference and thus the tympanic response (Pinder and Palmer 1983).

Because the response of the tympanic membrane depends on the relative amplitudes and phases of the sound pressure on each side, the membrane will act as an asymmetrical pressure-gradient receiver within an appropriate frequency range. Such devices are intrinsically directional, and it is believed that this confers the ability of frogs to localize sound at low frequencies (Vlaming et al. 1984; Aertsen et al. 1986). A detailed consideration of the models and experiments used to examine interaural coupling and sound localization in frogs is beyond the scope of this chapter; the interested reader is referred to Eggermont (1988) for a review.

Frog tympanic membranes respond to free-field sound at frequencies up to around 5 kHz (Pinder and Palmer 1983; Hetherington 1992a, 1994a; van Dijk et al. 2002). The response ranges of smaller frogs are shifted towards higher frequencies than those of larger frogs, and the peak sensitivities tend to be lower (Hetherington 1992a). The frequencies of peak response in frogs tend to coincide with regions of high sensitivity established neurophysiologically, and are typically frequencies in the basilar papilla range (Lombard and Straughan 1974; Anson et al. 1985). The velocity amplitude of the tympanic membrane of *Rana esculenta* (edible frog) at resonance frequency, around 1.6 kHz in this species, is similar to that of air particles, meaning that the acoustic impedance of the membrane at this frequency is close to that of air (Anson et al. 1985).

The tympanic membrane of frogs does not vibrate as a flat plate, as simple models of the middle ear impedance matching function tend to assume. The amplitude of motion at the point of attachment of the extrastapes has been found to be lower than the amplitude of the peripheral membrane (Moffat and Capranica 1978; Pinder and Palmer 1983; Anson et al. 1985), and the membrane vibrates in different modes according to frequency (Vlaming et al. 1984; Anson et al. 1985; Purgue 1997).

3.2 Ossicular Movements in Frogs

Saunders and Johnstone (1972) were among the first to measure stapes vibration in frogs, using the Mössbauer technique. In a more recent study using laser interferometry, Jørgensen and Kanneworff (1998) found that the footplate of *Rana temporaria* vibrated 180° out-of-phase with the center of the tympanic membrane at low frequencies, showing that the stapes rocks about a fulcrum rather than moving in and out of the oval window as a piston. The fulcrum was identified as the articulation between pars media and otic capsule, supporting anatomical observations (Bolt and Lombard 1985; Jaslow et al. 1988). These experimental findings were confirmed in *R. catesbeiana* (Mason and Narins 2002a,b).

Jørgensen and Kanneworff (1998) found the ratio of displacement amplitudes of the tympanic membrane to the center of the stapes footplate to be 20 in *Rana temporaria*. The equivalent ratios in *R. catesbeiana* were calculated to be 13.3 in males and 7.1 in females (Mason and Narins 2002a; Mason et al. 2003). These values are all considerably higher than the lever ratios of between 5.5 and 5.8 predicted from the anatomy, on the assumption that the extrastapes and stapes of these frogs operate as a single stiff unit (Jørgensen and Kanneworff 1998; Mason et al. 2003). Jørgensen and Kanneworff (1998) proposed that the reasons for this might include bending of the cartilaginous extrastapes, suggested by the fact that the phase difference between tympanic membrane and footplate increases with

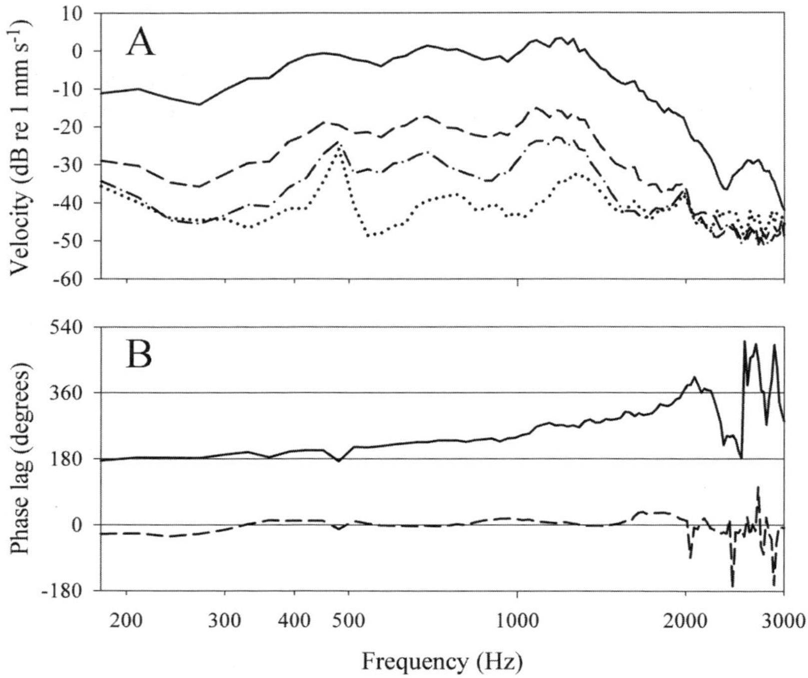

FIGURE 6.7. Vibrations of middle ear structures in a male *Rana catesbeiana* in response to airborne sound, as measured using laser interferometery. A: Velocity amplitudes. Solid line: tympanic membrane; dashed line: stapes footplate; dash-dotted line: operculum; dotted line: parotic crest (part of the skull in the auditory region, included as a control). Note that the responses of the footplate and operculum are of a similar shape to that of the tympanic membrane, but are of a lower amplitude. The responses of these structures are above the "background" response of the parotic crest at all but the lowest and highest frequencies. B: Phase lags between tympanic membrane and stapes footplate (solid line) and between footplate and operculum (dashed line). The footplate vibrates 180° out of phase with the tympanic membrane at low frequencies, this phase lag rising with frequency. However, the footplate and operculum vibrate in phase over a wide frequency range. After Mason and Narins (2002b), with permission of The Company of Biologists Ltd.

frequency (Fig. 6.7B). Mason and Narins (2002a) examined the movement of the extrastapes directly through laser interferometry and found that the body of the extrastapes is actually stiff, but its articulation with the pars media, where the cartilage of the extrastapes narrows considerably, is flexible. The ascending process connecting the extrastapes to the skull was found to be vital for ossicular function: when severed, the extrastapes rocks freely on its articulation with the pars media and the footplate movement is consequently minimal. Bending of the strap-like, cartilaginous ascending process allows the extrastapes to move

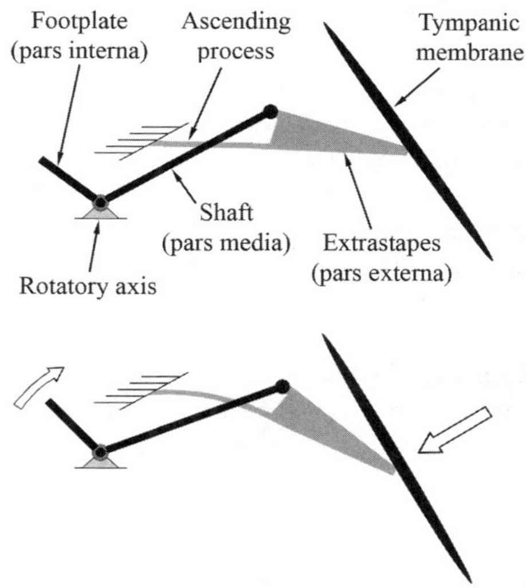

FIGURE 6.8. Diagrammatic representation of the (greatly exaggerated) movements of the middle ear structures in *Rana catesbeiana* in response to airborne sound, according to the interpretation of Mason and Narins (2002a). The lower diagram shows the effects of an inflection of the tympanic membrane. The stapes (pars interna and pars media) rocks about its rotatory axis, located ventrolateral to the footplate. Movement of the footplate relative to the tympanic membrane is reduced by flexion of the articulation between extrastapes and pars media. The ascending process is a vital part of this ossicular mechanism.

relative to the skull, such that pushing the tympanic membrane inwards is translated into a ventral displacement of the articulation between extrastapes and pars media, pulling the pars media shaft downwards (Fig. 6.8). The pars media rocks about its articulation with the otic capsule and the footplate is consequently pulled out of the oval window. Werner (2003) independently came to very similar conclusions, based on anatomical and electrophysiological studies of *R. catesbeiana*.

It has therefore been argued that the *Rana catesbeiana* middle ear apparatus contains two functional "ossicles," one being the extrastapes and the other the pars media and pars interna acting together as a unit (Mason and Narins 2002a; Werner 2003). The flexible articulation between extrastapes and pars media introduces an extrastapedial lever, which is probably responsible for the very high velocity ratios observed in frogs. Flexibility in the ossicular apparatus, apparently a universal feature of tympanic ears, presumably allows for a measure of protection against high-amplitude displacements of the tympanic membrane. In frogs, such displacements might be due to pressure applied to the external surface of the membrane as the frog dives or if the membrane is accidentally touched, or to the high pressures inside the middle ear cavity during breathing or vocalizing (see Section 5.5).

The discovery that the ascending process is a vital part of the *Rana catesbeiana* ossicular apparatus was unexpected: its function had been largely ignored in previous studies. However, several anuran species are said to lack this process (de Villiers 1931; Sedra and Michael 1959; Jaslow et al. 1988; Smirnov and Vorobyeva 1988), warning against the dangers of extrapolating results from one species to all.

As well as the force and velocity transformations achieved by the ossicular apparatus, the impedance-matching function of the middle ear should be aided by the area ratio between tympanic membrane and footplate. This ratio is 12.8 in *Bufo americanus* (American toad), 9.1 in *Hyla cinerea* (green treefrog), 20.0 in female *Rana catesbeiana*, and 50.6 in males of this species (Moffat and Capranica 1978; Mason et al. 2003). These values include the entire tympanic membrane area in the calculation. The fact that the tympanic membrane vibrates as a flexible membrane rather than a stiff plate suggests that its "effective" area will differ from this "anatomical" value (von Békésy 1960; Moffat and Capranica 1978). The area ratio in frogs might also be affected by the operculum, which appears to be coupled to the stapes footplate, but the contribution of the operculum to the total volume velocity at the oval window when the stapes is displaced was estimated to be quite small (Mason and Narins 2002b).

3.3 Sexual Dimorphism in the Middle Ear

The tympanic membrane of most frog species is either the same diameter in both sexes, or slightly larger in females (Duellman and Trueb 1986). Larger membranes in females of certain *Hyla* species might be responsible for lower auditory thresholds compared to males (McClelland et al. 1997). However, the tympanic membrane area of certain ranid frogs is considerably larger in males (Wright and Wright 1949; Shofner and Feng 1981; Boatright-Horowitz and Simmons 1995; Mason et al. 2003). The mass of the membrane is also increased in males by means of a large, connective tissue pad located centrally on the inner surface (Purgue 1997; Mason et al. 2003; Werner 2003). Purgue (1997) showed that the enlarged tympanic membrane of the male *Rana catesbeiana* (Fig 6.2B) is used to radiate high-frequency components of its vocalization, an insight that has widespread implications for the interpretation of middle ear structures in frogs.

Although sound radiation may well represent the primary selective pressure towards its enlargement, an enlarged tympanic membrane is also expected to affect hearing. The tympanic membrane of the male *Rana catesbeiana* has an augmented vibratory response to frequencies of around 200 Hz, coinciding with a low-frequency peak in the mating call (Hetherington 1994a). Mason et al. (2003) found that the stapes footplate area is little different in size between males and females of this species, and the operculum area does not differ at all. Another consequence of the marked dimorphism in the eardrum area is therefore that the anatomical area ratio between the tympanic membrane and stapes footplate is considerably higher in male than in female *R. catesbeiana*. The velocity ratio between the tympanic membrane and stapes was also found to be higher in males than in females (Mason and Narins 2002a). Mason et al. (2003) calculated that the overall impedance transform ratio is much higher in male *R. catesbeiana* than in females, and higher in females than in other vertebrates for which values are available. This might be required if the inner ear impedance is unusually high in *R. catesbeiana*, especially in males. There is, however, no reason to suppose that this is the case: these values more likely reflect the considerable oversimplifications in existing models of middle ear transformer function.

The tympanic disc in *Xenopus laevis* (African clawed frog) is relatively larger in males than in females (Elepfandt 1996b). The tympanic membrane is adorned with a fleshy papilla in males of certain *Petropedetes* species (Fig. 6.2C), the presence of which increases the frequency of peak vibratory amplitude of the membrane, perhaps by stiffening it (Narins et al. 2001). The adaptive significance of these examples of dimorphism are unclear.

4. Extratympanic Sound Transmission

Although many species of adult amphibians lack tympanic ears, all those tested retain some degree of sensitivity to airborne sound, especially at low frequencies (see, e.g., Wever 1985; Hetherington 1989; Lindquist and Hetherington 1996; Hetherington and Lindquist 1999). Some "earless" frogs actually have hearing comparable in range and sensitivity to species with tympanic ears (Loftus-Hills 1973; Jaslow and Lombard 1996; Lindquist et al. 1998). It is clear from comparisons between tympanic and neural responses to sound that even frogs possessing tympanic ears must have alternative routes by which airborne sound can reach their inner ear receptors at the low frequencies at which the tympanic membrane is relatively unresponsive (Chung et al. 1978; Pinder and Palmer 1983; Aertsen et al. 1986). Indeed, Lombard and Straughan (1974) found that bilateral removal of the tympanic membrane did not have a significant impact on midbrain responses to airborne sound in *Hyla* species at frequencies below around 1 kHz.

The transmission of airborne sound vibrations to the inner ear of amphibians by routes other than through the tympanic membrane and stapes is referred to as extratympanic transmission. Wilczynski et al. (1987) found that extratympanic transmission was more effective than tympanic transmission below around 200 Hz in *Rana pipiens* (northern leopard frog). From 200 Hz to 1 kHz, responses to sound could be induced almost as effectively by extratympanic as by tympanic routes; above 1 kHz the tympanic route was found to become relatively much more effective. Extratympanic transmission, by definition, does not involve the tympanic ear, and must therefore involve the excitation of other, less specialized parts of the body. Given that transmission of low-frequency sound is less impeded by mass than is transmission of higher frequencies, it is unsurprising that extratympanic transmission is most significant at low frequencies. Certain models of hearing in frogs are improved by the inclusion of an extratympanic route for low-frequency sound transmission (Aertsen et al. 1986; Eggermont 1988).

The potential role of the opercularis system in the detection of airborne sound is discussed in Section 5.3. Other possible mechanisms are described below.

4.1 Acoustic Pathways Involving the Lungs

Narins et al. (1988) showed that a portion of the lateral body wall overlying the lung of the leptodactylid frog *Eleutherodactylus coqui* (Puerto Rican coqui)

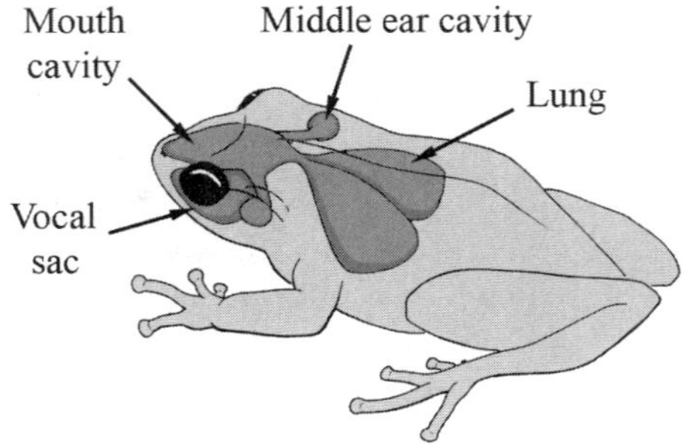

FIGURE 6.9. Contiguous air spaces within a frog with tympanic middle ears (based on *Eleutherodactylus coqui*). The pathway from lungs to middle ear cavity represents one possible route by which sound is transferred to the ear. Reprinted with permission from Peter M. Narins, The Journal of the Acoustical Society of America, 91, 3551 (1992). Copyright 1992, Acoustical Society of America.

vibrates in response to airborne sound. These authors suggested that sound vibrations might pass through the body wall, lung, and glottis to the mouth cavity, thereby reaching the tympanic membranes from the inside (Fig. 6.9). This was confirmed in subsequent studies (Ehret et al. 1990; Jørgensen 1991; Jørgensen et al. 1991; Ehret et al. 1994). The lateral body wall overlying the lungs is more responsive in smaller anuran species than in large ones; its peak vibratory amplitude in small species is often comparable to that of the tympanic membrane, but the peak tends to be at a lower frequency (Hetherington 1992a).

Because the lung-mouth-ear pathway in frogs like *Eleutherodactylus coqui* ultimately involves the excitation of the tympanic membrane, it is not strictly an extratympanic pathway. However, amphibian species lacking tympanic ears also detect air-borne sound using their lungs (Lindquist et al. 1998; Hetherington and Lindquist 1999; Hetherington 2001). The auditory midbrain response to airborne sound was reduced by 20 to 25 dB when the lateral body wall of the "earless" toad *Bombina orientalis* was covered with silicone grease, or when the lungs were filled with oxygenated saline solution (Hetherington and Lindquist 1999). In salamanders possessing lungs, the anterolateral body wall was found to be considerably more responsive to airborne sound over a wide frequency range than was the lateral head region (Hetherington 2001). The anterolateral body wall was not found to be responsive to airborne sound in lungless plethodontid salamanders (Hetherington 2001).

The routes by which sound vibrations in the lungs reach the inner ear in these species have not been established, but there are several theories. Hetherington and Lindquist (1999) suggested that sound vibrations reaching the mouth cavity of frogs could potentially enter the inner ear through the round window, which, although covered with muscle and connective tissue, is close to the roof of the mouth cavity. Alternatively, lung vibrations might cause the suprascapular cartilage to vibrate, its movement relative to the skull being conveyed to the inner ear via the opercular apparatus (Hetherington and Lindquist 1999). Another possible route involves the cranial cavity, discussed next.

4.2 Acoustic Pathways Between the Inner Ear and Cranial Cavity

Narins et al. (1988) hypothesized that lung vibrations might be able to excite the inner ear via the endolymphatic sacs, diverticula of the inner ear which pass into the cranial cavity and (in frogs) extend from there into the vertebral canal. The extensions of the perilymphatic system into the cranial cavity represent another possible route (Hetherington 2001). Experiments on urodeles suggest that fluid vibrations can indeed pass between the cranial cavity and inner ears, probably via the perilymphatic rather than the endolymphatic ducts (Smith 1968; Wever 1978, 1985). Seaman (2002) has demonstrated that vibrations applied to the exposed dura above the dorsal midbrain in anesthetized *Rana catesbeiana* evoke responses in the eighth nerve, the conclusion of this study being that intracranial pressure changes can stimulate the inner ear receptors.

Seismic vibrations are known to result in relative movement between head and body in frogs (Hetherington 1988a), and presumably generate pressure fluctuations within the cerebrospinal fluid. Smith (1968) observed vibrations of the fluid in the ear side of the perilymphatic foramen when the bodies of salamanders were exposed to vibrations between 50 and 250 Hz. The communication between cranial cavity and inner ear therefore represents a potential pathway for seismic vibratory stimulation of the amphibian ear that might not require the opercularis system (Mason and Narins 2002b); this remains to be tested experimentally.

4.3 Sound Transmission in Caecilians

Caecilians exposed to closed-field airborne sound were found to be most responsive to low frequencies, below around 1 to 2 kHz (Wever 1975, 1985; Wever and Gans 1976). The greatest response in terms of microphonic potentials (electrical responses to sound recorded from the inner ear) occurred when the sound source was applied to the region directly over the stapes footplate. The sound presumably travels directly through the head to the inner ear via the stapes, the impedance mismatch between air and tissue resulting in the relatively low sensitivity observed (Wever and Gans 1976). Wever (1975) argued that the articulation between stapes and quadrate might serve to reduce high-amplitude movement of the stapes, caused, for example, by a blow to the head, while exerting little influence on normal hearing.

Using a vibrating probe, Wever (1975) found that inner ear responses were greatest when the probe was applied to the skin directly over the stapes footplate. Responses obtained when the vibrations were applied directly to the skull were considerably smaller. This suggests that the primary mechanism of vibration detection in caecilians is direct, but the possibility of seismic sensitivity through other means, for example, a form of inertial bone conduction (Hetherington 1992b), remains to be examined.

5. The Opercularis System

The opercularis system, consisting of the operculum within the oval window and the opercularis muscle connecting it to the shoulder girdle (Fig. 6.10), is unique to amphibians. As discussed earlier, what is referred to as the "opercularis muscle" is not homologous among all amphibians, and the muscle may insert on a fenestral plate in urodeles rather than on a discrete operculum. The opercularis system *sensu lato* is present in most anurans and urodeles after metamorphosis, notable exceptions being species that remain strictly aquatic as adults. Consequently, it has long been considered to be associated with terrestrial life (Kingsbury and Reed 1909; de Villiers 1934; Eiselt 1941). The development of the opercularis system in frogs is complete at the time of initiation of terrestrial habits, whereas development of the tympanic system is more variable and in some species is not complete until some time afterwards (Hetherington 1987b). This again suggests that the opercularis system is in some way associated with terrestriality, and not necessarily with the other components of the middle ear apparatus.

The opercularis muscle contains a high proportion (80 to 100%) of tonic fibers in frogs of the genera *Hyla* and *Rana* (Becker and Lombard 1977; Hetherington

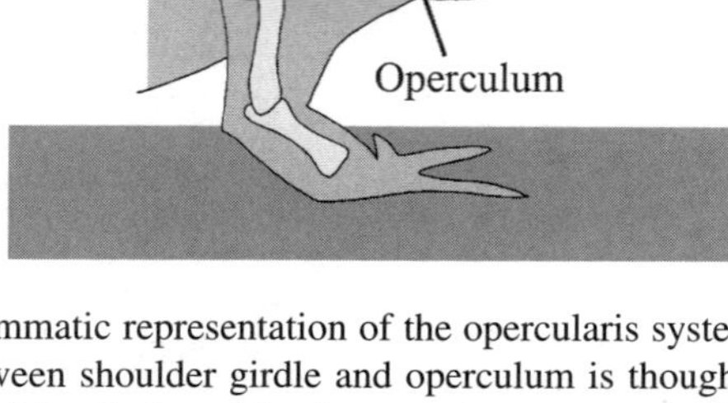

FIGURE 6.10. Diagrammatic representation of the opercularis system in a frog. The muscular connection between shoulder girdle and operculum is thought to allow differences between the motion of forelimb and head, occurring as a result of seismic vibrations in the substrate, to be translated into vibrations of the inner ear fluids. See text for further details. Adapted from Figure 21.2 in Hetherington TE (1992) The effects of body size on the evolution of the amphibian middle ear. In: Webster DB, Fay RR, Popper AN (eds) The Evolutionary Biology of Hearing, pp. 421–437. Copyright © 1992, Springer-Verlag New York Inc.

and Tugaoen 1990). In *Rana catesbeiana*, the muscle has a slow velocity of contraction and relaxation and was found to be fatigue-resistant (Hetherington 1987a). Turning to salamanders, Becker and Lombard (1977) found that the opercularis muscle contained around 90% tonic fibers in *Desmognathus* species, in which it is derived from the *m. cucullaris*, and 45% tonic fibers in *Ambystoma tigrinum* (tiger salamander), in which it is derived from the *m. levator scapulae*. However, Hetherington and Tugaoen (1990) found only 9% tonic fibers in the opercularis muscle of the latter species, the vast majority being fast oxidative-glycolytic or fast glycolytic twitch fibers. Although the opercularis is primarily a tonic muscle in some amphibians, this is clearly not universally the case.

The several proposed functions of the opercularis system are described below.

5.1 The Opercularis System and the Detection of Body Movements

Eiselt (1941) suggested that the opercular apparatus might function in the control of posture during terrestrial locomotion, by coupling the movements of the forelimb to the inner ear. Baker (1969) found that cutting the opercularis muscle led to an impairment of postural control in a frog placed on a tilting platform, supporting this hypothesis. However, Baker's results were challenged by Becker and Lombard (1977), who suggested that the entire *m. levator scapulae* might have been severed in that study rather than just the opercularis slip. The hypothesis that the opercularis system is used in postural control has not resurfaced in recent years.

5.2 The Opercularis System and Protection Against Intense Sounds

Wever (1979, 1985) hypothesized that the opercular apparatus forms part of an acoustic protective mechanism, based primarily on an examination of *Rana* species. Contraction of a "columellar muscle" extending between footplate and shoulder girdle was believed to pull the stapes footplate caudally, locking it together with the operculum and thus restraining its movements. Conversely, contraction of the *m. opercularis* and relaxation of the columellar muscle would free the stapes. Although details of the experiments were not given, Wever (1985) found that tension applied to the muscles resulted in changes in inner ear potential recordings in accordance with his theory, although the effects of anesthesia or sectioning of the muscles were apparently less clear.

Later authors have denied the existence of a discrete "columellar muscle" in the majority of frogs (see Section 2.1.4), and the slow velocity of contraction and relaxation of the anuran opercularis muscle appears unsuitable for rapidly modifying the movement of the stapes (Hetherington 1987a). Hetherington (1994b) found that neither removal nor artificial stimulation of the opercularis in lightly anesthetized *Rana catesbeiana* had any effect on the tympanic membrane response to airborne sound, even at high sound pressure levels. Wever's theory has therefore attracted little support.

Considering urodeles, Wever (1985) applied masses to the opercularis muscle of anesthetized *Ambystoma* salamanders and found a reduction of up to 12 dB in the inner ear responses to air-borne sound when the muscle was thus artificially tensed. He therefore proposed that the opercularis muscle has a protective function in urodeles. Wever (1985) also maintained that contractions of the quadrate muscle, which runs from fascia on the midline of the throat to the palatoquadrate, would move the palatoquadrates ventrally and medially, displacing the stapes footplate ventrally and dampening its movements. These ideas have apparently not been explored further.

5.3 The Opercularis System as a Mechanism to Enhance Perception of Airborne Sound

Lombard and Straughan (1974) measured midbrain responses to airborne sound in frogs of the genera *Hyla*, *Smilisca*, and *Leptodactylus*. Sectioning the opercularis muscle, transecting its nerve, or changing the frog's posture reduced hearing sensitivity by up to around 30 dB at frequencies below 1 kHz, but the responses to higher frequencies were hardly affected. The authors did not believe that their manipulations were affecting sensitivity to low-frequency substrate vibrations.

Lombard and Straughan (1974) hypothesized that contraction of the opercularis muscle would lock the operculum and stapes footplate together (the opposite of Wever's contention), effectively increasing the surface area of the footplate and improving force transfer efficiency at low frequencies. Perception of low frequencies would thus require sustained contraction of the opercularis muscle, a role for which the anuran muscle is apparently well-suited (Becker and Lombard 1977; Hetherington 1987a). Relaxation of the muscle would uncouple the stapes and operculum, decreasing the inertia of the middle ear apparatus and thereby increasing sensitivity to the higher frequencies transmitted through the tympanic system. Capranica (1976) disputed this functional scenario based on the fact that the area ratio of the middle ear should always be constant, so as to achieve optimum impedance-matching. However, this would only be true if the impedance of the inner ear is always constant, and if the "effective area" of the tympanic membrane does not change with frequency. Given that the tympanic membrane of frogs is known to vibrate in frequency-dependent modes, the latter, at least, is unlikely.

In *Rana catesbeiana*, removal of the opercularis muscle had only a very small effect on the response to airborne sound, decreasing microphonic responses by 1 to 2 dB at frequencies lower than 200 to 300 Hz (Hetherington 1989). Neither removing nor electrically stimulating the opercularis muscle had any effect on the tympanic membrane's response to airborne sound at any frequency (Hetherington 1994b). Hetherington concluded that Lombard and Straughan's

(1974) hypothesis is unlikely to be correct, inasmuch as one would predict that the increased mass at the stapes footplate associated with opercular coupling would affect the tympanic membrane response. In the salamander *Ambystoma tigrinum* (subspecies unspecified), removal of the opercularis muscle resulted in small decreases in microphonic responses to airborne sounds (averaging around 4 dB) across a wide frequency range from 100 Hz to 3 kHz, although the effects on different individuals were found to be variable (Hetherington 1989). Although *A. t. tigrinum* has a mobile stapes (Monath 1965), the opercularis system is present in many species of frogs lacking all components of a tympanic ear. Even if valid for some species, Lombard and Straughan's hypothesis clearly cannot be generalized to all amphibians (Hetherington et al. 1986; Hetherington 1989).

As an alternative hypothesis, low-frequency airborne sound might cause the whole body of a frog to vibrate, any amplitude or phase differences between the vibrations of the skull and shoulder region being communicated to the inner ear by the opercularis system in a manner analogous to that proposed for seismic vibrations (Hetherington 1985, 1987a, 1989). This would not require the involvement of the tympanic system, and is therefore applicable both to urodeles and to "earless" frogs. Indeed, Walkowiak (1980) found that midbrain responses to sound in the "earless" *Bombina bombina* were greatest when the closed-field sound source was applied over the shoulder and neck region (example cited by Hetherington 1992a). The shoulder region is more responsive to airborne sound in smaller frogs, so this mechanism might be more effective in the species studied by Lombard and Straughan than in *Rana catesbeiana*, explaining the differences in experimental findings (Hetherington 1989, 1992a).

5.4 The Opercularis System and the Detection of Substrate Vibrations

Kingsbury and Reed (1909) proposed that in terrestrial amphibians, the opercularis system could serve to communicate "jars and vibrations" from the forelimb to the inner ear via the opercularis muscle. This hypothesis was developed further and evidence supporting it collected in a series of papers by Hetherington and colleagues (Hetherington 1985, 1987a, 1988a; Hetherington et al. 1986). The functional scenario is as follows. Substrate vibration results in relative movement between the suprascapular cartilage and the skull, which is translated into relative movement of the operculum within the oval window by means of the opercularis muscle (Fig. 6.10). The muscle is believed to be in a state of constant tension during terrestrial situations, allowing it to pull on the operculum when the animal is exposed to such vibrations. It is unlikely that a muscle could transmit compressive forces to the operculum, so elasticity in the ligaments and cartilage connecting it to the oval window are probably responsible for its return to position. Vibrations of the operculum would be translated into vibrations of the fluid in the inner ear, which would stimulate receptor end-organs such as those in the sacculus or lagena. The opercularis system, used in the detection of seismic vibrations, is thus considered to be functionally independent of the tympanic system, used in the detection of airborne sound. The loss of the opercularis system in primarily aquatic frogs and urodeles is consistent with this scenario.

It has also been suggested that a massive operculum might act as an inertial element, able to function in vibration detection even without an associated muscle (Hetherington 1988a, 1992b). This seems unlikely, however, in the case of the many frogs and urodeles in which the operculum remains unossified and therefore of relatively low density.

Mechanical vibrations applied directly to the operculum do excite inner ear receptors in urodeles (Wever 1978; Ross and Smith 1982), but what is the evidence that the operculum is set into motion by seismic vibrations in vivo? Hetherington (1985) looked at inner ear microphonic responses to vibration in *Rana catesbeiana*, at frequencies from 20 to 250 Hz. The mean decreases in response following opercularis muscle section ranged from 45 to 74% (5 to 12 dB). Sectioning the nerve to the opercularis muscle caused only slight decreases in vibratory responses (mean <2 dB) over a similar frequency range (Hetherington 1987a). Artificial stimulation of the muscle resulted in small increases of up to 4.5 dB, although responses were sometimes depressed. In a more convincing series of experiments, Hetherington (1988a) found that the amplitude of inner ear microphonic responses to vibration was largest when there was the most relative movement between head and scapula. Removal of the opercularis muscle resulted in a reduction in response amplitude of up to 18 dB, depending on frequency and position of the frog. The largely vertical orientation of the opercularis muscle in anurans (Hetherington et al. 1986) suggests that the opercularis system would be most responsive to vertically oriented vibrations, which appeared to be the case. However, the largely horizontal orientation of the muscle in urodeles (Hetherington et al. 1986) suggests that horizontally oriented vibrations would be most effective here.

Although the hypothesis that the opercularis system of amphibians is involved in vibration reception is currently the most popular explanation, the experimental evidence cited above is not especially strong. Indeed, not all facts are consonant with this hypothesis. The opercularis muscle of the tiger salamander *Ambystoma tigrinum* is not a tonic muscle (Hetherington and Tugaoen 1990), which is inconsistent with the idea that the muscle should be in a permanent state of tension. Severing the opercularis muscle resulted in a small decrease in vibratory sensitivity, as measured by inner ear microphonic potentials, in adult *Notophthalmus viridescens* (red eft) salamanders, but it made no difference to juveniles, nor to adult *Plethodon cinereus* (eastern red-backed salamander; Ross and Smith 1980). Anesthetized urodeles, lying on their backs, can still respond to vibratory stimuli (Ross and Smith 1979), and the sensitivity of anesthetized frogs in a similar position is acute (Koyama et al. 1982; Narins and Lewis 1984; Yu et al. 1991). Under these conditions the forelimbs and opercularis system are presumably ineffective as a transmission path, so there must be another means of exciting the inner ear receptors (Yu et al. 1991). The saccular recess of amphibians is filled with a large otoconial mass coupled to the hair cells, which probably operates in vibration detection as an inertial sensor (Ross and Smith 1979, 1980; Lewis 1984; Lewis and Lombard 1988): this mechanism of vibratory sensitivity would not require the opercularis system (Ross and Smith 1979). Given that the inner ear is so sensitive to vibrations without opercular contribution, it is unclear why the opercular apparatus would be needed as a supplemental pathway.

The view that the opercularis system is functionally independent from the tympanic system (Hetherington et al. 1986; Hetherington 1988a, 1992b; Jaslow et al. 1988) has also been challenged. Mason and Narins (2002b) demonstrated that the operculum vibrates along with the stapes footplate in response to airborne sound (Fig. 6.7A). The two elements vibrate in phase (Fig. 6.7B) but about different axes; coupling is presumably achieved by means of the flange of the stapes that extends under the operculum, and the connective tissue joining them (Figs. 6.3, 6.11). Severing the extrastapes eliminates vibration of both footplate and operculum, showing that the stapes causes the operculum to move, rather than the operculum moving independently. The frogs examined were anesthetized, and it is possible that the stapes and operculum become uncoupled if the opercularis muscle is under tonic tension in an alert frog. However, tension applied manually to the opercularis muscle can be seen to move the stapes as well as the operculum outwards from the oval window (Mason and Narins 2002b). These

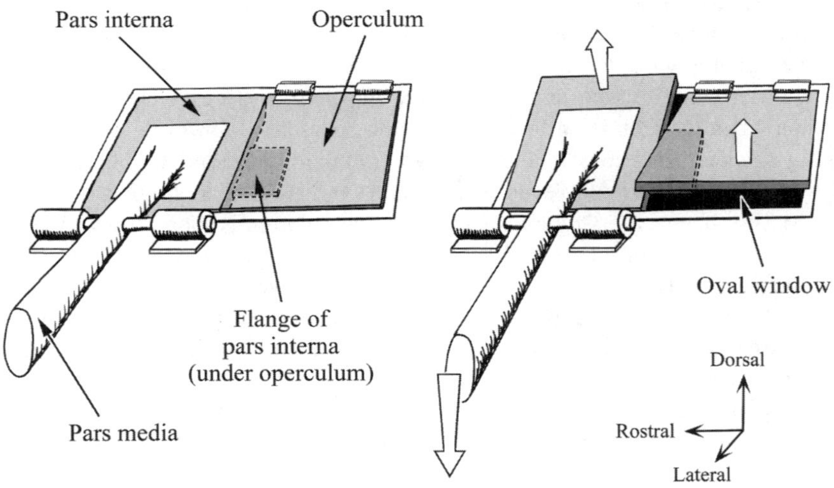

FIGURE 6.11. Diagrammatic representation of the movements of the stapes and operculum of *Rana catesbeiana* at low frequencies. The operculum is portrayed as semitransparent to reveal the flange of the pars interna underneath. When the pars media is pushed down, which happens when the tympanic membrane is inflected, the pars interna (footplate) is moved out of the oval window. Due to the coupling between footplate and operculum, the operculum also moves outwards. The footplate is hinged ventrolaterally but the operculum is hinged dorsomedially. From Mason and Narins (2002b), with permission of The Company of Biologists Ltd.

data suggest that the tympanic and opercular systems are not functionally independent, and the role of the opercularis system should be reconsidered in this light.

5.5 The Opercularis System and Pressure Buffering

Ventilation in frogs is by a "buccal force pump" mechanism (de Jongh and Gans 1969; Gans et al. 1969). The pressure within the buccal cavity rises sharply during the phase that air is forced into the elastic lungs, and this high pressure is communicated via the Eustachian tubes to the middle ear cavity: the eardrums visibly bulge outwards (Hetherington and Lombard 1983a; Narins 1992; Narins et al. 2001). Due to the lever action of the ossicular apparatus, the stapes footplate would be forced into the inner ear during these events. A pressure release pathway within the inner ear might help to prevent high-amplitude displacements associated with breathing and vocalizing from causing damage (Purgue and Narins 2000a,b). Protection from the superimposed higher-frequency vibrations during vocalization might be afforded through stiffening of the tympanic membrane concomitant with the bulging (Jaslow et al. 1988; Narins 1992).

In an electromyographic study of the opercularis muscle in *Rana catesbeiana*, rhythmic, low-amplitude activity was only seen when the frog was on land, or when in water if the nostrils were above the surface (Hetherington and Lombard 1983a). High-amplitude bursts of activity coincided with the phase of ventilation when air is forced into the lungs. The correlation between opercularis activity and ventilation was interpreted as an indirect means of linking opercularis activity with terrestriality, the result being that the muscle is in a state of sustained tension when the frog is on land. This would be suitable for a role in communicating seismic vibrations to the inner ear (Hetherington and Lombard 1983a; Hetherington 1987a). However, Mason and Narins (2002b) proposed that the relationship between opercularis activity and ventilation is more direct. Activity in the opercularis muscle would act to pull the operculum outwards, helping to restrain the inward movement of the stapes as the tympanic membranes bulge outwards. This would represent another mechanism to protect the inner ear from damage during ventilatory and vocalization events, explaining why the muscle is active in a floating frog when the legs are not touching the bottom (Hetherington and Lombard 1983b), circumstances in which the opercularis system could not be involved in the detection of seismic vibrations.

This protective mechanism clearly cannot be valid for urodeles and frogs that lack a stapes and/or a middle ear cavity, nor can it explain the convergent loss of the opercular apparatus in strictly aquatic species. A more general role of the opercularis system could be to buffer changes in inner ear fluid pressure associated with (terrestrial) locomotory movements (Schmalhausen 1957, 1968; Mason and Narins 2002b). Movements of the head relative to the body in urodeles result in fluid movement between the inner ear and the cranial cavity, via the perilymphatic foramen (Smith 1968). Displacements of the operculum also result in fluid movement between these two compartments (Smith 1968; Ross and Smith 1982).

Tensing and relaxing the opercularis muscle could potentially limit fluid movement out of and into the inner ear respectively, during relative movements of skull and spine. This hypothesis remains to be tested experimentally.

6. Underwater Hearing in Amphibians

Two basic modes of vertebrate hearing underwater have been described (Fay and Popper 1985; Rogers and Cox 1988). Because the acoustic impedance of the vertebrate body is very similar to that of water, the body is essentially transparent to underwater sound vibrations. A structure of higher density than the rest of the body, for example, a calcified otolith, will vibrate at a different amplitude and phase to the tissues surrounding it, due to its inertia. Sound can be detected if stereocilia of sensory hair-cells are attached to this otolith. This direct detection of the acoustic particle displacement is sometimes referred to as the "inertial mode" of sound detection. A second and indirect mode of hearing underwater relies on the presence of a bubble of air within the body, for example, trapped within the swim-bladder of a fish. Because air is compressible, the bubble will expand and contract in response to the sound-induced pressure fluctuations in the surrounding tissue. At frequencies lower than the resonant frequency of the bubble (usually the frequencies of biological interest), the bubble acts as a spring and displacement of its wall is proportional to incident pressure (Rogers and Cox 1988). The pulsation of the bubble generates considerable near-field particle displacement close to the ear of the animal, which can be of much higher amplitude than the particle displacement in the original far-field sound wave (van Bergeijk 1967). Sensitivity can be improved further if there is a special means of coupling the pulsations of the bubble to the inner ear, for example, the Weberian ossicles of otophysan fish. Because the signal transmitted to the ear is proportional to the incident sound pressure, this mode of underwater hearing is sometimes referred to as the "pressure mode." In fish, the pressure mode of hearing seems to be the more effective at frequencies above a couple of hundred Hertz, whereas the inertial mode is important at lower frequencies (Fay and Popper 1985).

The contribution of the amphibian lateral line system to sensory perception underwater is discussed by Russell (1976) and Elepfandt (1996a); it is not considered further here.

6.1 Underwater Hearing in Frogs with Tympanic Ears

Rana catesbeiana is known to hear well underwater (Lombard et al. 1981). Hetherington and Lombard (1982) demonstrated that *Rana* species are sensitive primarily to underwater sound pressure rather than particle displacement at frequencies from 200 to 3000 Hz. At frequencies near the estimated resonant frequency of the middle ear air space, pulsations of the middle ear cavity seemed to be able to stimulate the inner ear receptors directly, perhaps through shaking of the whole otic capsule. At frequencies farther away from the resonance frequency, the response of the inner ear could be significantly reduced by severing the stapes, suggesting that pulsations of the middle ear cavity stimulate the inner ear via excitation of the tympanic membrane and stapes. Air contained in the mouth cavity and lungs had little effect on the responses observed, although Hetherington and Lombard suspected that these additional "bubbles" might be more significant at frequencies lower than those tested. The frogs were found to be responsive to particle motion directly when the stimulus was presented at high levels, although this response tended to be masked by the more sensitive pressure detection response. Particle displacement detection might be mediated by the sacculus (Hetherington and Lombard 1982).

6.2 Underwater Hearing in *Xenopus laevis*

Christensen-Dalsgaard et al. (1990) found that the vibration velocities of the tympanic disc of *Xenopus* in response to underwater sound were up to 40 dB higher than water particle velocities at frequencies from 200 Hz to 4 kHz. The authors concluded that pulsations in the air filled middle ear cavity probably account for a large part of the vibrations observed, confirming and extending the results of an earlier study of this species (Hetherington and Lombard 1982). Christensen-Dalsgaard and Elepfandt (1995) found that the tympanic disk exhibited two velocity peaks in response to underwater sound. The higher peak, at 1.7 to 2.2 kHz, seemed to be due to pulsation of air in the middle ear cavities. The lower peak, at 0.6 to 1.2 kHz, was ascribed to pulsations of air in the lungs, communicated to the middle ear cavities by means of the larynx, which rests in a median recess in the roof of the mouth from where both Eustachian tubes originate. The air-space continuity between the middle ear cavities, which persists even when the frog is underwater, might represent a pressure-difference receiver mechanism allowing sound localization (Christensen-Dalsgaard and Elepfandt 1995).

Vibrations of the tympanic disk induced by pulsations of the middle ear air cavity are presumably communicated to the inner ear via the stapes. Christensen-Dalsgaard and Elepfandt (1995) concluded that although the tympanic disk of *Xenopus* probably does not exhibit higher velocity amplitudes than the tympanic membrane of *Rana catesbeiana* in response to underwater sound, it does have a lower response to airborne sound. The adaptive advantage of the pipid middle ear morphology, they speculate, might lie in the tight coupling between disc and stapes, resulting in more efficient transfer of vibrations to the inner ear. In accordance with this suggestion, the tympanic disc was found to respond as a rigid plate when exposed to vibrations, and there was no apparent lever ratio between disk and inner ear (Wever 1985).

6.3 Underwater Hearing in Urodeles

Kingsbury and Reed (1909) suggested that in urodele larvae, or in aquatic adults, vibrations might be transmitted from the floor of the mouth via the mandible to the suspensorium of the jaw, and from there to the columella through its articulation with the palatoquadrate or squamosal. In *Siren*, the columella was thought to be excited through its connection to the hyoid. Wever (1985) argued that underwater sound would actually set the whole of the side of the head into vibration, due to the similarities in acoustic impedance with that of the water, whereas the otic capsule would tend to vibrate with a slightly different amplitude and phase due to differences in inertia and articulations with surrounding structures. The relative movement between the two regions would be coupled to the inner ear via the stapes. Schmalhausen (1968) suggested that sound vibrations could be transmitted to the inner ear via the venous system in larval amphibians, adult urodeles, and adult aquatic frogs. None of these theories appear to have been tested experimentally.

Hetherington and Lombard (1983b) demonstrated that *Ambystoma tigrinum*, like frogs, responds primarily to the pressure component of underwater sound at frequencies from 200 to 1000 Hz. The bubble of air responsible is, in this case, contained within the mouth cavity, but the exact route of transmission between mouth and ear was not investigated. The air-filled lungs had little effect. Like frogs, salamanders were found to be responsive to particle displacement directly only at high sound levels: the mechanism was unclear.

6.4 Underwater Hearing in Tadpoles

In early postembryonic anuran larvae, the oval window is a wide opening in the lateral wall of the otic capsule, covered only with connective tissue (Hetherington 1987b; Horowitz et al. 2001). Hetherington (1987b) suggested that the open fenestra could act as a low-attenuation window through which sound waves penetrating the body could enter the inner ear directly. This could not be true of urodele larvae, which never possess an "empty" oval window: in urodeles, the expansion of the fenestra follows the expansion of the stapes and operculum inside it (Hetherington 1988b).

Witschi (1949, 1950, 1951) proposed that the air-filled lung of anuran tadpoles resonates underwater, and that the inner ear is excited by means of connections between the lung and the round window. In *Alytes*, the perilymphatic duct prolapses through the round window on each side to approach the ipsilateral bronchus (Witschi 1951), whereas in *Xenopus* a diverticulum of the bronchus reaches the round window (Witschi 1950, 1951). In the larvae of *Rana* species, the suspensory ligament of the seventh (pulmonary) pharyngeal pouch forms a fibrous connection between bronchus and round window, referred to as a "bronchial columella" (Witschi 1949, 1955; Hetherington 1987b; Horowitz et al. 2001). The bronchial columella (Fig. 6.12) is present in late hatchling and early post-embryonic tadpoles, disappearing at metamorphic climax (Horowitz et al. 2001). The idea that the coupling between lung and inner ear in tadpoles serves in acoustic transmission is attractive due to the analogy with the transmission of sound from swim-bladder to inner ear in fish. There have, however, been critics:

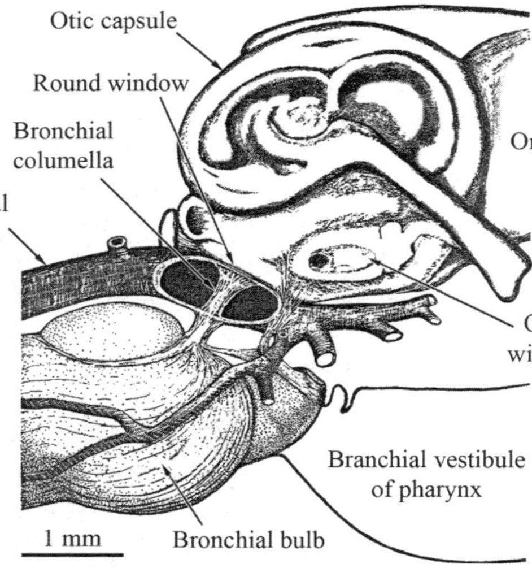

FIGURE 6.12. Lateral view of the caudal part of the skull and pharynx of a *Rana catesbeiana* tadpole. The dorsal aorta has been opened to reveal the bronchial columella, which connects the bronchus to the round window. From Witschi E (1955). The bronchial columella of the ear of larval Ranidae. J Morphol 96: 497–511. Copyright © 1955, The Wistar Institute of Anatomy and Biology. Reprinted with permission of Wiley-Liss, Inc., a subsidiary of John Wiley & Sons, Inc.

van Bergeijk (1959) argued that the low damping of the lung in *Xenopus* larvae would result in almost continuous "ringing" at the resonant frequency, which would be undesirable for an auditory organ. Horowitz et al. (2001) argued that the flexible bronchial columella in *Rana* is unsuitable for efficient sound transmission.

Boatright-Horowitz and Simmons (1997) found that stimulus-evoked midbrain potentials could be recorded in *Rana catesbeiana* tadpoles at the stage when the oval window is covered over only by connective tissue. Responses could be reduced by "earmuffs" covering the oval window, leading the authors to conclude that sound passed directly through the oval window at this stage, as Hetherington (1987b) originally suggested. The bronchial columellae were not considered to represent a primary route for sound transmission, but the lungs of the larvae were uninflated in these experiments. Late prometamorphic tadpoles were unresponsive to sound, probably due to a transient loss of neural connectivity between medullary and midbrain auditory nuclei. Animals in metamorphic climax regained responsiveness, which was not reduced by blockade of the oval window. Boatright-Horowitz and Simmons suggested that the opercularis system was responsible for sound transmission at this stage, but this is just one of many possibilities. It is clear that further experimental study is required to test rigorously the various hypotheses about tadpole hearing.

7. Summary

The role of the tympanic ear in tetrapods is fairly well understood: it has to cope with demands including impedance matching, sound localization, and protection from high sound levels and static pressures. Adaptations in frogs and other tetrapods might well be analogous rather than homologous (Lombard and Bolt 1979), allowing one to examine critically assumptions about what is required of an ear. Indeed, many amphibians lack a tympanic ear altogether yet are still capable of hearing in air. Hetherington and Lindquist (1999) suggest that audition in early tetrapods might have involved the lung.

Acoustic communication in frogs commonly relies on airborne sound, but it can also involve aquatic (Yager 1996) or seismic (Lewis and Narins 1985) channels. The mechanisms for sound transfer to the inner ear in these cases are far less clear, possibly involving structures unique to amphibians such as the bronchial columella or the opercularis system. The functions of these structures have still not been conclusively resolved: although the relevant anatomy is in many cases well described and there is a wealth of ingenious proposals for how sound transfer might be achieved, rigorous experimental work that might distinguish between the competing hypotheses is generally lacking. The need for further study is especially acute in the case of caecilians and urodeles.

Their small size, poikilothermy, and ability to breathe cutaneously make amphibians ideal subjects for physiological experiments. Many of the unanswered questions about amphibian hearing are therefore eminently tractable, and are, it is hoped, to be resolved in the near future.

Acknowledgments. Some of the research for this chapter was undertaken in the Department of Physiological Science, UCLA, and was supported by grant no. R01 DC00222 from the NIDCD, National Institutes of Health, to Peter M. Narins. The author wishes to thank Professor Narins for his help and support during the preparation of this chapter.

References

Aertsen AMHJ, Vlaming MSMG, Eggermont JJ, Johannesma PIM (1986) Directional hearing in the grassfrog (*Rana temporaria* L.). II. Acoustics and modelling of the auditory periphery. Hear Res 21:17–40.

Anson M, Pinder AC, Keating MJ, Chung SH (1985) Acoustic vibration of the amphibian eardrum studied by white noise analysis and holographic interferometry. J Acoust Soc Am 78:916–923.

Baker MC (1969) The effect of severing the opercularis muscle on body orientation of the leopard frog, *Rana pipiens*. Copeia 1969:613–616.

Becker RP, Lombard RE (1977) Structural correlates of function in the "opercularis" muscle of amphibians. Cell Tissue Res 175:499–522.

Boatright-Horowitz SS, Simmons AM (1995) Postmetamorphic changes in auditory sensitivity of the bullfrog midbrain. J Comp Physiol [A] 177:577–590.

Boatright-Horowitz SS, Simmons AM (1997) Transient "deafness" accompanies auditory development during metamorphosis from tadpole to frog. Proc Natl Acad Sci USA 94:14877–14882.

Bolt JR, Lombard RE (1985) Evolution of the amphibian tympanic ear and the origin of frogs. Biol J Linn Soc Lond 24:83–99.

Brand DJ (1956) On the cranial morphology of *Scolecomorphus uluguruensis* (Barbour & Loveridge). Annals Univ Stellenbosch A 32:1–25.

Cannatella DC, Trueb L (1988) Evolution of pipoid frogs: Morphology and phylogenetic relationships of *Pseudhymenochirus*. J Herpetol 22:439–456.

Capranica RR (1976) Morphology and physiology of the auditory system. In: Llinás R, Precht W (eds) Frog Neurobiology: A Handbook. Berlin: Springer Verlag, pp. 551–557.

Carroll RL (2001) The origin and early radiation of terrestrial vertebrates. J Paleontol 75: 1202–1213.

Christensen-Dalsgaard J, Elepfandt A (1995) Biophysics of underwater hearing in the clawed frog, *Xenopus laevis*. J Comp Physiol [A] 176:317–324.

Christensen-Dalsgaard J, Breithaupt T, Elepfandt A (1990) Underwater hearing in the clawed frog, *Xenopus laevis*. Tympanic motion studied with laser vibrometry. Naturwissenschaften 77:135–137.

Chung S-H, Pettigrew A, Anson M (1978) Dynamics of the amphibian middle ear. Nature 272:142–147.

Dallos P (1973) The Auditory Periphery: Biophysics and Physiology. New York: Academic.

de Jager EFJ (1939a) The gymnophione quadrate and its processes, with special reference to the processus ascendens in a juvenile *Ichthyophis glutinosus*. Anat Anz 88:223–232.

de Jager EFJ (1939b) Contributions to the cranial anatomy of the Gymnophiona. Further points regarding the cranial anatomy of the genus *Dermophis*. Anat Anz 88:193–222.

de Jager EFJ (1947) Some points in the development of the stapes of *Ichthyophis glutinosus*. Anat Anz 96:203–210.

de Jongh HJ, Gans C (1969) On the mechanism of respiration in the bullfrog, *Rana catesbeiana*: A reassessment. J Morphol 127:259–290.

de Villiers CGS (1931) The cranial characters of the brevicipitid genus *Cacosternum*. Q J Microsc Sci 74:275–302.

de Villiers CGS (1932) Über das Gehörskelett der aglossen Anuren. Anat Anz 74:33–55.

de Villiers CGS (1934) Studies of the cranial anatomy of *Ascaphus truei* Stejneger, the American "liopelmid". Bull Am Mus Comp Zool Harvard 77:3–38.

de Villiers CGS (1938) A comparison of some cranial features of the East African gymnophiones *Boulengerula boulengeri*, Tornier and *Scolecomorphus ulugurensis* Boulenger. Anat Anz 86:1–26.

Duellman WE, Trueb L (1986) Biology of Amphibians. Baltimore, MD: The Johns Hopkins University Press.

Dunn ER (1941) The "opercularis" muscle of salamanders. J Morphol 69:207–216.

Ecker A (1889) The Anatomy of the Frog (translated by George Haslam). Amsterdam: A. Asher.

Eggermont JJ (1988) Mechanisms of sound localization in anurans. In: Fritzsch B, Ryan MJ, Wilczynski W, Hetherington TE, Walkowiak W (eds) The Evolution of the Amphibian Auditory System. New York: Wiley, pp. 307–336.

Ehret G, Keilwerth E, Kamada T (1994) The lung-eardrum pathway in three treefrog and four dendrobatid frog species: Some properties of sound transmission. J Exp Biol 195: 329–343.

Ehret G, Tautz J, Schmitz B, Narins PM (1990) Hearing through the lungs: lung-eardrum transmission of sound in the frog *Eleutherodactylus coqui*. Naturwissenschaften 77: 192–194.

Eiselt J (1941) Der Musculus opercularis und die mittlere Ohrsphäre der anuren Amphibien. Arch Naturgesch 10:179–230.

Elepfandt A (1996a) Sensory perception and the lateral line system in the clawed frog, *Xenopus*. In: Tinsley RC, Kobel HR (eds) The Biology of *Xenopus*. Oxford: Clarendon, pp. 97–120.

Elepfandt A (1996b) Underwater acoustics and hearing in the clawed frog, *Xenopus*. In: Tinsley RC, Kobel HR (eds) The Biology of *Xenopus*. Oxford: Clarendon, pp. 177–193.

Fay RR, Popper AN (1985) The octavolateralis system. In: Hildebrand M, Bramble DM, Liem KF, Wake DB (eds) Functional Vertebrate Morphology. London: Belknap, pp. 291–316.

Feng AS, Narins PM, Xu C-H (2002) Vocal acrobatics in a Chinese frog, *Amolops tormotus*. Naturwissenschaften 89:352–356.

Fox JH (1995) Morphological correlates of auditory sensitivity in anuran amphibians. Brain Behav Evol 45:327–338.

Gans C, de Jongh HJ, Farber J (1969) Bullfrog (*Rana catesbeiana*) ventilation: how does the frog breathe? Science 163:1223–1225.

Goodrich ES (1930) Studies on the Structure and Development of Vertebrates. New York: Dover.

Hetherington TE (1985) Role of the opercularis muscle in seismic sensitivity in the bullfrog *Rana catesbeiana*. J Exp Zool 235:27–43.

Hetherington TE (1987a) Physiological features of the opercularis muscle and their effects on vibration sensitivity in the bullfrog *Rana catesbeiana*. J Exp Biol 131:189–204.

Hetherington TE (1987b) Timing of development of the middle ear of Anura (Amphibia). Zoomorphology 106:289–300.

Hetherington TE (1988a) Biomechanics of vibration reception in the bullfrog, *Rana catesbeiana*. J Comp Physiol [A] 163:43–52.

Hetherington TE (1988b) Metamorphic changes in the middle ear. In: Fritzsch B, Ryan MJ, Wilczynski W, Hetherington TE, Walkowiak W (eds) The Evolution of the Amphibian Auditory System. New York: Wiley, pp. 339–357.

Hetherington TE (1989) Effect of the amphibian opercularis muscle on auditory responses. Prog Zool 35:356–359.

Hetherington TE (1992a) The effects of body size on functional properties of middle ear systems of anuran amphibians. Brain Behav Evol 39:133–142.

Hetherington TE (1992b) The effects of body size on the evolution of the amphibian middle ear. In: Webster DB, Fay RR, Popper AN (eds) The Evolutionary Biology of Hearing. New York: Springer-Verlag, pp. 421–437.

Hetherington TE (1994a) Sexual differences in the tympanic frequency responses of the American bullfrog (*Rana catesbeiana*). J Acoust Soc Am 96:1186–1188.

Hetherington TE (1994b) The middle ear muscle of frogs does not modulate tympanic responses to sound. J Acoust Soc Am 95:2122–2125.

Hetherington TE (2001) Laser vibrometric studies of sound-induced motion of the body walls and lungs of salamaders and lizards: Implications for lung-based hearing. J Comp Physiol [A] 187:499–507.

Hetherington TE, Lindquist ED (1999) Lung-based hearing in an “earless” anuran amphibian. J Comp Physiol [A] 184:395–401.

Hetherington TE, Lombard RE (1982) Biophysics of underwater hearing in anuran amphibians. J Exp Biol 98:49–66.
Hetherington TE, Lombard RE (1983a) Electromyography of the opercularis muscle of *Rana catesbeiana*: An amphibian tonic muscle. J Morphol 175:17–26.
Hetherington TE, Lombard RE (1983b) Mechanisms of underwater hearing in larval and adult tiger salamanders *Ambystoma tigrinum*. Comp Biochem Physiol 74A:555–559.
Hetherington TE, Tugaoen JR (1990) Histochemical studies on the amphibian opercularis muscle (Amphibia). Zoomorphology 109:273–279.
Hetherington TE, Jaslow AP, Lombard RE (1986) Comparative morphology of the amphibian opercularis system: I. General design features and functional interpretation. J Morphol 190:43–61.
Horowitz SS, Chapman JA, Kaya U, Megela Simmons A (2001) Metamorphic development of the bronchial columella of the larval bullfrog (*Rana catesbeiana*). Hear Res 154:12–25.
Inger RF (1966) The systematics and zoogeography of the Amphibia of Borneo. Fieldiana: Zoology 52:1–402.
Jaslow AP, Lombard RE (1996) Hearing in the neotropical frog, *Atelopus chiriquiensis*. Copeia 1996:428–432.
Jaslow AP, Hetherington TE, Lombard RE (1988) Structure and function of the amphibian middle ear. In: Fritzsch B, Ryan MJ, Wilczynski W, Hetherington TE, Walkowiak W (eds) The Evolution of the Amphibian Auditory System. New York: Wiley, pp. 69–91.
Jørgensen MB (1991) Comparative studies of the biophysics of directional hearing in anurans. J Comp Physiol [A] 169:591–598.
Jørgensen MB, Kanneworff M (1998) Middle ear transmission in the grass frog, *Rana temporaria*. J Comp Physiol [A] 182:59–64.
Jørgensen MB, Schmitz B, Christensen-Dalsgaard J (1991) Biophysics of directional hearing in the frog *Eleutherodactylus coqui*. J Comp Physiol [A] 168:223–232.
Kingsbury BF, Reed HD (1909) The columella auris in Amphibia. J Morphol 20:549–627.
Koyama H, Lewis ER, Leverenz EL, Baird RA (1982) Acute seismic sensitivity in the bullfrog ear. Brain Res 250:168–172.
Lewis ER (1984) Inertial motion sensors. In: Bolis L, Keynes RD, Maddrell SHP (eds) Comparative Physiology of Sensory Systems. Cambridge, UK: Cambridge University Press, pp. 587–610.
Lewis ER, Lombard RE (1988) The amphibian inner ear. In: Fritzsch B, Ryan MJ, Wilczynski W, Hetherington TE, Walkowiak W (eds) The Evolution of the Amphibian Auditory System. New York: Wiley, pp. 93–123.
Lewis ER, Narins PM (1985) Do frogs communicate with seismic signals? Science 227:187–189.
Lindquist ED, Hetherington TE (1996) Field studies on visual and acoustic signalling in the "earless" Panamanian golden frog, *Atelopus zeteki*. J Herpetol 30:347–354.
Lindquist ED, Hetherington TE, Volman SF (1998) Biomechanical and neurophysiological studies on audition in eared and earless harlequin frogs (*Atelopus*). J Comp Physiol [A] 183:265–271.
Loftus-Hills JJ (1973) Neural mechanisms underlying acoustic behaviour of the frog *Pseudophryne semimarmorata* (Anura: Leptodactylidae). Anim Behav 21:781–787.
Lombard RE (1977) Comparative morphology of the inner ear in salamanders (*Caudata: Amphibia*). Basel: S. Karger.
Lombard RE, Bolt JR (1979) Evolution of the tetrapod ear: An analysis and reinterpretation. Biol J Linn Soc Lond 11:19–76.

Lombard RE, Straughan IR (1974) Functional aspects of anuran middle ear structures. J Exp Biol 61:71–93.
Lombard RE, Fay RR, Werner YL (1981) Underwater hearing in the frog, *Rana catesbeiana*. J Exp Biol 91:57–71.
Mason MJ, Narins PM (2002a) Vibrometric studies of the middle ear of the bullfrog *Rana catesbeiana* I. The extrastapes. J Exp Biol 205:3153–3165.
Mason MJ, Narins PM (2002b) Vibrometric studies of the middle ear of the bullfrog *Rana catesbeiana* II. The operculum. J Exp Biol 205:3167–3176.
Mason MJ, Lin CC, Narins PM (2003) Sex differences in the middle ear of the bullfrog (*Rana catesbeiana*). Brain Behav Evol 61:91–101.
McClelland BE, Wilczynski W, Rand AS (1997) Sexual dimorphism and species differences in the neurophysiology and morphology of the acoustic communication system of two neotropical hylids. J Comp Physiol [A] 180:451–462.
McDiarmid RW (1971) Comparative morphology and evolution of frogs of the Neotropical genera *Atelopus*, *Dendrophryniscus*, *Melanophryniscus*, and *Oreophrynella*. Bull Los Angeles County Mus Nat Hist Science 12:1–66.
Moffat AJM, Capranica RR (1978) Middle ear sensitivity in anurans and reptiles measured by light scattering spectroscopy. J Comp Physiol [A] 127:97–107.
Monath T (1965) The opercular apparatus of salamanders. J Morphol 116:149–170.
Narins PM (1992) Reduction of tympanic membrane displacement during vocalization of the arboreal frog, *Eleutherodactylus coqui*. J Acoust Soc Am 91:3551–3557.
Narins PM, Lewis ER (1984) The vertebrate ear as an exquisite seismic sensor. J Acoust Soc Am 76:1384–1387.
Narins PM, Ehret G, Tautz J (1988) Accessory pathway for sound transfer in a neotropical frog. Proc Natl Acad Sci USA 85:1508–1512.
Narins PM, Lewis ER, Purgue AP, Bishop PJ, Minter LR, Lawson DP (2001) Functional consequences of a novel middle ear adaptation in the central African frog *Petropedetes parkeri* (Ranidae). J Exp Biol 204:1223–1232.
Noble GK (1931) The Biology of the Amphibia. New York: McGraw-Hill.
Overstreet EH, Ruggero MA (2001) Development of wide-band middle ear transmission in the Mongolian gerbil. J Acoust Soc Am 111:261–270.
Pinder AC, Palmer AR (1983) Mechanical properties of the frog ear: Vibration measurements under free- and closed-field acoustic conditions. Proc R Soc Lond B Biol Sci 219:371–396.
Purgue AP (1997) Tympanic sound radiation in the bullfrog *Rana catesbeiana*. J Comp Physiol [A] 181:438–445.
Purgue AP, Narins PM (2000a) A model for energy flow in the inner ear of the bullfrog (*Rana catesbeiana*). J Comp Physiol [A] 186:489–495.
Purgue AP, Narins PM (2000b) Mechanics of the inner ear of the bullfrog (*Rana catesbeiana*): The contact membranes and the periotic canal. J Comp Physiol [A] 186:481–488.
Reed HD (1920) The morphology of the sound-transmitting apparatus in caudate Amphibia and its phylogenetic significance. J Morphol 33:325–387.
Relkin EM (1988) Introduction to the analysis of middle-ear function. In: Jahn AF, Santos-Sacchi J (eds) Physiology of the Ear. New York: Raven, pp. 103–123.
Rogers PH, Cox M (1988) Underwater sound as a biological stimulus. In: Atema J, Fay RR, Popper AN, Tavolga WN (eds) Sensory Biology of Aquatic Animals. New York: Springer-Verlag.
Ross RJ, Smith JJB (1979) Detection of substrate vibrations by salamanders: Eighth cranial nerve activity. Can J Zool 57:368–374.

Ross RJ, Smith JJB (1980) Detection of substrate vibrations by salamanders: Frequency sensitivity of the ear. Comp Biochem Physiol 65A:167–172.

Ross RJ, Smith JJB (1982) Responses of the salamander inner ear to vibrations of the middle ear. Can J Zool 60:220–226.

Russell IJ (1976) Amphibian lateral line receptors. In: Llinás R, Precht W (eds) Frog Neurobiology: A Handbook. Berlin: Springer Verlag, pp. 513–550.

Saunders JC, Johnstone BM (1972) A comparative analysis of middle-ear function in non-mammalian vertebrates. Acta Otolaryngol (Stockh) 73:353–361.

Schmalhausen II (1957) The sound-transmitting mechanism of amphibians [in Russian]. Zool Zh 36:1044–1063.

Schmalhausen II (1968) The Origin of Terrestrial Vertebrates. New York: Academic.

Seaman RL (2002) Non-osseous sound transmission to the inner ear. Hear Res 166: 214–215.

Sedra SN, Michael MI (1957) The development of the skull, visceral arches, larynx and visceral muscles of the South African clawed toad, *Xenopus laevis* (Daudin) during the process of metamorphosis (from stage 55 to stage 66). Verh K Acad Wetenschappen, Afdeeling Natuurkunde, 2e Reeks 51:1–80.

Sedra SN, Michael MI (1959) The ontogenesis of the sound conducting apparatus of the Egyptian toad, *Bufo regularis* Reuss, with a review of this apparatus in Salientia. J Morphol 104:359–375.

Shofner WP, Feng AS (1981) Post-metamorphic development of the frequency selectivities and sensitivities of the peripheral auditory system of the bullfrog, *Rana catesbeiana*. J Exp Biol 93:181–196.

Smirnov SV (1991) The anuran middle ear: Developmental heterochronies and adult morphology diversification. Belg J Zool 121:99–110.

Smirnov SV, Vorobyeva EI (1988) Morphological grounds for diversification and evolutionary change in the amphibian sound-conducting apparatus. Anat Anz 166: 317–322.

Smith JJB (1968) Hearing in terrestrial urodeles: A vibration-sensitive mechanism in the ear. J Exp Biol 48:191–205.

Spannhof L (1954) Die Entwicklung des Mittelohres und des schalleitenden Apparates bei *Xenopus laevis* Daudin. Z Wiss Zool 158:1–30.

Stebbins RC, Cohen NW (1995) A Natural History of Amphibians. Princeton, NJ: Princeton University Press.

Stephenson NG (1951) On the development of the chondrocranium and visceral arches of *Leiopelma archeyi*. Trans Zool Soc Lond 27:203–253.

Taylor EH (1969) Skulls of Gymnophiona and their significance in the taxonomy of the group. Univ Kansas Sci Bull 48:585–687.

van Bergeijk WA (1959) Hydrostatic balancing mechanism of *Xenopus* larvae. J Acoust Soc Am 31:1340–1347.

van Bergeijk WA (1967) The evolution of vertebrate hearing. In: Neff WD (ed.) Contributions to Sensory Physiology, vol 2. New York: Academic, pp. 1–49.

van Dijk P, Mason MJ, Narins PM (2002) Distortion product otoacoustic emissions in frogs: Correlation with middle and inner ear properties. Hear Res 173:100–108.

van Seters WH (1922) Le développement du chondrocrâne d'*Alytes obstetricans* avant la métamorphose. Arch Biol (Liege) 32:373–491.

Vlaming MSMG, Aertsen AMHJ, Epping WJM (1984) Directional hearing in the grass frog (*Rana temporaria* L.): I. Mechanical vibrations of tympanic membrane. Hear Res 14:191–201.

von Békésy G (1960) Experiments in Hearing. New York: McGraw Hill.
Wagner DS (1934) The structure of the inner ear in relation to the reduction of the middle ear in the Liopelmidae (Noble). Anat Anz 79:20–36.
Walkowiak W (1980) The coding of auditory signals in the torus semicircularis of the fire-bellied toad and the grass frog: Responses to simple stimuli and to conspecific calls. J Comp Physiol 138:131–148.
Werner YL (2003) Mechanical leverage in the middle ear of the American bullfrog, *Rana catesbeiana*. Hear Res 175:54–65.
Wever EG (1975) The caecilian ear. J Exp Zool 191:63–72.
Wever EG (1978) Sound transmission in the salamander ear. Proc Natl Acad Sci USA 75:529–530.
Wever EG (1979) Middle ear muscles of the frog. Proc Natl Acad Sci USA 76:3031–3033.
Wever EG (1985) The Amphibian Ear. Princeton, NJ: Princeton University Press.
Wever EG, Gans C (1976) The caecilian ear: Further observations. Proc Natl Acad Sci USA 73:3744–3746.
Wilczynski W, Resler C, Capranica RR (1987) Tympanic and extratympanic sound transmission in the leopard frog. J Comp Physiol [A] 161:659–669.
Witschi E (1949) The larval ear of the frog and its transformation during metamorphosis. Z Naturforsch 4b:230–242.
Witschi E (1950) The bronchial diverticula of *Xenopus laevis* Daudin. Anat Rec 108:590.
Witschi E (1951) The development of the bronchial columella in the aquatic ear of Ranidae. Anat Rec 109:359–360.
Witschi E (1955) The bronchial columella of the ear of larval Ranidae. J Morphol 96: 497–511.
Wright AH, Wright AA (1949) Handbook of Frogs and Toads of the United States and Canada, 3rd ed. Ithaca, NY: Comstock.
Yager DD (1996) Sound production and acoustic communication in *Xenopus borealis*. In: Tinsley RC, Kobel HR (eds) The Biology of Xenopus. Oxford: Clarendon, pp. 121–141.
Yang D-T (1991) Phylogenetic systematics of the *Amolops* group of ranid frogs of Southeastern Asia and the Greater Sunda Islands. Fieldiana: Zoology 63:1–42.
Yu X, Lewis ER, Feld D (1991) Seismic and auditory tuning curves from bullfrog saccular and amphibian papillar axons. J Comp Physiol [A] 169:241–248.

7 Anatomy, Physiology, and Function of Auditory End-Organs in the Frog Inner Ear

Dwayne D. Simmons, Sebastiaan W.F. Meenderink, and Pantelis N. Vassilakis

1. Overview

The vertebrate ear is a highly sensitive frequency analyzer that receives sound through a specialized accessory apparatus (the external and middle ears) prior to its transmission to discrete end-organs containing sensory hair cells (the inner ear). Although there are significant differences in the structures used to receive and analyze sound, amphibian and mammalian ears function very similarly to each other. With few exceptions, the amphibian ear consists of a middle ear and an inner ear, but no external ear. As schematized in Figure 7.1, the amphibian middle ear has an exposed eardrum (tympanic membrane) overlying a funnel-shaped tympanic cavity that connects to the inner ear near the base of the skull (see Mason, Chapter 6, for a review of the amphibian middle ear). The amphibian inner ear or otic labyrinth is unique among vertebrate animals in that it has two sensory organs specialized for the reception of airborne sound, the amphibian papilla (AP) and the basilar papilla (BP). These sensory papillae reside within the posterior portion of the otic labyrinth and are contained in ventrally located recesses of the large, fluid-filled saccular chamber shared with two vibration-sensitive macular organs, the sacculus and lagena (Fig. 7.1). Both the AP and BP chambers have a thin contact membrane that separates periotic perilymph from the endolymph fluid of the saccular chamber.

Sound energy captured by the eardrum as well as other areas along the body of a frog is converted into fluid displacements and travels along pathways of the otic labyrinth that lead into the endolymphatic spaces of the inner ear (Hetherington et al. 1986; Lewis and Lombard 1988; Purgue and Narins 2000a). The sound path eventually leads into the AP and BP recesses before exiting into the caudal portion of the periotic canal and the round window (Purgue and Narins 2000a). Similar to the mammalian ear, the amphibian ear demonstrates exquisite intensity sensitivity and sharp frequency selectivity that are likely to arise from nonlinear, active amplification processes.

How theories of mammalian auditory function apply to amphibian hearing is not known. Mechanisms of tuning and sensitivity have been extensively studied in the mammalian cochlea. It is generally agreed that the initial stage of inner ear

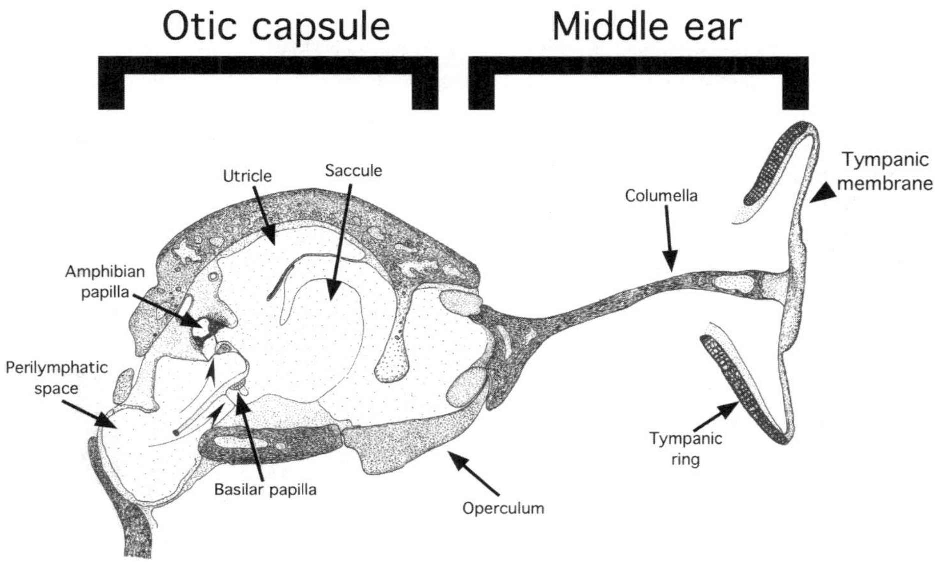

FIGURE 7.1. A schematic representation of the frog's ear. Modified from Wever, Ernest Glen; *the Amphibian Ear*. © 1985 Princeton University Press. Reprinted by permission of Princeton University Press.

frequency selectivity is achieved through the specialized mechanical properties of the basilar membrane, giving rise to a traveling wave (Von Békésy 1960; West 1985). Such traveling waves may be enhanced by electrically driven somatic movements of specialized outer hair cells that provide the work required for the active process (Dallos 1992; Nobili et al. 1998; Ashmore et al. 2000). Although outer hair cell-like motility has not been demonstrated in reptiles and birds, their papillae contain structures analogous to those in the mammalian cochlea such as a basilar membrane. However, one of the most striking anomalies concerning the frog AP is that it lacks a basilar membrane, and yet it clearly demonstrates sharp frequency resolution and sensitivity as well as otoacoustic emissions (OAEs) that are comparable to those found in mammals. In amphibians, reptiles, and birds, the best candidate for an active process may be the active motility of their mechanically sensitive hair bundles (Hudspeth et al. 2000; Fettiplace et al. 2001; Bozovic and Hudspeth 2003).

Although amphibian auditory organs may (Wever 1973) or may not (Will and Fritzsch 1988) have arisen independently and separately from those of other vertebrates, studies of these organs provide an opportunity to explore the details of what may be convergent design and function. Do analogous physiological responses arise from similar or different anatomical features? Does the sensory end-organs' fine structure and innervation follow common principles across species? Although there have been extensive investigations into the physiology of these organs, much less is known about the detailed structure of anatomical correlates of this physiology such as the relationship between conduction velocities and response latencies. Much of the physiology of hair cell and eighth-nerve responses of the AP and BP has been reviewed previously (Lewis and Narins 1999; Smotherman and Narins 2000). Sections 2 and 3 focus on the basic structural correlates of physiological responses of ranid frogs within the auditory papillae and nerve, respectively, and Section 4 focuses on what is known about amphibian OAEs, and discusses the relationship of their properties to the underlying hearing mechanisms as well as the differences between sensory organs.

2. AP and BP Sensory Organs

2.1 Structural Organization of the AP and BP

Acoustic sense organs attach to the dorsal wall (ceiling) of the AP and BP chambers. In ranid frogs, the AP chamber is oblong and roughly cylindrical. At its lateral end, it opens broadly to the saccule and, towards its medial end, it is closed by a thin contact membrane. As illustrated in Figure 7.2, the hatchet-shaped AP sensory organ can be divided into a rostral patch of hair cells (representing the hatchet blade) up to 30 hair cells wide, an omega-shaped, central curved section (representing the handle neck) with 6 to 10 hair cell rows, and a caudal extension (representing the handle base) that flares out at its caudalmost pole to accommodate 20 rows of hair cells (Wever 1973; Lewis et al. 1985; Simmons et al.

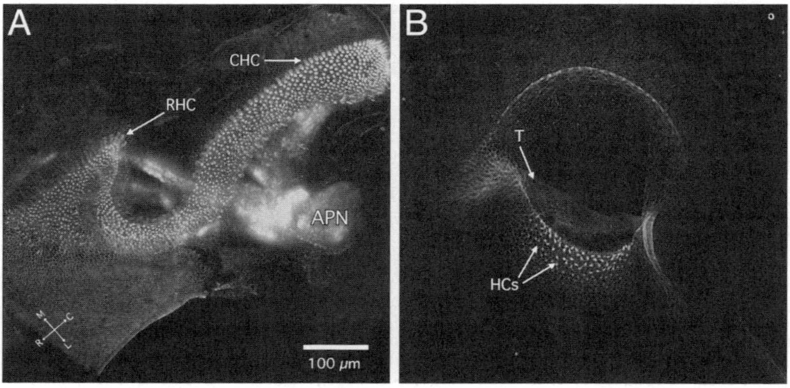

FIGURE 7.2. Confocal images of the amphibian and basilar papillae in the bullfrog. (A) Hair cells in the amphibian papilla (AP) are labeled for myosin VI. A transmission image of the AP nerve is superimposed onto the confocal image of the AP. Rostral (R), lateral (L), and caudal (C) orientations are as indicated. Courtesy of Simmons, Burton, and Baird. (B) Hair cells in the basilar papilla (BP) are labeled with phalloidin. The BP tectorial covering can also be seen. The scale bar in (A) is the same as in (B).

1994b, 2004). Overlying the sensory surface of the AP is a honeycomb meshwork of gelatinous tissue called the tectorium. A thin net of tectorial tissue (tectorial curtain) is suspended across the AP chamber between the AP nerve and the middle region of the AP tectorium. In contrast, the organization of the BP is quite simple. The cylindrical and much smaller BP chamber also opens to the saccule at its lateral end and is likewise closed by a thin contact membrane towards its medial end. Hair cells in the BP are organized into five or six rows containing up to 20 hair cells each (Wever 1985). The BP has a thin, semi-lunar tectorium stretching across a short tubular passage with an underlying sensory epithelium (Fig. 7.2B). In contrast to the mammalian cochlea, hair cells in both the AP and the BP are rigidly fixed to the cartilaginous wall of the respective papillar recesses, rather than over a flexible membrane. In mammals, traveling waves on the flexible basilar membrane are thought to contribute importantly to the functionality of the cochlea.

There are many features of the frog auditory system that may influence its sensitivity and tuning. The tectorial structures of each papilla, the size of the hair cell, its position within the papilla, as well as its stereovillar morphology all contribute to the thresholds and tuning characteristics of afferent nerve fibers. In addition, the synaptic area and number of synapses per hair cell, the region and number of hair cells innervated, and the diameter and length of each afferent fiber also contribute to the fiber's threshold, latency, and/or tuning responses. However, how the tectorium, hair cells, and synaptic architecture impose limits on eighth-nerve responses is not at all well understood.

For the gross tuning of frog auditory organs, Purgue and Narins (2000a,b) suggest that the frequency difference in neural responses between the AP and BP may be largely explained on the basis of the mechanical tuning of their respective contact membranes. However, only a few studies have attempted to characterize the micromechanics of the tectorium (Lewis and Leverenz 1983; Shofner and Feng 1983; Lewis et al. 1992). With the exception of Wever's hypothesis that the tectorial curtain intercepts the motion of the fluid in the AP chamber and transfers it to the tectorium (Wever 1985), very little is known or has been hypothesized about the dynamics between the point of auditory interception and the initiation of action potentials in the auditory nerve fiber (Smotherman and Narins 2000). In general, the degree to which the fine structure of the auditory end-organ and its nerve fibers are related to physiological parameters such as acoustic tuning, threshold responses, response latencies, and spontaneous activity is not as well understood in frogs as it is in the mammalian cochlea.

2.2 Hair Cell Stereovilli and the Tectorium

In frogs, auditory hair cells exhibit a variety of distinctive hair bundle morphologies that contain both stereovilli and a true cilium (kinocilium). The hair bundle consists of stereovilli having a graded series of heights with the tallest ones adjacent to the kinocilium. Deflection of the hair bundle towards the kinocilium opens tip-linked transduction channels, resulting in a depolarization of the hair cell. Deflection of the hair bundle away from the kinocilium hyperpolarizes the hair cell. Lewis and colleagues (Lewis and Li 1975; Lewis 1976; Baird and Lewis 1986; Lewis and Lombard 1988) have studied amphibian hair bundles extensively. In general, hair cells on one or more margins of these organs display bundles with an immature or juvenile pattern, suggesting that they make up a marginal growth zone that develops hair cells (Corwin and Warchol 1991; Baird et al. 2000). In contrast to marginal growth zones, hair cells within central regions of the papillae have mature-looking hair bundles. Lewis and colleagues further distinguished at least two categories of mature bundles: those in which the kinocilium extends just beyond the tips of the tallest stereovilli (class 1) and those in which the kinocilium extends well beyond the tallest stereovilli (class 2). Class 1 bundles can be further classified by the presence (1a) or absence (1b) of a conspicuous enlargement or bulb at the tip of the kinocilium and by the bundle height (tall, type E or short, type D).

Hair bundle morphology varies along the transverse axis of the AP, but has not been observed to vary significantly along the (rostrocaudal) axis of tonotopy (Lewis 1976). However, the hair cell polarization (i.e., the orientation of the tallest stereovilli within the bundle) varies along the tonotopic axis, rather than along the transverse direction. Hair bundles rostral to the tectorial curtain (or AP nerve branchlet) are oriented along the rostrocaudal axis, whereas hair bundles caudal to the tectorial curtain are oriented along the transverse (mediolateral) axis of the caudal extension (Lewis and Li 1975; Lewis 1976). Because maximum hair cell response occurs when hair bundles are displaced along their axis of orientation (Flock 1965; Hudspeth and Corey 1977), the optimal stimulus for hair cells requires the tectorium to vibrate along the mediolateral axis in caudal regions and along the rostrocaudal axis in more rostral regions (Lewis and Lombard 1988; Smotherman and Narins 2000). These polarization patterns suggest that the acoustically induced motion of the AP tectorium is complex.

In analogy with the mammalian cochlea, it has been proposed that traveling waves are supported in the AP tectorium (Fig. 7.3). This structure demonstrates a conspicuous tapering in its width and thickness (Shofner and Feng 1984; Lewis and Leverenz 1983). The difference in apparent mass along the tectorium suggests that the thickest portion overlying the rostral AP patch would exhibit larger response latency than the thinnest portion over the caudal AP extension, which would result in a traveling wave. Neurophysiological evidence has suggested that a low-velocity traveling wave exists in the AP (Hillery and Narins 1984, 1987; Lewis 1984). However, a number of studies do not support the idea of a traveling wave along the AP tectorium. For example, Lewis (1988) suggests that the tectorium lacks the appropriate mass and stiffness gradients necessary to support a mammalianlike traveling wave. Anuran DPOAE measurements also argue against the presence of a traveling wave in the AP (Meenderink et al. 2005a) but do support the presence of tectorial disturbances that may lead to OAE generation (Vassilakis et al. 2004).

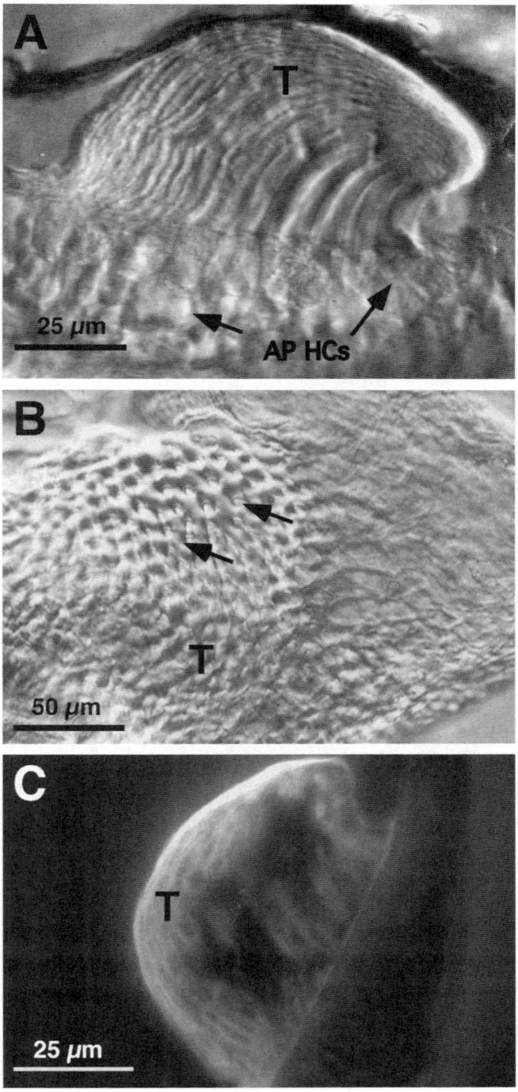

FIGURE 7.3. (A) Light-micrograph of a transverse section through the tectorial membrane, from a celloidin-embedded section using DIC (Nomarski) optics. (B) Light-micrograph of a horizontal section through the tectorial membrane as it is situated just above the hair bundles. (C) Same section as in A, but with celloidin removed and stained with wheat-germ agglutinin (WGA)-rhodamine and imaged with an epi-fluorescent microscope.

2.3 Hair Cell Morphometry

In Figure 7.4, characteristic frequencies (CFs) derived from the bullfrog by Lewis et al. (1982b) and hair cell lengths from studies of the leopard frog by Simmons et al. (1994b) are plotted as a function of normalized distance from the caudal AP pole. In rostral and middle AP regions, hair cell lengths are inversely correlated with a fiber's best frequencies. When moving rostrally, the anterior portion of the rostral AP patch begins around the 80% distance location where hair cell

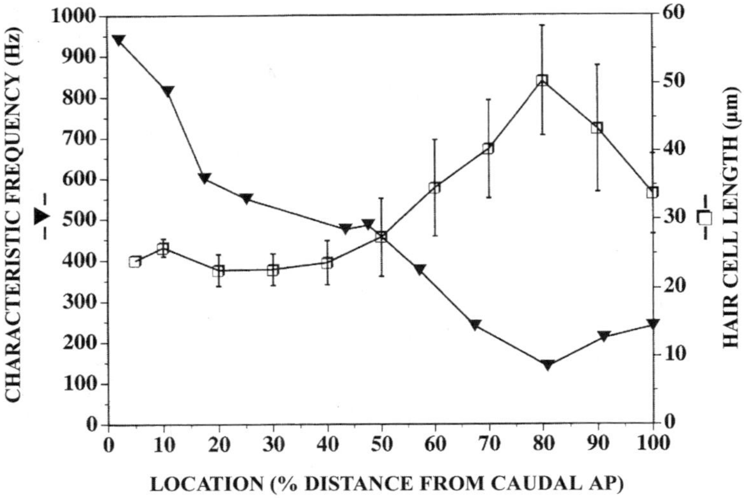

FIGURE 7.4. A plot superimposing the gradient of mean hair cell length in the AP of the leopard frog on the characteristic frequency (CF) tuning gradient in the AP of the bullfrog as determined by Lewis et al. 1982b. From Simmons et al. (1994b).

lengths begin to decrease and where frequencies begin to increase. The relationship between frequencies and hair cell length is less well correlated in the caudal extension. In other vertebrates, inner ear organs also demonstrate an association between hair cell dimensions and tuning. For example, studies of goldfish saccular hair cells reveal morphological gradients similar to those found in the frog AP and further provide evidence for the possibility of hair cell length being related to intrinsic tuning (Sugihara and Furukawa 1989). The pattern of changes in hair cell morphology along the AP rostrocaudal axis correlates well with its observed tonotopy and may have implications for the intrinsic tuning of the AP (Smotherman and Narins 2000). In a study of isolated AP hair cells, Smotherman and Narins (1999a) demonstrated that whole-cell capacitances vary predictably with hair cell body length and thus provided further evidence that the resonant frequency of a hair cell is generally inversely proportional to cell size.

2.4 Tuning Properties and Ion Channels

In addition to any mechanical tuning, many nonmammalian hair cells exhibit electrical tuning in which the hair cell's basolateral membrane shows bandpass filtering capabilities. In the frog, hair bundle morphology correlates with a unique set of physiological properties, including transduction sensitivity and rate of adaptation. However, neither bundle height nor bundle type co-varies with CF in any frog auditory organ. Several ionic currents that participate in the electrical

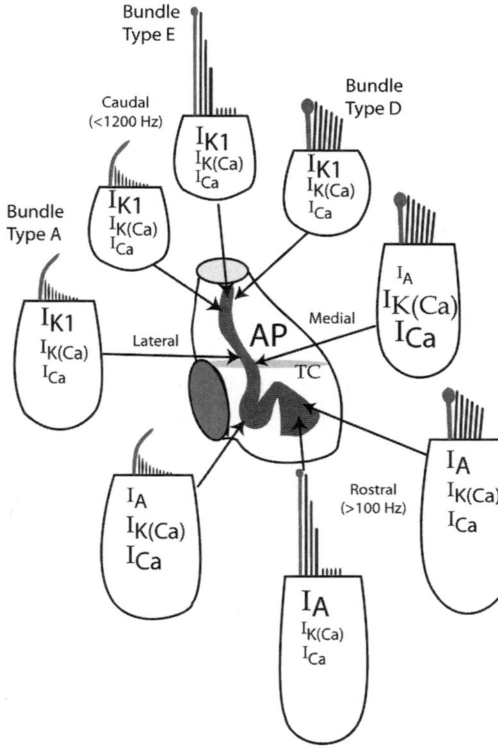

FIGURE 7.5. A schematic diagram of the amphibian papilla chamber in the leopard frog. Within the AP, hair cells show systematic variations in their size, shape, bundle architecture, and membrane electrical properties. The variations in cell size are consistent with cell capacitance changes and relative current amplitudes (represented here by font size). The three principal hair bundle types (A, E, D) are taken from Lewis and Li (1975). The tectorial curtain (TC) is also represented. Reprinted from J Exp Biol, 203, MS Smotherman and PM Narins Hair cells, hearing and hopping: a field guide to hair cell physiology in the frog 2237–2246, 2000, with permission from Company of Biologists Ltd.

tuning of AP hair cells have been identified from whole-cell recordings (Fig. 7.5). These include, among others, inward calcium currents and outwardly rectifying potassium currents (Smotherman and Narins 2000). BK channels conduct the major potassium current in hair cells in the bullfrog (Smotherman and Narins 1999b). As a result of these calcium and potassium currents, electrical resonances can be observed in isolated AP hair cells (Pitchford and Ashmore 1987; Smotherman and Narins 1999b) and occur at frequencies very close to the best frequencies of corresponding afferent fibers innervating the AP. Smotherman and Narins (1999a) found that there are two electrically distinct populations of hair cells (rostral and caudal). Within each population, the electrical properties and ionic currents vary along a rostrocaudal gradient. However, electrical tuning mechanisms are only well supported in AP hair cells from the rostral patch. Hair cells from the caudal extension do not appear to be electrically tuned (Smotherman and Narins 1999a). Another related feature of rostral AP hair cells is that they do not possess an inward rectifying K+ current (I_{K1} in Fig. 7.5). The unique absence of these channels in the rostral AP patch is consistent with the role of this ionic current in elevating auditory nerve thresholds by a reduction in neurotransmitter release (Smotherman and Narins 2000).

2.5 Synaptic Ultrastructure

The ultrastructure of synapses on hair cells in the AP is, according to several studies, consistent with other rostrocaudal gradients. Early ultrastructural studies predicted there might be a rostrocaudal gradient in synaptic architecture but provided no detailed evidence for such (Flock and Flock 1966; Frishkopf and Flock 1974). Studies of the ultrastructural characteristics of synapses and nerve fibers in both rostral and caudal regions of the AP have shown that hair cells in the rostral patch region differ significantly in their synaptic architecture from hair cells in the caudal extension (Simmons et al. 1995; Fig. 7.6A). Additionally, reconstructions of synaptic complexes suggest significant differences in the number of synaptic sites between rostral and caudal locations on tall and short hair cells, respectively. Both the number of synaptic sites and the amount of neurotransmitter released have been associated with the efficacy of synaptic transmission and subsequent nerve fiber responses. Thus, the ultrastructural data are consistent with the hypothesis that a hair cell's innervation pattern is related to specific features of auditory nerve fiber responses such as their spontaneous activity, thresholds, and their ability to follow rapid temporal variations in the stimulus.

Hair cell afferent synapses are known as ribbon synapses, and their structure has been well documented across most vertebrates (Smith and Sjostrand 1961; Gleisner et al. 1973; Liberman 1980; Sobkowicz et al. 1986; Chang et al. 1992; Simmons et al. 1994a, 1995; Lenzi et al. 1999). Similar to descriptions in other inner ear organs (Gleisner et al. 1973; Frishkopf and Flock 1974; Liberman et al. 1990; Chang et al. 1992), AP hair cell ribbon synapses contain a dense, spherical-to-elliptical presynaptic body, surrounded by a halo of tethered synaptic vesicles (Fig. F.6B). The ribbon or presynaptic body is typically seen in close apposition to the synaptic active zone that contains docked vesicles and Ca^{2+} entry sites for Ca^{2+}-induced exocytosis (Zenisek et al. 2003).

The most complete ultrastructural studies of the hair cell ribbon synapse in amphibians are from the bullfrog saccule (Lenzi et al. 1999) and the leopard frog AP (Simmons et al. 1995). In the leopard frog, each hair cell makes a synapse with several terminals, and afferent fibers appear to innervate several hair cells via *en passant* contacts. Simmons et al. (1995) reconstructed afferent ribbon synapses from serial thin sections and demonstrated a clear rostrocaudal gradient in the size of the presynaptic ribbon. In this study, significant differences were found in the innervation and synaptic ultrastructure between hair cells with tall (type E) bundles and those with short (type D) bundles in the rostral and caudal portions of the AP.

Studies at the neuromuscular junction suggest that the number and distribution of active zone particles at an active zone determines the probability of transmitter release at that site (Walrond and Reese 1985; Lnenicka et al. 1986). However, observations in the mammalian inner ear suggest just the opposite may be the case (Liberman et al. 1990). Presynaptic body size appears to be inversely related to fiber diameter and spontaneous rate. This discrepancy may simply reflect an

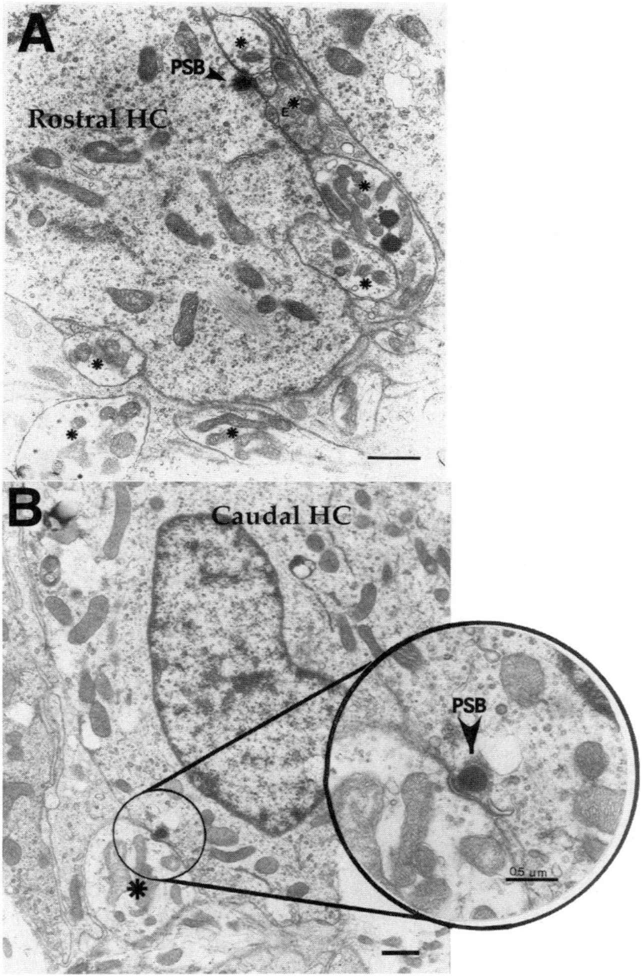

FIGURE 7.6. (A) Transmission electron micrograph (TEM) of a cross-section through the synaptic pole of a rostrally located tall hair cell (HC). Both afferent and efferent terminals are identified with asterisks. A hair cell presynaptic body (PSB) is opposite one of the afferent terminals. In the region shown, there are five afferent terminals and one efferent (E) terminal contacting the basal portion of the HC. (B) TEM of a cross-section through the synaptic pole of a caudally located short HC. In this micrograph, the caudal HC has just one afferent terminal (asterisk) in synaptic contact. This section is midway through the synaptic site. The inset shows the synaptic site with its PSB at higher magnification. Adapted from Simmons et al. (1995). Scale bars in (A) and (B) represent 10 µm.

inadequate understanding of the presynaptic body and its functional significance. Although the exact role of a morphological parameter such as synaptic ultrastructure in sculpting the physiological responses of the auditory nerve fibers is unknown, it is likely to correlate with the spatial and temporal analysis resolution applied to the incoming stimulus.

Few hair cell organs lack efferent synapses, the basilar papilla of the frog being one (Frishkopf and Flock 1974). Flock and Flock (1966) reported that efferent-type terminals were more abundant on rostral AP hair cells than on caudal AP hair cells. In a more detailed study of hair cell synapses, Simmons et al. (1995) found densely packed, vesiculated efferent terminals mostly on the tall hair cells in the rostral AP patch (Fig. 7.6A). Rostral hair cells are contacted by at least one and as many as three efferent terminals. Efferent terminals have punctate swellings rather than being *en passant* fibers, characteristic of the majority of the afferent terminals. Although generally smaller, efferent terminals can be as large as afferent terminals with diameters generally ranging from 0.2 to 2.0 μm.

3. AP and BP Nerve Fibers

3.1 AP and BP Nerve Fiber Responses

By selectively lesioning the eighth-nerve branchlets innervating individual sensory maculae within the otic capsule of the bullfrog inner ear, it was demonstrated that fibers tuned to low- and mid-frequencies innervate the AP and fibers tuned to higher frequencies innervate the BP (Feng et al. 1975). Subsequent studies assumed that similar organization held for other frog species (Capranica and Moffat 1975, 1983; Capranica 1978; Zakon and Capranica 1981; Narins 1983; Hödl et al. 2004). The only neurophysiologically derived tonotopic map available for any amphibian is for the bullfrog, *R. catesbeiana* (Lewis et al. 1982a,b). In the bullfrog AP, nerve fibers innervating the rostral AP patch exhibit CFs between 100 and 300 Hz. Fibers that innervate the caudal extension exhibit CFs roughly between 400 and 1000 Hz. Nerve fibers innervating the bullfrog BP have CFs between 1000 and 3000 Hz without any apparent tonotopy.

Typically, frog auditory nerve fibers are characterized in the frequency domain by frequency tuning curves (FTC), that is, a plot of their response threshold to a sinusoidal input (typically in decibels SPL, or re 0.0002 dynes/cm^2) as a function of the input frequency. Each fiber has its lowest threshold to one (characteristic) frequency (CF) and behaves as a bandpass filter; that is, its threshold rises as the stimulus frequency deviates in either direction from CF. The tuning curves of frog auditory nerve fibers are V-shaped when plotted on a log–log scale. The FTC shape (i.e., the relative degree of sharpness) reflects the underlying tuning mechanisms. In the frog, the sharpness of tuning as measured by the Q_{10dB} decreases with CF (Ronken 1990, 1991; Stiebler and Narins 1990). In the AP, afferent fibers with low CFs (rostral regions) have Q_{10dB} values as high as 5. Afferent fibers with high CFs from either caudal AP regions or the BP are more broadly tuned and rarely exceed a Q_{10dB} value of 2. Auditory nerve fibers in frogs exhibit similar frequency selectivity to that in mammals for the same CF despite clear anatomical differences between the ears of the two groups.

In addition to characterizing auditory nerve fiber responses by their frequency functions (i.e., tuning curves), a linear systems approach can be used to characterize nerve fiber responses to continuous broadband noise in the time domain by applying Wiener kernel analysis. Basically, a series of Wiener kernels characterizes a linear system that describes the (nonlinear) signal path between auditory input and neural output. This series can be determined experimentally by calculating the reverse correlation between a continuous, Gaussian noise stimulus and the ear's response to this stimulus, that is, the occurrence of spikes in the auditory nerve fiber (Schetzen 1989). Wiener kernel analysis reveals amplitude and phase characteristics of single nerve fiber responses (Eggermont 1993; van Dijk et al. 1994; Lewis et al. 2002; Lewis and van Dijk 2004). The first-order Weiner kernel is obtained by averaging the stimulus windows preceding each spike in the nerve fiber response. It is proportional to the reverse correlation (REVCOR) function of auditory nerve fibers and reveals the linear response of a system such as the ear (De Boer 1967; van Dijk et al. 1994). The second-order Wiener kernel is also calculated from the stimulus windows preceding neural spikes and provides (1) a two-dimensional visual image of a fiber's second-order nonlinear dynamics and (2) information about a fiber's tuning and timing of excitation, adaptation, and suppression responses (Lewis et al. 2002; Lewis and van Dijk 2004). Further analysis of Wiener kernels has provided qualitative information on the auditory filters and the interaction (tuning and timing) between suppression and excitation within the auditory nerve (Lewis and van Dijk, 2004). Most studies have been limited to the calculation of the first- and/or second-order Wiener kernel, although higher-order kernels may exist (e.g., van Dijk 1995), thus reflecting the presence of higher-order nonlinearities within the anuran inner ear.

In ranid frogs, AP and BP nerve fibers not only differ in their frequency tuning, but also in their temperature dependence, spontaneous activity, threshold and intensity responses, and OAEs. Table 7.1 summarizes many of the physiological characteristics of auditory nerve fibers found in anuran species. Although the extent of the frequency range in the AP varies with species, its properties are relatively uniform across groups (Zakon and Wilczynski 1988). The range of frequencies (at or below 100 Hz up to 1.0–1.4 kHz) represented along the AP sensory organ is correlated with the length of the caudal extension across species (Feng and Shofner 1981; Lewis 1981). Additionally, low-CF fibers show two-tone suppression, that is, reduction of a fiber's response to a tone by another tone of higher frequency (Liff and Goldstein 1970; Ehret et al. 1983; Rose and Capranica 1985; Christensen-Dalsgaard et al. 1998), or of lower frequency (Benedix et al. 1994). Two-tone suppression is not found in mid- or high-CF fibers. However, one-tone suppression (spontaneous activity suppressed by a single tone) may be a more general phenomenon within the AP (Christensen-Dalsgaard and Jørgensen 1996). The CFs of auditory nerve fibers from the AP also show a temperature

TABLE 7.1. Summary of peripheral auditory nerve physiology.

Parameter	AP rostral	AP caudal	BP	Possible anatomical correlate(s)	References
FREQUENCY CHARACTERISTICS					
Frequency tuning (Hz)	~100–400 Hz	~400–1000 Hz up to 1.4 kHz	>1000 Hz	Tectorium, hair bundle, hair cell locations	Feng et al. 1975 Feng and Shofner 1981 Lewis 1981 Zakon and Wilczynski 1988
Tonotopic organization	Decreasing (300 to 100 Hz) Increasing (100 to 400 Hz)	Increasing (400 to 1000 Hz)	None	Tectorium, hair bundle, hair cell locations	Lewis et al. 1982a,b
Tuning sharpness, $Q_{10\,dB}$	Broad (as high as 4–5)	Sharp (typically ≤2)	Typically ≤2	Tectorium, Innervation pattern	Ronken 1990, 1991 Stiebler and Narins 1990
Minimum threshold sensitivity to tones (dB SPL)	~30–90 dB SPL	~40–90 dB SPL	~50–80 dB SPL	Hair cell innervation density	Capranica and Moffat 1983
Tone suppression	Present	None	None	Tectorium, hair bundle	Ehret et al. 1983 Rose and Capranica 1985 Christensen-Dalsgaard and Jørgensen 1996 Christensen-Dalsgaard et al. 1998

Table 7.1. *Continued*

Parameter	AP rostral	AP caudal	BP	Possible anatomical correlate(s)	References
Phase locking	Strong	Moderate	Weak	Tectorium, hair bundle	Dunia and Narins 1989 Simmons et al. 1993
TEMPORAL CHARACTERISTICS					
Spontaneous rates (spikes/s)	0–80/s	0–50/s	0–40/s	Hair bundle, number of synapses	Ronken 1990 Christensen-Dalsgaard et al. 1998
Dynamic range (dB)	≤50 dB	≤30 dB	20–30 dB	Hair bundle, number of synapses	Capranica and Moffat 1983
Response (click) latency (ms)	Up to 8 ms	2–5 ms	2–3 ms	Tectorium, hair bundle, fiber diameter	Hillery and Narins 1984, 1987
Amplitude modulation response	Present	Present	Present	Tectorium, hair bundle	Simmons 1988 Simmons et al. 1993
Temperature sensitivity	Present	Present in lower frequencies	None	Hair cell (electrical tuning)	Stiebler and Narins 1990 van Dijk et al. 1990
Sexual dimorphism	None	None	Present		Narins and Capranica 1976
Size differences	None	None	Present		Wilczynski et al. 1992

dependence: lowering the animal's core temperature decreases a fiber's CF (Stiebler and Narins 1990; van Dijk et al. 1990). Although BP nerve fibers are typically tuned to frequencies >1.0 kHz and to a narrow range of CFs, tuning characteristics of nerve fibers are species-specific and, within a species, can vary with size, sex, or geographical location (Narins and Capranica 1976; Capranica 1978; Capranica and Moffat 1983). As a rule, the BP of large-sized species is tuned to lower frequencies than that of smaller-sized species and males have BPs tuned to generally higher frequencies than females (Narins and Capranica 1976; Wilczynski et al. 1992, 1993). Auditory nerve fibers of the BP show little or no change in their physiological properties when the animal's temperature is experimentally varied (Stiebler and Narins 1990; van Dijk et al. 1990, 1997; Carey and Zelick 1993; Benedix et al. 1994).

Auditory nerve fibers in the frog exhibit spontaneous activity, although reports as to the exact level vary widely based on the methods and species used. Earlier reports of spontaneous activity suggested that low-CF fibers generally have the lowest spontaneous rates, typically less than five spikes per second, whereas higher-CF fibers generally have higher spontaneous rates (Feng et al. 1975; Capranica 1978; Capranica and Moffat 1980, 1983; Zelick and Narins 1985; Ronken 1990; Ronken et al. 1993). However, more recent reports (Christensen-Dalsgaard et al. 1998) find no correlation between spontaneous activity and CF, but between spontaneous activity and threshold (fibers with low threshold responses to sounds tend to have higher spontaneous rates). Interestingly, low auditory thresholds in the frog have also been correlated with high DPOAE amplitudes (van Dijk et al. 2002; Vassilakis et al. 2004; and see Section 4.3.).

3.2 Fiber Morphometry, Conduction Velocity, and Response Latency

The physiological parameters of amphibian auditory nerve fibers can be defined by a structural unit that includes the tectorial membrane, the hair cells innervated by a single neuron, and the synaptic architecture and pattern of afferent innervation. However, how the anatomical patterns of hair cells and synaptic arrangements affect physiological responses is poorly understood. It has been tempting to associate such patterns with the observed physiological distribution of primary axons: low frequency, suppressible units, and mid-frequency, nonsuppressible units (Feng et al. 1975; Lewis 1976; Smotherman and Narins 2000).

In auditory organs, one of the best-characterized examples of structure–function relationships has been in the mammalian cochlea, where different spontaneous rate categories of auditory afferent fibers have been correlated with the size, location on the hair cell, synaptic arrangement, and mitochondrial content of nerve fiber terminals (Liberman 1980; Liberman and Simmons 1985; Gleich and Wilson 1993). To date, structure–function relationships have not been investigated at the same level of detail in the frog.

The innervation patterns of individual auditory neurons within AP and BP organs have been reconstructed in both leopard frogs and bullfrogs (Lewis et al.

1982b; Simmons et al. 1992). Most notably, the innervation of both AP and BP organs is highly convergent; that is, individual neurons make synaptic contact with multiple hair cells. Nerve fibers in the AP terminate on varying numbers of receptor hair cells (Simmons et al. 1992), with the physiological gradients following roughly the rostrocaudal gradient of hair cell innervation (Lewis et al. 1982b; Simmons et al. 1994b, 1995). Nerve fibers that respond best to lower frequencies contact more hair cells than fibers that respond best to higher frequencies. Retrograde-labeling studies suggest that (1) response latencies and conduction velocities may be correlated with the papillar location as well as the number of hair cells innervated and afferent fiber size (Lewis et al. 1982b; Simmons et al. 1992) and (2) the pattern of hair cell innervation may in part determine physiological parameters of auditory nerve fiber responses such as tuning sharpness and sensitivity. For example, large innervation areas (i.e., dendritic fields) might be correlated with broad tuning observed in a given auditory nerve fiber and quite possibly with poor frequency resolution.

3.3 AP Nerve Structural Correlates

Nerve fibers terminating in the caudal (high-frequency) end of the AP have smaller diameters than fibers terminating in the rostral (low frequency) end of the AP (Simmons et al. 1992; Simmons and Narins 1995). Both caudal and rostral AP fibers have nearly equal lengths from the ganglion to their terminations and virtually all of the fibers are myelinated (average ± s.d. myelin thickness of $0.13 \pm 0.05\,\mu m$; Simmons and Narins 1995). Larger diameter fibers appear more heavily myelinated and are situated within the center of the nerve bundle. These data suggest that fiber size is directly correlated with the area of innervation. In combination with previous neurophysiological results (Hillery and Narins 1984, 1987), such differences in fiber size may also relate to the differences found in response latencies to click stimuli (Hau et al. 2004).

The relationship between fiber morphometry and conduction velocity may play a significant role in understanding response latencies. Auditory nerve latency is an important feature of the acoustic transformation process and may encode additional information about the acoustic stimuli, including the directionality and periodicity of the source of sound (Bleeck and Langner 2001). In the frog AP as well as the mammalian cochlea, auditory nerve fibers that respond best to high-frequency stimuli have short response latencies and nerve fibers that respond best to low-frequency stimuli have longer response latencies (Kiang et al. 1965; Hillery and Narins 1987; Fitzgerald et al. 2001; Hau et al. 2004). In mammals, this observation has been linked to the traveling wave transit time along the basilar membrane in addition to the differences in path length between high-frequency (with short distances to the brain) and low-frequency (with longer distances to the brain) fibers. Although a similar latency–frequency relationship exists in the frog AP, this organ neither has a basilar membrane nor exhibits a significant difference in path length between high- and low-frequency fibers. However, an observation not suspected to play a significant role in mammalian response latencies is that frog auditory nerve fibers responding to high frequencies are typically thin and fibers responding to lower frequencies are thicker.

Previous theoretical studies of conduction velocities in nerve fibers (Rushton 1951; Deutsch 1969; Smith and Koles 1970) indicate that myelin thickness in nerve fibers is such that, for a fiber of given external diameter, conduction velocity is maximized. For fibers with the same myelin thickness, fibers with large diameters conduct impulses faster. Based on preliminary estimates of theoretical conduction velocities [velocity = k (diameter)], rostral AP fibers should have the highest conduction velocity, caudal AP fibers should have low conduction velocities, and BP fibers should have a range of conduction velocities. The difference between theoretical estimates and measured latencies suggests that other variables may be involved in determining response latencies. Two such variables are the time it takes an acoustic signal to stimulate hair cell stereovilli and the time it takes hair cells, once stimulated, to give rise to action potentials in auditory nerve fibers (Hau, Simmons, and Narins 2005, unpublished data).

The tectorium may also contribute to the response latencies observed in AP nerve fibers, as it is a complex gelatinous structure with its most massive segment overlying the rostral AP patch (Shofner and Feng 1983; Hau et al. 2004). A traveling wave across the tectorium would be expected to stimulate higher-frequency regions before lower-frequency regions by as much as a factor of 10 and would be consistent with the observed longer latencies of rostrally innervating fibers.

4. Otoacoustic Emissions

4.1 Introduction

One of the remarkable properties of the vertebrate inner ear is its great sensitivity. From early on it was recognized that such sensitivity could not arise solely from passive responses to sound. Rather, some active amplification mechanism would be required to enhance the vibration of inner ear structures in response to low-level acoustic stimuli (Gold 1948). In mammals, this mechanism has been linked to outer hair cell (OHC) motility and receptor potential (see reviews by Probst et al. 1991 and Robles and Ruggero 2001) and is termed the "cochlear amplifier" (Davis 1983). It utilizes bioelectrochemical energy (Zheng et al. 2000; Liberman et al. 2002), which may propagate through the middle ear to the outside in the form of vibrations in the hearing frequency range. The discovery of low-level sounds corresponding to such vibrations (Kemp 1978) provided the first evidence for the presence of an active amplification mechanism within the inner ear. These sounds are now known as OAEs and can be measured by placing a sensitive microphone in the ear canal.

Since their discovery in mammals, OAEs have been reported to be present in all classes of terrestrial vertebrates (Köppl 1995; Köppl and Authier 1995), suggesting that they reflect a fundamental property of normal hearing. Because OAEs can be recorded noninvasively, they provide an important tool to hearing researchers. OAEs are often used for diagnostic screening of frequency-dependent cochlear function (Lonsbury-Martin et al. 1993; Stover et al. 1996; Wagner and Plinkert 1999; Abdala 2000; Emmerich et al. 2000) or middle ear damage (Owens et al. 1992, 1993; Zhang and Abbas 1997). In addition, OAEs provide a noninvasive means of examining inner ear mechanisms in mammals (Kössl and Vater 1996; Maison et al. 1997; Shera and Guinan 1999) and other vertebrates (Rosowski et al. 1984; Vassilakis et al. 2004), as well as in insects (Kössl and Boyan 1998).

Traditionally, OAEs have been classified into spontaneous OAEs (SOAEs) and evoked OAEs (EOAEs), based on whether an external stimulus is required for their generation. Furthermore, based on the type of stimulus required for their generation, EOAEs are classified into (1) stimulus-frequency OAEs or SFOAEs (evoked by single sinusoidal stimuli), (2) transient-evoked OAEs or TEOAEs (evoked by short-duration, low-level signals of various spectra), and (3) distortion-product OAEs or DPOAEs (evoked by two continuous sinusoidal stimuli) (Probst et al. 1991). It has been argued that the above classifications do not do justice to the mechanisms underlying OAE generation (Shera and Guinan 1999; Kalluri and Shera 2001). However for the sake of consistency, the traditional classification scheme is presented. Both the AP and BP exhibit OAEs but their properties differ considerably.

4.2 Stimulus-Frequency and Transient-Evoked Otoacoustic Emissions

Upon stimulation with single continuous sinusoids or short signals of various spectral distributions (e.g., clicks), the inner ear generates SFOAEs or TEOAEs, respectively, which coincide in frequency with the component(s) of the external stimulus. It is currently believed that both SFOAEs and TEOAEs arise from the same inner ear mechanism. Unfortunately, the number of studies exploring these types of OAEs in frogs is limited.

Whitehead et al. (1986) examined SFOAEs and their relationship to body temperature in *Rana temporaria*. The authors were able to evoke SFOAEs only for frequencies below 1 kHz, suggesting that in this species, SFOAEs may arise only from the AP. Palmer and Wilson (1982) were able to evoke measurable SFOAEs in *Rana esculenta* for frequencies up to 2 kHz, implying that in this species, both papillae generate SFOAEs. The effect of body temperature on SFOAE frequencies is very pronounced, suggesting that SFOAE generation may involve an electrochemical tuning mechanism along, possibly, with some form of mechanical tuning. In addition, SFOAE levels increase just before the start of the frog breeding season (Whitehead et al. 1986), suggesting that OAE generation in frogs may be seasonal and under hormonal control.

TEOAEs were reported by Palmer and Wilson (1982) but not in enough detail to support an account of TEOAE-properties.

4.3 Distortion-Product Otoacoustic Emissions (DPOAEs)

Upon stimulation with two continuous sinusoids (often termed primaries) of appropriately chosen frequencies (f_1, f_2, where $f_2 > f_1$) and intensities (L_1, L_2), the inner ear generates DPOAEs. DPOAE frequencies coincide with the intermodulation distortion frequencies of the two primaries, whereas the most-studied DPOAEs coincide in frequency with the third-order intermodulation distortion frequencies ($2f_1$–f_2 and $2f_2$–f_1). As such, DPOAEs are the only type of EOAEs that do not match in frequency any of the frequencies present in the external stimulus, a fortunate property that simplifies their study.

With the exception of early reports (Baker et al. 1989), DPOAEs have been recorded in frog species from the families Ranidae, Hylidae, and Pipidae (van Dijk and Manley 2001; van Dijk et al. 2002, 2003; Meenderink and van Dijk 2004, 2005a; Vassilakis et al. 2004; Meenderink et al. 2005a,b). No DPOAEs were found in species from the families Bombinatoridae and Pelobatidae (van Dijk et al. 2002). It is thought this apparent difference between frogs from different families is related to the overall sensitivity of the hearing apparatus (van Dijk et al. 2002).

The amplitude and phase of DPOAEs depends on the parameters that describe the two primaries. These parameters outline a four-dimensional parameter space usually defined by f_1, f_2/f_1, L_1, and L_1–L_2. In general, individual studies limit their exploration to isolated slices/planes within this space. When the two primary frequencies are systematically varied (i.e., changes only in the f_1 dimension of the parameter space), plotting DPOAE amplitude versus frequency results in a DPOAE-audiogram. In frogs, DPOAE-audiograms typically exhibit two maxima; one below and one above approximately 1 kHz (Fig. 7.7). It is currently believed that the bimodal shape of frog DPOAE-audiograms reflects emission generation from the two papillae, with the low- and high-frequency peaks representing DPOAEs generated within the AP and the BP, respectively. This bimodality seems to be independent of both the primary frequency ratio (van Dijk and Manley 2001; Meenderink et al. 2005a) and the absolute levels of the primaries (Meenderink and van Dijk 2004). DPOAEs from the AP (AP-DPOAEs) and the BP (BP-DPOAEs) differ in several respects and are referred to separately.

DPOAE-audiogram data can be extended by recording multiple DPOAE audiograms from the same animal, each with a different f_2/f_1 (Fig. 7.8). The resulting (f_1, f_2) area maps reveal DPOAE amplitude and phase patterns that are different from those obtained from the mammalian cochlea (Knight and Kemp 2000; Schneider et al. 2003). A transmission-line model that incorporates cochlear properties such as traveling waves may explain the mammalian DPOAE patterns (Knight and Kemp 2001). In contrast, the frog patterns obtained from both the AP and the BP may be modeled by a single nonlinearity, suggesting the absence of mammalianlike traveling waves (Meenderink et al. 2005a). Several mammalian studies (e.g., Shera and Guinan 1999; Faulstich and Kössl 2000) have argued that f_2/f_1 must be ≥ 1.15, to avoid phase complications (e.g., rapidly rotating phase and regular amplitude variations of the DPOAEs, due to beating

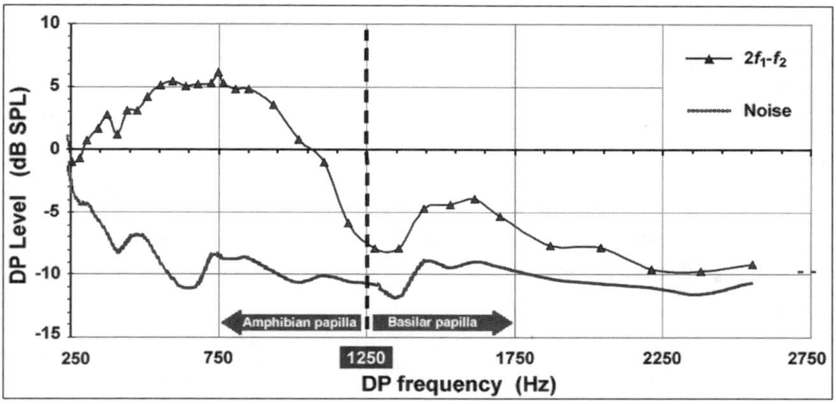

FIGURE 7.7. Average levels of the $2f_1$–f_2 (-▲-) DPOAE from *R. p. pipiens* (20 ears: both ears of five males and five females) for $f_2/f_1 = 1.15$, $L_1 = L_2 = 60$ dB SPL (±2.5 dB), and $240 \geq f_1 \geq 3000$ Hz as a function of DP frequency. The dashed vertical line marks the "break" in the frequency coverage of the AP and the BP (1250 Hz). Adapted from Vassilakis et al. (2004).

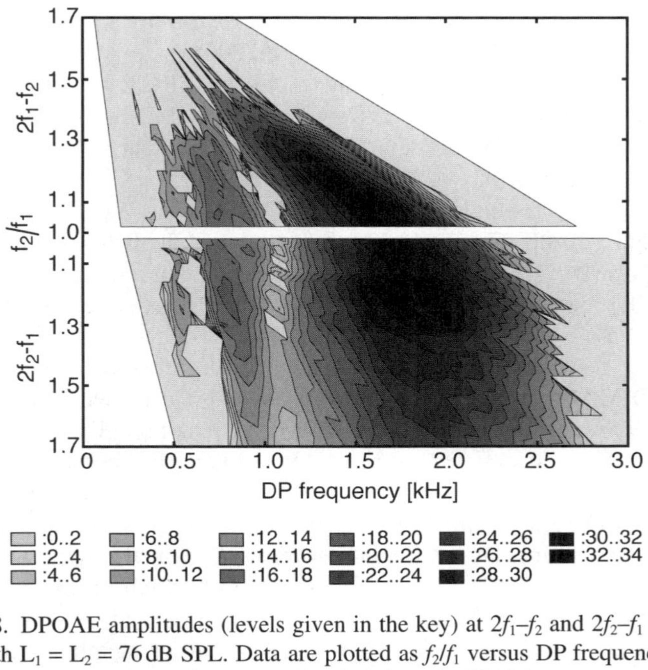

FIGURE 7.8. DPOAE amplitudes (levels given in the key) at $2f_1$–f_2 and $2f_2$–f_1 in dB SPL, evoked with $L_1 = L_2 = 76$ dB SPL. Data are plotted as f_2/f_1 versus DP frequency. Contour lines are drawn at 2 dB intervals. From Meenderink et al. (2005a).

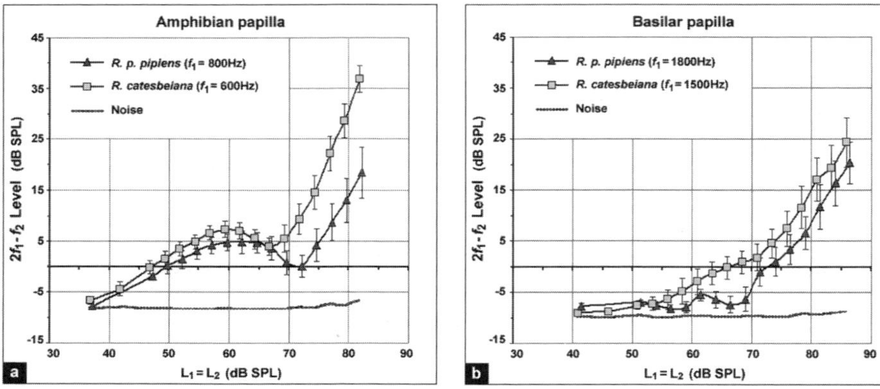

FIGURE 7.9. DPOAE I/O curves for $f_2/f_1 = 1.15$ from *R. p. pipiens* (-▲-, $f_1 = 800$ Hz, 1800 Hz) and *R. catesbeiana* (-□-, $f_1 = 600$ Hz, 1500 Hz). Average DPOAE levels ($2f_1$–f_2) and standard errors (20 ears each species: both ears of five males and five females). (A) AP; (B) BP. Adapted from Vassilakis et al. (2004).

between the primaries) that arise as f_2/f_1 drops below 1.1. Consistent with filtering mechanism and two-tone suppression differences between the mammalian and the frog inner ears, no such complications arise for low f_2/f_1 values in the frog (van Dijk and Manley 2001; Meenderink et al. 2005a).

The relationship of DPOAE amplitude to the absolute levels of the primaries (i.e., to changes in both stimulus levels with L_1–$L_2 = 0$) is most commonly visualized in the form of DPOAE input/output curves (I/O curves). For $L_1 = L_2$ <~ 70 dB SPL, the growth rate of the AP-DPOAE I/O curves is compressive (i.e., ≤1 dB/dB), whereas for higher primary levels ($L_1 = L_2$ >~ 70 dB SPL) it is expansive (i.e., 2 to 3 dB/dB; Fig. 7.9A). The growth rate of the BP-DPOAE I/O curves does not appear to vary with absolute primary levels, but remains expansive (i.e., 2 to 3 dB/dB) over the entire stimulus level range tested (Meenderink and van Dijk 2004; Vassilakis et al. 2004; Fig. 7.9B).

Recording DPOAEs for unequal primary levels provides a more complete picture of DPOAE-dependence on L_1 and L_2. For low-to-moderate primary levels and relatively small frequency ratios ($f_2/f_1 \leq 1.15$), maximum DPOAE amplitudes are obtained for L_1–$L_2 \approx 0$ in both the AP and the BP (Vassilakis et al. 2004; Meenderink and van Dijk 2005a; Fig. 7.10). For larger frequency ratios (e.g., $f_2/f_1 = 1.3$), the primary level difference (L_1–L_2) resulting in maximum BP-DPOAE amplitude (i.e., optimal level difference) varies with primary frequencies (Meenderink and van Dijk 2005a). In mammals, the relationship of this optimal level difference (L_1–L_2) to the absolute primary levels (L_1, L_2) and the relative primary frequencies (f_2/f_1) may be explained in terms of basilar membrane disturbance envelopes (e.g., Kummer et al. 2000). It has been argued that the difference in these relationships between mammals and frogs may be related to differences in

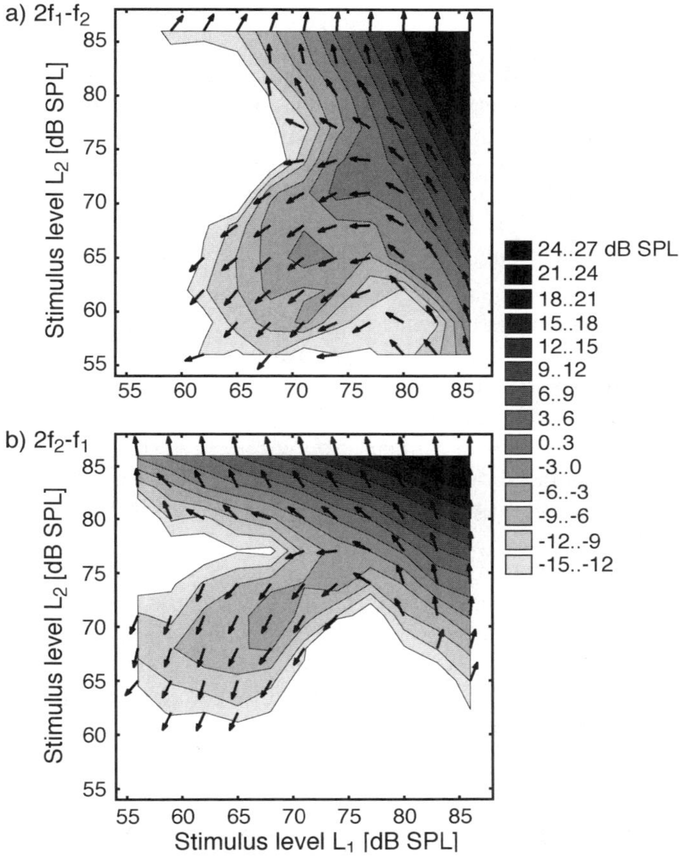

FIGURE 7.10. Amplitude and phase of DPOAEs at (A) $2f_1$–f_2 and (B) $2f_2$–f_1 as a function of stimulus levels (L_1, L_2). Due to the stimulus parameters used (f_1 = 1913 Hz; f_2/f_1 = 1.1), the resulting DPOAEs originate in the BP. Contour lines (drawn at 3 dB intervals) represent amplitude (levels given in the key) and arrows represent phase. Phase is relative to the phase at $L_1 = L_2$ = 86 dB SPL. From Meenderink and van Dijk (2005).

the mechanical tuning properties of the mammalian basilar membrane and the frog tectorium (Vassilakis et al. 2004).

AP- and BP-DPOAEs arising in response to low-level primaries also differ in their vulnerability to physiological insults (van Dijk et al. 2003) and to changes in body temperature (Meenderink and van Dijk 2005b). AP-DPOAEs rapidly disappear when oxygen supply is hindered or when body temperature is decreased. In contrast, anoxia takes much longer to influence BP-DPOAEs, whereas temperature changes have no clear effect.

It is currently believed that DPOAEs result from the combination of two level-dependent DPOAE components (Whitehead et al. 1992). The first component dominates for low-level primaries, saturating as primary levels increase. The second component dominates for high-level primaries, with its amplitude dropping below that of the low-level component as primary levels decrease. A phase difference of ~π rad between these two components is believed to be responsible for the DPOAE I/O-curve notches observed for intermediate primary levels (Mills and Rubel 1994). In contrast to mammalian data, this phase difference increases systematically from 0 to π rad with increasing primary frequencies in both frog papillae (Meenderink and van Dijk 2004).

Vassilakis et al. (2004) compared the frequency ranges of the AP and BP response with DPOAE-audiograms plotted as functions of f_1, f_2, and DP frequency, and interpreted their degree of correlation as providing evidence for the site of DPOAE-generation. Opposite to mammals, they found that the generation of the $2f_1$–f_2 DPOAE in the frog may primarily occur at or near the DPOAE frequency place, whereas the generation of the $2f_2$–f_1 DPOAE may primarily occur at a frequency place between the two primaries. This opposite behavior may be related to various anatomical differences between the mammalian and frog ears, including the absence, in the frog ear, of a basilar membrane, and may be a manifestation of differences regarding the possibility of traveling wave development.

The single study exploring sex differences in frog DPOAEs (Vassilakis et al. 2004) indicates that, although overall frog DPOAE generation mechanisms may be independent of sex, females produce stronger emissions than males, from both the AP and the BP (Fig. 7.11). This difference is consistent with mammalian results and may be related to sex differences in frog ear frequency tuning, sensitivity (Narins and Capranica 1976), and middle- and inner-ear physiology (Mason et al. 2003).

4.4 Spontaneous Otoacoustic Emissions (SOAEs)

As is clear from their name, SOAEs arise spontaneously, without the need of an external evoking stimulus. The peaks in the frequency spectrum corresponding to SOAEs can be identified by (a) their consistent presence in the average of a large number of spectra, (b) their susceptibility to changes in temperature, and (c) their modification (suppression and frequency modulation) by external tones. SOAEs have been reported for frog species in the families Ranidae, Hylidae, and Leptodactylidae (Baker et al. 1989; van Dijk et al. 1996), but not for species in the families Pipidae and Bombinatoridae (van Dijk et al. 1996). They have been recorded in approximately 70% of the subjects tested within the species exhibiting these emissions, with approximately one to four SOAEs per ear. On average, the frequency spacing between consecutive SOAEs from the same ear is ~0.5 octave, exceeding the frequency spacing found in humans (~0.1 octave; reviewed in van Dijk et al. 1989). SOAE spectral peaks are limited within the 450 to 1350 Hz frequency range (van Dijk et al. 1989), suggesting that SOAEs may only arise from the caudal portion of the AP (van Dijk and Manley 2001).

Similar to SFOAEs and AP-DPOAEs, SOAEs are sensitive to changes in temperature. Lowering (increasing) the subjects' body temperature results in lower

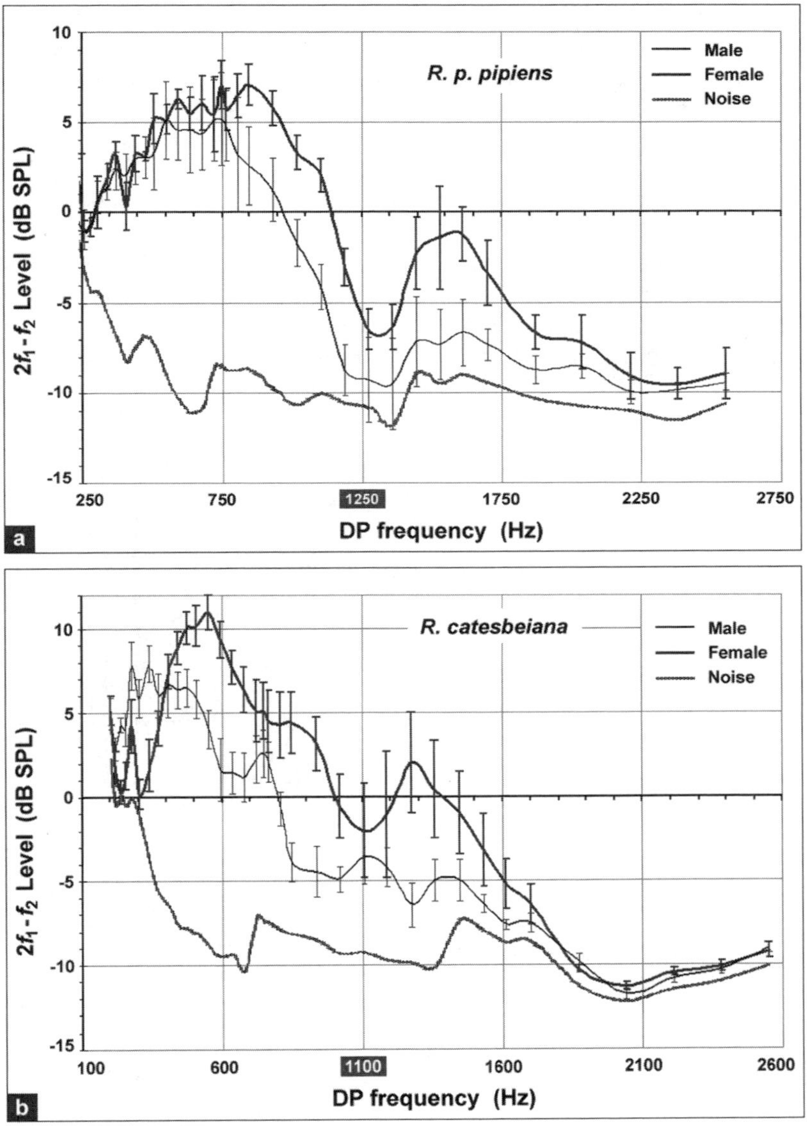

FIGURE 7.11. Comparison between DPOAE ($2f_1$–f_2) audiograms from both ears of five male and five female *R. p. pipiens* (A) and *R. catesbeiana* (B). DPOAE levels are plotted as a function of DP frequency. Female subjects exhibit stronger emissions than males, especially from the BP (from Vassilakis et al. 2004).

(higher) SOAE frequencies (van Dijk and Wit 1987; van Dijk et al. 1989; Long et al. 1996). The rate of this change appears to depend on emission frequency, with SOAEs <600 Hz exhibiting less sensitivity to temperature changes than SOAEs >600 Hz (Long et al. 1996; van Dijk et al. 1996; Fig. 7.12). Temperature changes also affect SOAE amplitudes, but in an on–off rather than continuous manner (van Dijk and Wit 1987).

Both the frequency and amplitude of SOAEs change in the presence of an external sinusoidal stimulus. If the external stimulus is both close in frequency to an SOAE and sufficiently strong, the SOAE frequency will shift to match the frequency of the stimulus (van Dijk and Wit 1990). On the contrary, if the frequency separation between an SOAE and the external stimulus is large, the SOAE frequency will shift away from that of the external stimulus (Baker et al. 1989;

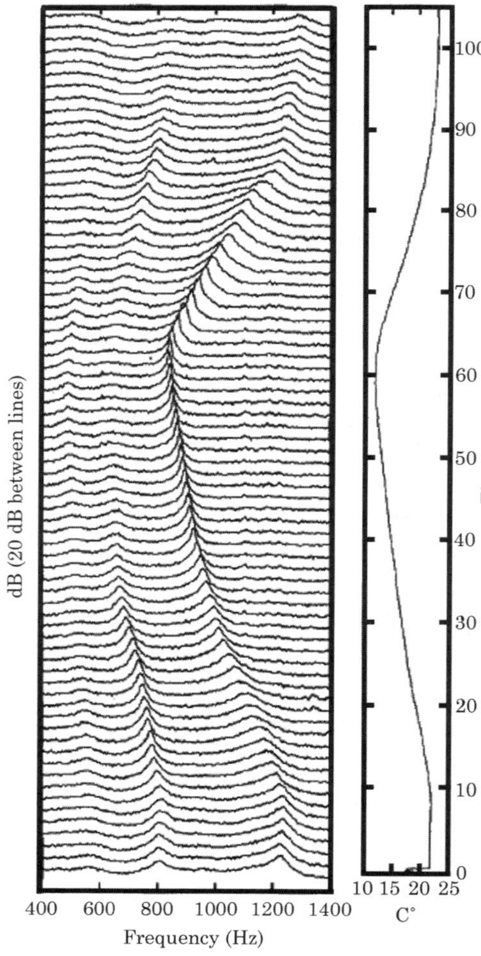

FIGURE 7.12. SOAE spectra (64 averages per spectrum) and temperature changes plotted over time, illustrating the dependence of SOAE frequencies on temperature. From Long et al. (1996).

Long et al. 1996). The presence of an external stimulus may also decrease the amplitude of an SOAE. The amount of suppression depends on the frequency of the external stimulus, as seen in isosuppression contours (Baker et al. 1989). Suppression is most prominent when the SOAE and the external stimulus coincide in frequency, and decreases as the frequency separation between the two increases. The amount of suppression is highest for external stimulus frequencies below the SOAE frequency, resulting in asymmetric isosuppression contours that are similar in shape to neural tuning curves (Baker et al. 1989).

Although SOAEs could be easily recorded in the summer, they were not detected in the winter (van Dijk et al. 1989), confirming the earlier observation that OAE generation may be seasonal.

4.5 OAEs Suggest the BP Acts as a Single Auditory Filter

Due to the BP's relatively simple anatomy and physiology, it has been suggested that this papilla may be functioning as a bandpass filter with a single CF. In ranid frogs, for example, all BP hair bundles follow the same orientation (Lewis 1978), and almost all BP nerve fibers are tuned to the same frequency and have identically shaped tuning curves (Ronken 1990). Based on such observations, several frog DPOAE studies have used the Duffing oscillator as a model for the BP (van Dijk and Manley 2001; Meenderink et al. 2005a,b), successfully predicting the dependence of DPOAE amplitudes and phases on the relative and absolute primary frequencies, as well as the observed correlations between DPOAEs and neural tuning curves. These results support the notion of the BP as a single, broadly tuned auditory filter. The uniqueness of this property within vertebrate hearing makes the frog an excellent subject for DPOAE studies, inasmuch as BP-DPOAEs are not influenced by auxiliary structures (in contrast to the mammalian ear or the frog AP) and only reflect the properties of the nonlinearities directly involved in OAE generation.

4.6 OAEs and Inner Ear Amplification in the Frog

OAE data obtained from the frog support several speculations regarding the functioning of the amphibian inner ear. It is generally accepted that mammalian OAE generation is closely linked to the cochlear amplifier, a term introduced by Davis (1983) to denote the summed contribution of those mechanisms that appear to enhance the movement of mammalian inner ear structures. Besides the presence of OAEs, several observations support the involvement of an amplifier in the ear's transduction of sound. First, the auditory system exhibits higher sensitivity and frequency selectivity than expected by its passive mechanical properties alone. Second, the system's response grows compressively, being highly amplified for low-intensity stimuli and less so as stimulus intensity increases. Third, the system is highly susceptible to various forms of physical, physiological, and chemical insults, consistent with the fact that the cochlear amplifier depends on biochemical sources for its energy.

SOAEs are the only type of OAEs whose mere occurrence may be interpreted as evidence for the presence of an inner ear amplifier. In the frog, SOAEs have been measured only within a frequency range corresponding to the frequency response of the AP, suggesting that an amplification process may be present in this papilla but not in the BP. Several differences between AP- and BP-DPOAEs support this observation. Only DPOAE I/O curves from the AP exhibit compressive growth at low primary levels. Furthermore, low-level AP-DPOAEs are much more sensitive to physiological insults and changes in temperature than low-level BP-DPOAEs, suggesting the presence of biochemical energy sources in the AP that would power an amplifier.

An amplification process may therefore be present in the frog AP but not in the BP. This potential difference is consistent with and may be related to several physiological differences between the two papillae. In the *R. catesbeiana* ear, for example, the AP has approximately fifteen times as many hair cells and is innervated by approximately three times as many afferent nerve fibers as the BP (Geisler et al. 1964; Lewis et al. 1985). At the same time, efferent nerve fibers, in most cases, innervate the AP but not the BP (Robbins et al. 1967; Simmons et al. 1995).

It has been suggested that, because the AP is tonotopically organized (Lewis et al. 1982b; Simmons et al. 1992) and the BP responds as a bandpass filter with a single characteristic frequency (van Dijk and Manley 2001), the presence of an amplifier might play the role of increasing the sharpness of tuning of the AP fibers (Vassilakis et al. 2004). The absence of an amplifier from the BP may be beneficial as well. Due to their ectothermic physiology, frog bodies may undergo relatively large temperature fluctuations. Given the lack of temperature sensitivity of the BP fibers (Stiebler and Narins 1990; van Dijk et al. 1990), such fluctuations do not result in the loss of frequency-specific information from this sensory organ.

Although specifying the exact location and the underlying molecular motor of a possible AP amplifier is not essential to understanding whether amplification occurs in the frog ear, some relevant speculations can be made. According to general consensus, the inner ear amplifier is located within the sensory hair cells. In mammals, a likely candidate for the cochlear amplifier has been identified in the OHCs. These cells contain within their lateral membrane a protein (prestin) that undergoes voltage-mediated conformational changes (Santos-Sacchi 1991). This deformation alters the OHC shape in synchrony with an incoming stimulus, providing a cycle-by-cycle amplification of the stimulus-induced OHC movement (Zheng et al. 2000). Nonmammalian vertebrates, on the other hand, lack OHCs and there is no evidence of or likelihood for fast somatic motility of their hair cells (Manley 2001). Oscillation of the hair bundles has been proposed as an alternative amplification mechanism. Such oscillations have been observed in vitro in several nonmammalian vestibular organs (Hudspeth 1997) and in the hearing organ of turtles (Crawford and Fettiplace 1985), and in vivo in the bobtail lizard (Manley et al. 2001). Great interest has been generated from recent reports (Chan and Hudspeth 2005; Jia and He 2005) documenting active hair bundle motion in mammalian hair cell bundles, requiring a reassessment of the role of evolution in the genesis of motility in auditory hearing organs.

5. Summary

The sensitivity (e.g., spontaneous activity and thresholds) and tuning (e.g., frequency responses) of the frog auditory periphery is determined in large part by its unique anatomical and physiological features, not the least of which is that there are two organs specialized for the reception of airborne sounds. The frog ear relies heavily on an elaborate infrastructure, which subserves conduction, filtering, and transduction of auditory information by both the AP (low to mid-frequencies) and the BP (high frequencies).

As the input to the central auditory system, auditory nerve fiber responses are determined by the arrangement and number of their connections to hair cells, by the events at the hair cell synapse, by the intrinsic properties of hair cells and their hair bundles, as well as by the bandpass spectral filtering of the overlying tectorial structures. The frog AP is a uniquely organized tonotopic end-organ whereas the frog BP is a broadly tuned organ that acts as a single auditory filter. In both the AP and BP, hair cells are rigidly fixed to a cartilaginous wall and lack a basilar membrane, a structure common to auditory organs of reptiles, birds, and mammals. The frog AP has a complicated frequency-related distribution of hair bundle types with polarization patterns that grossly correspond to other rostrocaudal gradients in the ear, such as hair cell height and tectorial membrane mass.

In the frog inner ear, frequency selectivity may depend on the mechanics of the tectorium and the fluids as well as their interaction with the hair cells. Movement of the tectorium presumably produces deflection of the stereovilli that leads to the release of transmitter, activating auditory nerve fibers. Thus, the inertial lag times associated with tectorium mechanics along with differences in stereovillar lengths or stiffness may be the primary determinants of response latency. Determination of the actual travel times of action potentials in auditory nerve fibers from high- and low-frequency locations should give a better understanding of how the motion of the tectorium is coupled to the stereovillar bundle.

Rostrocaudal variations in hair cell height are inversely related to tonotopy such that the tallest hair cells are found in the lowest-frequency regions and shorter hair cells are found in higher-frequency regions. Whole-cell capacitances also covary with hair cell body length. Thus, basic morphological features (e.g., hair cell height) can be related to the presence of intrinsic electrical tuning mechanisms (e.g., capacitance and frequency tuning). The variation of synaptic architecture is consistent with innervation patterns: rostrally located hair cells have a greater number of synapses and nerve fiber contacts than caudally located hair cells. Similarly, efferent synapses are predominant in rostral areas and not detectable in caudal areas. Frog auditory nerve fibers also have anatomical and physiological gradients such as larger diameters and longer response latencies from low-frequency rostral fibers, and thinner diameters and shorter response latencies from higher-frequency caudal fibers.

It is currently well-established that the mammalian cochlea exhibits tonotopic organization (Von Békésy, 1960). This tonotopy arises from mass and stiffness gradients along the basilar membrane and can be observed by measuring the mechanical response of the basilar membrane (Robles and Ruggero 2001). One of the most intriguing manifestations of this gradient-induced tonotopy is the presence of traveling waves on the basilar membrane. Conversely, tonotopic organization that does not move gradually from low to high frequencies (or vice versa) cannot support a traveling wave. Because the AP is also tonotopically organized, with highest sensitivity to different frequencies distributed at different locations along the sensory epithelium, the question arises whether this papilla may also support traveling waves. Neurophysiological evidence (Hillery and Narins 1984) suggests that this may indeed be the case. However, if there is an absence of appropriate mass and stiffness gradients within the tectorium (or any other frog inner ear structure) then the presence of mammalianlike traveling waves in the AP may be excluded (Lewis et al. 1985). This exclusion is further supported by the patterns present in DPOAE (f_1, f_2) area maps (Meenderink et al. 2005a). Rather, it seems more plausible that the tectorium overlying the AP sensory epithelium functions as a broadband filter, with tonotopy originating in additional filtering mechanisms (mechanical and/or electrical) that may be closely associated with the hair cells. Data on the dependence of DPOAE amplitude on primary level difference (L_1–L_2) also suggest that the tectorium may function as a broadband filter (Vassilakis et al. 2004; Meenderink and van Dijk 2005a), outlining the boundaries of the AP frequency response range. A possible difference in DPOAE generation sites between mammals and frogs observed by Vassilakis et al. (2004) also questions the development of a mammalianlike traveling wave in the frog ear.

OAEs are indirect manifestations of the vibration of structures within the AP and, as such, they can only provide indirect evidence for the types of available spectral filtering (i.e., the presence or absence of traveling waves). Conclusive evidence for the presence of a traveling wave can best be obtained by directly recording the vibration of the tectorial membrane or of other traveling-wave relevant AP structures.

Acknowledgments. The authors are indebted to Peter M. Narins for his superb editorship, guidance, and assistance. His gentle spirit, encouraging words, and persistent intellectual rigor were invaluable to the writing of this chapter.

References

Abdala C (2000) Distortion product otoacoustic emission (2f1–f2) amplitude growth in human adults and neonates. J Acoust Soc Am 107:446–456.

Ashmore JF, Geleoc GS, Harbott L (2000) Molecular mechanisms of sound amplification in the mammalian cochlea. Proc Natl Acad Sci USA 97:11759–11764.

Baird RA, Burton MD, Fashena DS, Naeger RA (2000) Hair cell recovery in mitotically blocked cultures of the bullfrog saccule. Proc Natl Acad Sci USA 97:11722–11729.

Baird RA, Lewis ER (1986) Correspondences between afferent innervation patterns and response dynamics in the bullfrog utricle and lagena. Brain Res 369:48–64.

Baker RJ, Wilson JP, Whitehead ML (1989) Otoacoustic evidence for nonlinear behavior in frog hearing: suppression but no distortion products. In: Wilson J, Kemp DT (eds.) Cochlear Mechanisms: Structure, Function and Models. Plenum, New York, pp. 349–356.

Benedix JH, Jr., Pedemonte M, Velluti R, Narins PM (1994) Temperature dependence of two-tone rate suppression in the northern leopard frog, *Rana pipiens pipiens*. J Acoust Soc Am 96:2738–2745.

Bleeck S, Langner G (2001) Functional significance of latencies. In: Greenberg S, Slaney M (eds.) Computational Models of Auditory Function (NATO Science Series). IOS Press, Burke, VA, pp. 205–220.

Bozovic D, Hudspeth AJ (2003) Hair-bundle movements elicited by transepithelial electrical stimulation of hair cells in the sacculus of the bullfrog. Proc Natl Acad Sci USA 100:958–963.

Capranica RR (1978) Auditory processing in anurans. Fed Proc 37:2324–2328.

Capranica RR, Moffat AJM (1975) Selectivity of the peripheral auditory system of spadefoot toads (*Scaphiopus couchi*) for sounds of biological significance. J Comp Physiol 100.

Capranica RR, Moffat AJM (1980) Nonlinear properties of the peripheral auditory system of anurans. In: Popper AN, Fay RR (eds.) Comparative Studies of Hearing in Vertebrates. Springer, New York, pp. 139–165.

Capranica RR, Moffat AJM (1983) Neurobehavioral correlates of sound communication in anurans. In: Ewert J, Capranica R, Ingle D (eds.) Advances in Vertebrate Neuroethology. Plenum, New York, pp. 701–730.

Carey MB, Zelick R (1993) The effect of sound level, temperature and dehydration on the brainstem auditory evoked potential in anuran amphibians. Hear Res 70:216–228.

Chan DK, Hudspeth AJ (2005) Ca2+ current-driven nonlinear amplification by the mammalian cochlea in vitro. Nat Neurosci 8:149–155.

Chang JS, Popper AN, Saidel WM (1992) Heterogeneity of sensory hair cells in a fish ear. J Comp Neurol 324:621–640.

Christensen-Dalsgaard J, Jørgensen MB (1996) One-tone suppression in the frog auditory nerve. J Acoust Soc Am 100:451–457.

Christensen-Dalsgaard J, Jørgensen MB, Kanneworff M (1998) Basic response characteristics of auditory nerve fibers in the grassfrog (*Rana temporaria*). Hear Res 119:155–163.

Corwin JT, Warchol ME (1991) Auditory hair cells: Structure, function, development, and regeneration. Ann Rev Neurosci 14:301–333.

Crawford AC, Fettiplace R (1985) The mechanical properties of ciliary bundles of turtle cochlear hair cells. J Physiol 364:359–379.

Dallos P (1992) The active cochlea. J Neurosci 12:4575–4585.

Davis H (1983) An active process in cochlear mechanics. Hear Res 9:79–90.

De Boer E (1967) Correlation studies applied to the frequency resolution of the cochlea. J Aud Res 7:209–217.

Deutsch S (1969) The maximization of nerve conduction velocity. IEEE Trans Syst Sci Cybern 5:86–91.

Dunia R, Narins PM (1989) Temporal resolution in frog auditory-nerve fibers. J Acoust Soc Am 85:1630–1638.

Eggermont JJ (1993) Wiener and Volterra analyses applied to the auditory system. Hear Res 66:177–201.

Ehret G, Moffat AJM, Capranica RR (1983) Two-tone suppression in auditory nerve fibers of the green treefrog (*Hyla cinerea*). J Acoust Soc Am 73:2093–2095.

Emmerich E, Richter F, Reinhold U, Linss V, Linss W (2000) Effects of industrial noise exposure on distortion product otoacoustic emissions (DPOAEs) and hair cell loss of the cochlea—long term experiments in awake guinea pigs. Hear Res 148:9–17.

Faulstich M, Kössl M (2000) Evidence for multiple DPOAE components based upon group delay of the 2f(1)-f(2) distortion in the gerbil. Hear Res 140:99–110.

Feng AS, Narins PM, Capranica RR (1975) Three populations of primary auditory fibers in the bullfrog (*Rana catesbeiana*): Their peripheral origins and frequency sensitivities. J Comp Physiol 100:221–229.

Feng AS, Shofner WP (1981) Peripheral basis of sound localization in anurans. Acoustic properties of the frog's ear. Hear Res 5:201–216.

Fettiplace R, Ricci AJ, Hackney CM (2001) Clues to the cochlear amplifier from the turtle ear. Trends Neurosci 24:169–175.

Fitzgerald JV, Burkitt AN, Clark GM, Paolini AG (2001) Delay analysis in the auditory brainstem of the rat: Comparison with click latency. Hear Res 159:85–100.

Flock A (1965) Transducing mechanisms in the lateral line canal organ receptors. Cold Spring Harb Symp Quant Biol 30:133–145.

Flock A, Flock B (1966) Ultrastructure of the amphibian papilla in the bullfrog. J Acoust Soc Am 40:1262.

Frishkopf LS, Flock A (1974) Ultrastructure of the basilar papilla, an auditory organ in the bullfrog. Acta Otolaryngol 77:176–184.

Geisler CD, Van Bergeijk W, Frishkopf LS (1964) The inner ear of the bullfrog. J Morphol 114:43–57.

Gleich O, Wilson S (1993) The diameters of guinea pig auditory nerve fibres: Distribution and correlation with spontaneous rate. Hear Res 71:69–79.

Gleisner L, Flock A, Wersall J (1973) The ultrastructure of the afferent synapse on hair cells in the frog labyrinth. Acta Otolaryngol 76:199–207.

Gold T (1948) Hearing. II. The physical basis of the action of the cochlea. Proc Roy Soc London B, Biological Sciences 135:492–498.

Hau LWT, Simmons DD, Narins PM (2004) Frequency-dependence of auditory-nerve latency in the northern leopard frog, *Rana pipiens pipiens*. Assoc for Research in Otolaryngology Abstract 380.

Hetherington TE, Jaslow AP, Lombard RE (1986) Comparative morphology of the amphibian opercularis system: I. General design features and functional interpretation. J Morphol 190:43–61.

Hillery CM, Narins PM (1984) Neurophysiological evidence for a traveling wave in the amphibian inner ear. Science 225:1037–1039.

Hillery CM, Narins PM (1987) Frequency and time domain comparison of low-frequency auditory fiber responses in two anuran amphibians. Hear Res 25:233–248.

Hödl W, Amezquita A, Narins PM (2004) The role of call frequency and the auditory papillae in phonotactic behavior in male dart-poison frogs *Epipedobates femoralis* (Dendrobatidae). J Comp Physiol A 190:823–829.

Hudspeth AJ (1997) How hearing happens. Neuron 19:947–950.
Hudspeth AJ, Corey DP (1977) Sensitivity, polarity, and conductance change in the response of vertebrate hair cells to controlled mechanical stimuli. Proc Natl Acad Sci USA 74:2407–2411.
Hudspeth AJ, Choe Y, Mehta AD, Martin P (2000) Putting ion channels to work: mechanoelectrical transduction, adaptation, and amplification by hair cells. Proc Natl Acad Sci USA 97:11765–11772.
Jia S, He DZ (2005) Motility-associated hair-bundle motion in mammalian outer hair cells. Nat Neurosci 8:1028–1034.
Kalluri R, Shera CA (2001) Distortion-product source unmixing: A test of the two-mechanism model for DPOAE generation. J Acoust Soc Am 109:622–637.
Kemp DT (1978) Stimulated acoustic emissions from within the human auditory system. J Acoust Soc Am 64:1386–1391.
Kiang NYS, Watanabe T, Thomas L, Clark L (1965) Discharge Patterns of Single Fibers in the Cat's Auditory Nerve. MIT Press, Cambridge, MA.
Knight RD, Kemp DT (2000) Indications of different distortion product otoacoustic emission mechanisms from a detailed f1, f2 area study. J Acoust Soc Am 107:457–473.
Knight RD, Kemp DT (2001) Wave and place fixed DPOAE maps of the human ear. J Acoust Soc Am 109:1513–1525.
Köppl C (1995) Otoacoustic emissions as an indicator for active cochlear mechanics: a primitive property of vertebrate auditory organs. In: Manley GA, Klump GM, Köppl C, Fastl H, Oeckinghaus H (eds.), Advances in Hearing Research, World Scientific, Singapore, pp. 207–216.
Köppl C, Authier S (1995) Quantitative anatomical basis for a model of micromechanical frequency tuning in the Tokay gecko, *Gekko gecko*. Hear Res 82:14–25.
Kössl M, Boyan G (1998) Acoustic distortion products from the ear of a grasshopper. J Acoust Soc Am 104:326–335.
Kössl M, Vater M (1996) Further studies on the mechanics of the cochlear partition in the mustached bat. II. A second cochlear frequency map derived from acoustic distortion products. Hear Res 94:78–86.
Kummer P, Janssen T, Hulin P, Arnold W (2000) Optimal L(1)–L(2) primary tone level separation remains independent of test frequency in humans. Hear Res 146:47–56.
Lenzi D, Runyeon JW, Crum J, Ellisman MH, Roberts WM (1999) Synaptic vesicle populations in saccular hair cells reconstructed by electron tomography. J Neurosci 19:119–132.
Lewis ER (1976) Surface morphology of the bullfrog amphibian papilla. Brain Behav Evol 13:196–215.
Lewis ER (1978) Comparative studies of the anuran auditory papillae. Scan Electron Microsc 11:633–642.
Lewis ER (1981) Suggested evolution of tonotopic organization in the frog amphibian papilla. Neurosci Lett 21:131–136.
Lewis ER (1984) On the frog amphibian papilla. Scan Electron Microsc Pt 4:1899–1913.
Lewis ER (1988) Tuning in the bullfrog ear. Biophys J 53:441–447.
Lewis ER, Leverenz EL (1983) Morphological basis for tonotopy in the anuran amphibian papilla. Scan Electron Microsc Pt 1:189–200.
Lewis ER, Li CW (1975) Hair cell types and distributions in the otolithic and auditory organs of the bullfrog. Brain Res 83:35–50.

Lewis ER, Lombard R (1988) The amphibian inner ear. In: Fritzsch B, Ryan M, Wilcynski W, Hetherington T, Walkowiak W (eds.) The Evolution of the Amphibian Auditory System. Wiley Interscience, New York, pp. 93–124.
Lewis ER, Narins PM (1999) The acoustic periphery of amphibians: Anatomy and physiology. In: Fay R, Popper A (eds.) Comparative Hearing: Fish and Amphibians. Springer, New York, pp. 101–154.
Lewis ER, van Dijk P (2004) New variation on the derivation of spectro-temporal receptive fields for primary auditory afferent axons. Hear Res 189:120–136.
Lewis ER, Baird RA, Leverenz EL, Koyama H (1982a) Inner ear: Dye injection reveals peripheral origins of specific sensitivities. Science 215:1641–1643.
Lewis ER, Hecht EI, Narins PM (1992) Diversity of form in the amphibian papilla of Puerto Rican frogs. J Comp Physiol [A] 171:421–435.
Lewis ER, Henry KR, Yamada WM (2002) Tuning and timing of excitation and inhibition in primary auditory nerve fibers. Hear Res 171:13–31.
Lewis ER, Leverenz EL, Bialek WS (1985) The Vertebrate Inner Ear. CRC, Boca Raton, FL.
Lewis ER, Leverenz EL, Koyama H (1982b) The tonotopic organization of the bullfrog amphibian papilla, an auditory organ lacking a basilar membrane. J Comp Physiol 145:437–445.
Liberman MC (1980) Morphological differences among radial afferent fibers in the cat cochlea: an electron-microscopic study of serial sections. Hear Res 3:45–63.
Liberman MC, Simmons DD (1985) Applications of neuronal labeling techniques to the study of the peripheral auditory system. J Acoust Soc Am 78:312–319.
Liberman MC, Dodds LW, Pierce S (1990) Afferent and efferent innervation of the cat cochlea: Quantitative analysis with light and electron microscopy. J Comp Neurol 301:443–460.
Liberman MC, Gao J, He DZ, Wu X, Jia S, Zuo J (2002) Prestin is required for electromotility of the outer hair cell and for the cochlear amplifier. Nature 419:300–304.
Liff HJ, Goldstein MH, Jr. (1970) Peripheral inhibition in auditory fibers in the frog. J Acoust Soc Am 47:1538–1547.
Lnenicka GA, Atwood HL, Marin L (1986) Morphological transformation of synaptic terminals of a phasic motoneuron by long-term tonic stimulation. J Neurosci 6:2252–2258.
Long GR, van Dijk P, Wit HP (1996) Temperature dependence of spontaneous otoacoustic emissions in the edible frog (*Rana esculenta*). Hear Res 98:22–28.
Lonsbury-Martin BL, McCoy MJ, Whitehead ML, Martin GK (1993) Clinical testing of distortion-product otoacoustic emissions. Ear Hear 14:11–22.
Maison S, Micheyl C, Collet L (1997) Medial olivocochlear efferent system in humans studied with amplitude-modulated tones. J Neurophysiol 77:1759–1768.
Manley GA (2001) Evidence for an active process and a cochlear amplifier in nonmammals. J Neurophysiol 86:541–549.
Manley GA, Kirk DL, Koppl C, Yates GK (2001) In vivo evidence for a cochlear amplifier in the hair-cell bundle of lizards. Proc Natl Acad Sci USA 98:2826–2831.
Mason MJ, Lin CC, Narins PM (2003) Sex differences in the middle ear of the bullfrog (*Rana catesbeiana*). Brain Behav Evol 61:91–101.
Meenderink SWF, van Dijk P (2004) Level dependence of distortion product otoacoustic emissions in the leopard frog, *Rana pipiens pipiens*. Hear Res 192:107–118.
Meenderink SWF, van Dijk P (2005a) Characteristics of distortion product otoacoustic emissions in the frog from L1, L2 maps. J Acoust Soc Am 118:279–286.

Meenderink SWF, van Dijk P (2005b) Temperature dependence of distortion product otoacoustic emissions in the frog. *in preparation.*

Meenderink SWF, Narins PM, van Dijk P (2005a) Detailed f1, f2 area study of distortion product otoacoustic emissions in the frog. J Assoc Res Otolaryngol 6:37–47.

Meenderink SWF, van Dijk P, Narins PM (2005b) Comparison between distortion product otoacoustic emissions and nerve fiber responses from the basilar papilla of the frog. J Acoust Soc Am 117:3165–3173.

Mills DM, Rubel EW (1994) Variation of distortion product otoacoustic emissions with furosemide injection. Hear Res 77:183–199.

Narins PM (1983) Synchronous vocal response mediated by the amphibian papilla in a neotropical treefrog: Behavioral evidence. J Exp Biol 105:95–105.

Narins PM, Capranica RR (1976) Sexual differences in the auditory system of the treefrog, *Eleutherodactylus coqui.* Science 192:378–380.

Nobili R, Mammano F, Ashmore J (1998) How well do we understand the cochlea? Trends Neurosci 21:159–167.

Owens JJ, McCoy MJ, Lonsbury-Martin BL, Martin GK (1992) Influence of otitis media on evoked otoacoustic emission in children. Seminars in Hearing 13:53–65.

Owens JJ, McCoy MJ, Lonsbury-Martin BL, Martin GK (1993) Otoacoustic emissions in children with normal ears, middle ear dysfunction, and ventilating tubes. Am J Otol 14:34–40.

Palmer AR, Wilson JP (1982) Spontaneous and evoked acoustic emissions in the frog *Rana esculenta.* J Physiol 324:66.

Pitchford S, Ashmore JF (1987) An electrical resonance in hair cells of the amphibian papilla of the frog *Rana temporaria.* Hear Res 27:75–83.

Probst R, Lonsbury-Martin BL, Martin GK (1991) A review of otoacoustic emissions. J Acoust Soc Am 89:2027–2067.

Purgue AP, Narins PM (2000a) Mechanics of the inner ear of the bullfrog (Rana catesbeiana): The contact membranes and the periotic canal. J Comp Physiol [A] 186: 481–488.

Purgue AP, Narins PM (2000b) A model for energy flow in the inner ear of the bullfrog (*Rana catesbeiana*). J Comp Physiol [A] 186:489–495.

Robbins RG, Bauknight RS, Honrubia V (1967) Anatomical distribution of efferent fibers in the 8th cranial nerve of the bullfrog (*Rana catesbeiana*). Acta Otolaryngol 64:436–448.

Robles L, Ruggero MA (2001) Mechanics of the mammalian cochlea. Physiol Rev 81: 1305–1352.

Ronken DA (1990) Basic properties of auditory-nerve responses from a "simple" ear: The basilar papilla of the frog. Hear Res 47:63–82.

Ronken DA (1991) Spike discharge properties that are related to the characteristic frequency of single units in the frog auditory nerve. J Acoust Soc Am 90:2428–2440.

Ronken DA, Bosch WR, Molnar CE (1993) Effects of spike discharge history on discharge probability and latency in frog basilar papilla units. Hear Res 69:55–75.

Rose GJ, Capranica RR (1985) Sensitivity to amplitude modulated sounds in the anuran auditory nervous system. J Neurophysiol 53:446–465.

Rosowski JJ, Peake WT, White JR (1984) Cochlear nonlinearities inferred from two-tone distortion products in the ear canal of the alligator lizard. Hear Res 13:141–158.

Rushton WA (1951) A theory of the effects of fibre size in medullated nerve. J Physiol 115:101–122.

Santos-Sacchi J (1991) Reversible inhibition of voltage-dependent outer hair cell motility and capacitance. J Neurosci 11:3096–3110.
Schetzen M (1989) The Volterra and Weiner Theories of Nonlinear Systems. Krieger, Malabar, FL.
Schneider S, Prijs VF, Schoonhoven R (2003) Amplitude and phase of distortion product otoacoustic emissions in the guinea pig in an (f1,f2) area study. J Acoust Soc Am 113:3285–3296.
Shera CA, Guinan JJ, Jr. (1999) Evoked otoacoustic emissions arise by two fundamentally different mechanisms: A taxonomy for mammalian OAEs. J Acoust Soc Am 105: 782–798.
Shofner WP, Feng AS (1983) A quantitative light microscopic study of the bullfrog amphibian papilla tectorium: Correlation with the tonotopic organization. Hear Res 11:103–116.
Shofner WP, Feng AS (1984) Quantitative light and scanning electron microscopic study of the developing auditory organs in the bullfrog: Implications on their functional characteristics. J Comp Neurol 224:141–154.
Simmons AM (1988) Masking patterns in the bullfrog (*Rana catesbeiana*). I: Behavioral effects. J Acoust Soc Am 83:1087–1092.
Simmons AM, Reese G, Ferragamo M (1993) Periodicity extraction in the anuran auditory nerve. II: Phase and temporal fine structure. J Acoust Soc Am 93:3374–3389.
Simmons DD, Narins PM (1995) Conduction velocity, fiber diameter and response latency in auditory nerve fivers of *Rana pipiens pipiens*: Toward temporal separation of coincidence? Proceedings of the 4th International Congress of Neuroethology. 347.
Simmons DD, Bertolotto C, Leong M (1994a) Ultrastructural reconstruction of auditory hair cells and their synapses in low and high frequency regions of the frog inner ear. Proc Int'l Cong Elctr Micrs 13:629–630.
Simmons DD, Bertolotto C, Narins PM (1994b) Morphological gradients in sensory hair cells of the amphibian papilla of the frog, *Rana pipiens pipiens*. Hear Res 80:71–78.
Simmons DD, Bertolotto C, Leong M (1995) Synaptic ultrastructure within the amphibian papilla of *Rana pipiens pipiens*: rostrocaudal differences. Auditory Neurosci 1:183–193.
Simmons DD, Bertolotto C, Narins PM (1992) Innervation of the amphibian and basilar papillae in the leopard frog: Reconstructions of single labeled fibers. J Comp Neurol 322:191–200.
Simmons DD, Burton MD, Hooper RN, Baird RA (2004) Hair cell damage and recovery from high-level noise exposures in the amphibian papilla of the bullfrog. Association for Research in Otolaryngology Abstract #365.
Smith CA, Sjostrand FS (1961) Structure of the nerve endings on the external hair cells of the guinea pig cochlea as studied by serial section. J Ultrastruc Res 5:184–192.
Smith RS, Koles ZJ (1970) Myelinated nerve fibers: Computed effect of myelin thickness on conduction velocity. Am J Physiol 219:1256–1258.
Smotherman MS, Narins PM (1999a) The electrical properties of auditory hair cells in the frog amphibian papilla. J Neurosci 19:5275–5292.
Smotherman MS, Narins PM (1999b) Potassium currents in auditory hair cells of the frog basilar papilla. Hear Res 132:117–130.
Smotherman MS, Narins PM (2000) Hair cells, hearing and hopping: A field guide to hair cell physiology in the frog. J Exp Biol 203:2237–2246.
Sobkowicz HM, Rose JE, Scott GL, Levenick CV (1986) Distribution of synaptic ribbons in the developing organ of Corti. J Neurocytol 15:693–714.

Stiebler IB, Narins PM (1990) Temperature-dependence of auditory nerve response properties in the frog. Hear Res 46:63–81.

Stover L, Gorga MP, Neely ST, Montoya D (1996) Toward optimizing the clinical utility of distortion product otoacoustic emission measurements. J Acoust Soc Am 100: 956–967.

Sugihara I, Furukawa T (1989) Morphological and functional aspects of two different types of hair cells in the goldfish sacculus. J Neurophysiol 62:1330–1343.

van Dijk P (1995) Polynomial correlation used to estimate the degree of nonlinearity of the frog inner ear system. J Acoust Soc Am 97:3414.

van Dijk P, Manley GA (2001) Distortion product otoacoustic emissions in the tree frog *Hyla cinerea*. Hear Res 153:14–22.

van Dijk P, Wit HP (1987) Temperature dependence of frog spontaneous otoacoustic emissions. J Acoust Soc Am 82:2147–2150.

van Dijk P, Wit HP (1990) Synchronization of spontaneous otoacoustic emissions to a 2f1–f2 distortion product. J Acoust Soc Am 88:850–856.

van Dijk P, Wit HP, Segenhout JM (1989) Spontaneous otoacoustic emissions in the European edible frog (*Rana esculenta*): Spectral details and temperature dependence. Hear Res 42:273–282.

van Dijk P, Lewis ER, Wit HP (1990) Temperature effects on auditory nerve fiber response in the American bullfrog. Hear Res 44:231–240.

van Dijk P, Mason MJ, Narins PM (2002) Distortion product otoacoustic emissions in frogs: correlation with middle and inner ear properties. Hear Res 173:100–108.

van Dijk P, Narins PM, Mason MJ (2003) Physiological vulnerability of distortion product otoacoustic emissions from the amphibian ear. J Acoust Soc Am 114:2044–2048.

van Dijk P, Narins PM, Wang J (1996) Spontaneous otoacoustic emissions in seven frog species. Hear Res 101:102–112.

van Dijk P, Wit HP, Segenhout JM (1997) Dissecting the frog inner ear with Gaussian noise. II. Temperature dependence of inner ear function. Hear Res 114:243–251.

van Dijk P, Wit HP, Segenhout JM, Tubis A (1994) Wiener kernel analysis of inner ear function in the American bullfrog. J Acoust Soc Am 95:904–919.

Vassilakis PN, Meenderink SWF, Narins PM (2004) Distortion product otoacoustic emissions provide clues to hearing mechanisms in the frog ear. J Acoust Soc Am 116:3713–3726.

Von Békésy G (1960) Experiments in Hearing. Acoustical Society of America Press, New York.

Walrond JP, Reese TS (1985) Structure of axon terminals and active zones at synapses on lizard twitch and tonic muscle fibers. J Neurosci 5:1118–1131.

West CD (1985) The relationship of the spiral turns of the cochlea and the length of the basilar membrane to the range of audible frequencies in ground dwelling mammals. J Acoust Soc Am 77:1091–1101.

Wever EG (1973) The ear and hearing in the frog, *Rana pipiens*. J Morphol 141:461–477.

Wever EG (1985) The Amphibian Ear. Princeton University, Princeton, NJ.

Whitehead ML, Lonsbury-Martin BL, Martin GK (1992) Evidence for two discrete sources of 2f1-f2 distortion-product otoacoustic emission in rabbit. II: Differential physiological vulnerability. J Acoust Soc Am 92:2662–2682.

Whitehead ML, Wilson JP, Baker RJ (1986) The effects of temperature on otoacoustic emission tuning properties. In: Moore B, Patterson R (eds.) Auditory Frequency Selectivity. Plenum, New York, pp. 39–48.

Wilczynski W, Keddy-Hector AC, Ryan MJ (1992) Call patterns and basilar papilla tuning in cricket frogs. I. Differences among populations and between sexes. Brain Behav Evol 39:229–237.

Wilczynski W, McClelland BE, Rand AS (1993) Acoustic, auditory, and morphological divergence in three species of neotropical frog. J Comp Physiol 172:425–438.

Will U, Fritzsch B (1988) The eighth nerve of amphibians. In: Fritzsch B, Ryan M, Walczynski W, Hetherington T, Walkowiak W (eds.) The Evolution of the Amphibian Auditory System. Wiley-Interscience, New York, pp. 159–183.

Zakon HH, Capranica RR (1981) An anatomical and physiological study of regeneration of the eighth nerve in the leopard frog. Brain Res 209:325–338.

Zakon HH, Wilczynski W (1988) The physiology of the anuran eighth nerve. In: Fritzsch B, Ryan M, Walczynski W, Hetherington T, Walkowiak W (eds.) The Evolution of the Amphibian Auditory System. Wiley-Interscience, New York, pp. 125–155.

Zelick R, Narins PM (1985) Temporary threshold shift, adaptation, and recovery characteristics of frog auditory nerve fibers. Hear Res 17:161–176.

Zenisek D, Davila V, Wan L, Almers W (2003) Imaging calcium entry sites and ribbon structures in two presynaptic cells. J Neurosci 23:2538–2548.

Zhang M, Abbas PJ (1997) Effects of middle ear pressure on otoacoustic emission measures. J Acoust Soc Am 102:1032–1037.

Zheng J, Shen W, He DZ, Long KB, Madison LD, Dallos P (2000) Prestin is the motor protein of cochlear outer hair cells. Nature 405:149–155.

8
Central Auditory Pathways in Anuran Amphibians: The Anatomical Basis of Hearing and Sound Communication

WALTER WILCZYNSKI AND HEIKE ENDEPOLS

1. Introduction

The auditory system is more clearly tied physiologically and behaviorally to social communication in anuran amphibians than in any other vertebrate. This, coupled with the phylogenetic position of amphibians within vertebrates, their developmental transition from aquatic to terrestrial hearing, and the curious mixture of primitive and derived characters marking anurans has made the central auditory system a target of many neuroanatomical investigations. The results have shown that auditory connections within the anuran central nervous system are very widespread, a characteristic consistent with the importance of acoustic signals in guiding all aspects of anuran social behavior. Paradoxically, as auditory pathways move to higher brain areas, they progressively lose the purely auditory nature of their anatomy. The result is a gradual change in the nature of central auditory structures, moving from the familiar sensory organization of brainstem auditory areas to a forebrain system of widespread auditory connections to motor, endocrine, and a variety of limbic structures without an obvious functional analogue to telencephalic pallial sensory areas seen in mammals and birds. The largest component of the auditory system, the midbrain torus semicircularis, serves as a key point in the central auditory pathways integrating ascending auditory and descending forebrain inputs as a transition from the lower brainstem auditory areas and its forebrain targets, and as an audiomotor interface. Sections through the frog brain illustrating auditory and other nuclei are shown in Figure 8.1, and components of the ascending and descending auditory pathways are diagrammed in Figure 8.2. In this chapter, we treat three levels of the central auditory system—lower brainstem, midbrain, and forebrain—separately as we review the structure of the central auditory system.

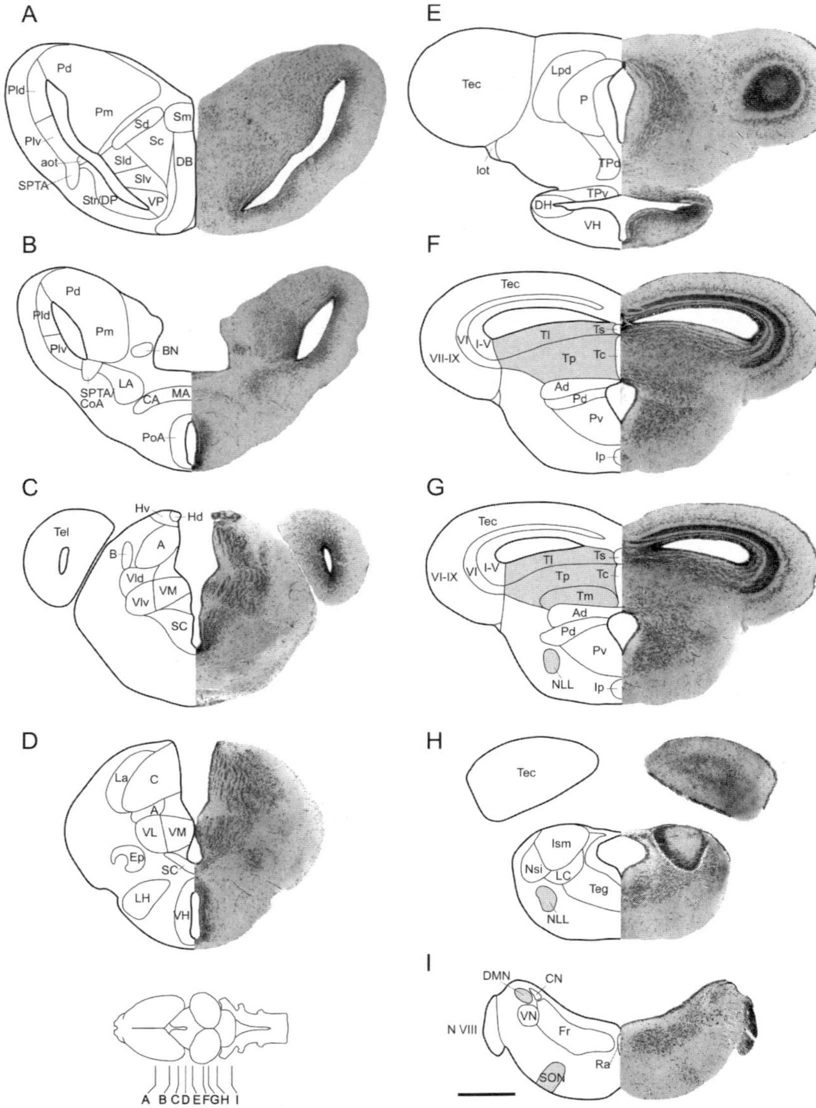

FIGURE 8.1. Frontal sections through the *Hyla versicolor* brain; Nissl stained photographs are on the right of each section, with the nuclei outlined and labeled on the left. Brainstem auditory nuclei are shaded. Nearly all diencephalic and telencephalic areas received some auditory input, hence none are distinguished by shading. Scale bar: 500 µm.

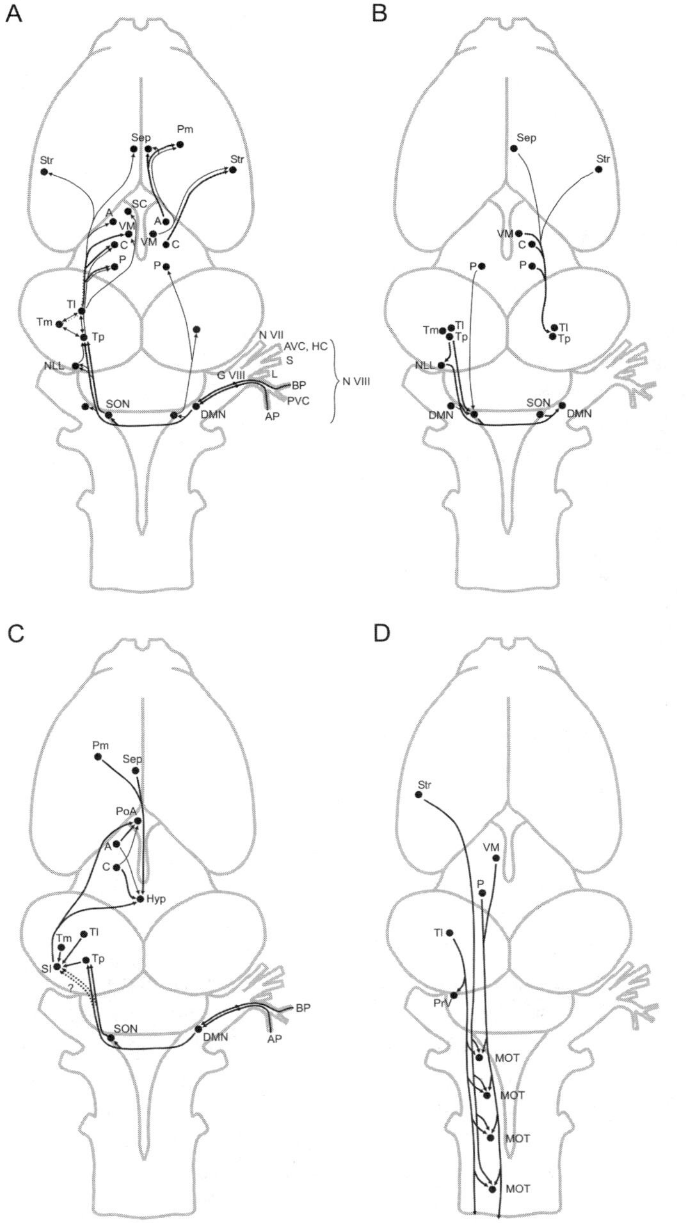

FIGURE 8.2. Schematic drawing of auditory connections. A. Ascending auditory pathway. B. Descending auditory pathway. C. Audioendocrine connections. D. Descending audiomotor connections.

2. Lower Brainstem Auditory Regions

The organization of brainstem auditory systems is similar to that of other vertebrates, although anurans present several special problems for understanding the organization of auditory centers. Some anurans possess a mechanical lateral line system, with nuclei and central tracts that ascend to the midbrain in parallel with the auditory fibers. A second problem is that anurans have several inner ear auditory papillae that are sensitive to acoustic stimuli of different, but overlapping, frequencies, the amphibian papilla (AP), basilar papilla (BP), sacculus, and lagena. The latter two are also sensitive to vibratory stimuli and their central projections tie them to the vestibular system as well as to auditory areas. This complicates deciding what constitutes an "auditory nucleus," especially at the level of the first central synapse, and therefore how many primary auditory nuclei should be recognized. Nevertheless, the basic organization of amphibian brainstem auditory pathways reflects that of a conserved tetrapod octavolateralis system that contains auditory, vestibular, and lateral line primary nuclei in a dorsolateral region of the medulla, with each forming separate efferent pathways. For the auditory component, afferents traveling in the eighth cranial nerve terminate ipsilaterally on neurons in the upper medulla, as do the vestibular fibers, which target their own set of primary nuclei.

2.1 Primary Nuclei

Anurans are most generally recognized as having a single primary auditory nucleus in the dorsal lateral medulla at the entrance of the eighth cranial nerve (Fig. 8.2). This nucleus has been variously termed the dorsal nucleus (Larsell 1934; Gregory 1974; Feng and Lin 1996; now not used so as to avoid confusion with an electrosensory nucleus of the same name in fish; Will and Fritzsch 1988), the dorsolateral nucleus (Wilczynski 1988; Will et al. 1988), and the dorsal medullary nucleus (DMN; Will et al. 1985a,b; Feng 1986a), which is the term we use here.

The DMN sits within a complex of nuclei, the octavolateralis area, in the dorsolateral medulla. Vestibular nuclei lie below it (the ventral nucleus, sometimes subdivided into three nuclei; Gregory 1972; Will et al. 1985b, Will 1988), rostral to it (the anterior nucleus; Nikundiwe and Nieuwenhuys 1983; Will et al. 1985b, Will 1988) and medial and caudal to it (caudal nucleus; Opdam et al. 1976; Will 1988). Lateral line nuclei are present in the medial and rostral areas when a lateral line system is preserved, for example, in the pipids (e.g., *Xenopus*) and the genus *Bombina* (Fritzsch et al. 1984; Will et al. 1985a,b; Will and Fritzsch 1988). Dendrites of DMN cells are mainly oriented in the rostrocaudal direction (Gregory 1974; Will et al. 1985b). Feng and Lin (1996) have published the only comprehensive cytoarchitecture study of the DMN, reporting a large number of distinct cell types based on soma size and dendritic morphology. They concluded that many types were similar to those found in mammalian cochlear nuclei, such as "bushy," "octopus," "radiate," and "giant" cells, although there is no indication that these are connectionally or physiologically equivalent to the similarly named mammalian cells. The larger cells appear to be the efferent neurons of the DMN (Will et al. 1985b; Feng and Lin 1996). Smaller cells are uniformly distributed throughout the DMN, but the various large cell types are not (Fig. 8.3). For example, around the boundaries of the DMN, Feng and Lin (1996) found fusiform (or bipolar) cells mainly in the ventrolateral portion of the nucleus, octopus cells medially, and bushy cells both laterally and medially, whereas stellate, giant, and small round neurons occupied the central region of the nucleus. Histochemical features also differentiate parts of the DMN (Fig. 8.3). Gamma-aminobutyric acid- (GABA-) immunoreactive cells are concentrated in the medial and ventral areas of the DMN and terminal puncta are seen throughout it (Simmons and Chapman 2002). NADPH-diaphorase positive cells are concentrated mainly in the medial parts of the DMN (Muñoz et al. 1996), whereas acetylcholinesterase positive cells are found throughout the nucleus (Hall and Bunker 1994).

The variation in cell types within the DMN is not reflected in any obvious way in the termination of eighth nerve auditory afferents. Incoming eighth nerve fibers distribute throughout the DMN (Gregory 1972; Matesz 1979; Aitken 1981; Will et al. 1985a), with most fibers bifurcating into ascending and descending branches (Will and Fritzsch 1988). The branching in amphibians is not functionally equivalent to the trifurcation of mammalian eighth nerve fibers, which ultimately construct three distinct tonotopic maps in three cochlear nuclei. Rather, there appears to be a single auditory projection field in frogs constructed from the combined inputs from the peripheral end organs, forming a ventrolateral-to-dorsomedial tonotopic map (Fig. 8.4A). Afferents from the AP (low and mid frequencies) and

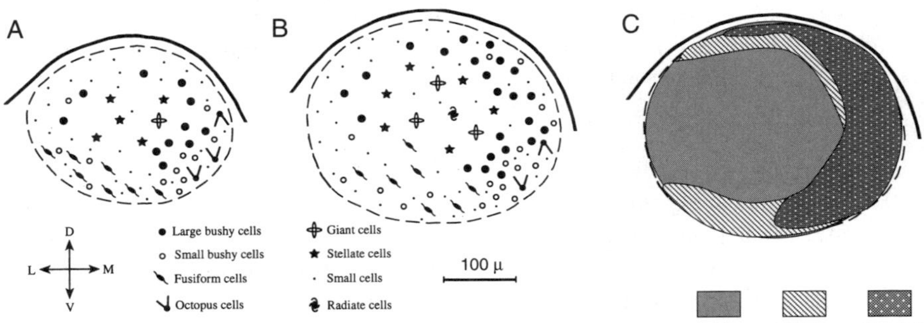

FIGURE 8.3. Cytoarchitecture and chemoarchitecture of the anuran DMN. A, B. Distribution of cell types in the rostral and caudal DMN (from Feng and Lin 1996). C. Location of neurochemically defined cell bodies as described by various authors: acetylcholinesterase (AchE) cells are distributed throughout the nucleus (described in Hall and Bunker 1994); GABA (illustrated in Simmons and Chapman 2002) and NADPH-diaphorase (illustrated in Muñoz et al. 1996) are more restricted (fibers and terminals of both are distributed throughout the nucleus).

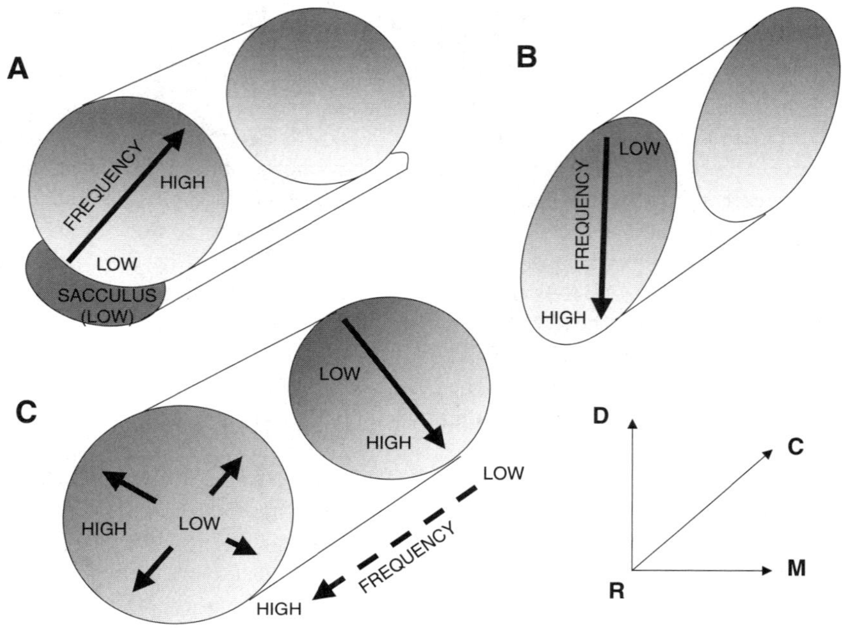

FIGURE 8.4. Tonotopic organization of brainstem auditory areas. A. DMN map from combined AP and BP input, with low-frequency saccular input ventral to the DMN's low-frequency representation. The tonotopic map appears to extend in the caudal–rostral plane as isofrequency contours. B. SON map; as for the DMN, the map extends in the caudal–rostral plane as isofrequency contours. C. Torus frequency organization is more complex. Rostrally, high frequencies predominate although a core area of low-frequency representation is present; caudally, low frequencies predominate, although high-frequency sensitivity is found in the ventromedial area. Note, however, that multiple frequency sensitivities are found in every toral region, and the frequency representations do not correspond clearly to the boundaries of the toral nuclei.

BP (high frequencies) enter with AP fibers more dorsal than BP fibers in the nerve (Fuzessery and Feng 1981; Will and Fritzsch 1988). BP fibers terminate in a restricted dorsomedial region of the nucleus, whereas the far more numerous AP fibers spread throughout the DMN including dorsal and ventral to the BP terminal area (Lewis et al. 1980; Fuzessery and Feng 1981; Will et al. 1985a). There is some evidence that the tonotopic map of AP hair cells is preserved in its primary fiber projections onto the DMN (Lewis et al. 1980; Fuzessery and Feng 1981; Fuzessery 1988). The various descriptions of physiological tonotopy and AP/BP terminations suggest that the topographic organization in the mediolateral–dorsoventral plane extends in the rostrocaudal axis as isofrequency slabs or bands. As incoming AP afferents travel in the rostrocaudal direction, they periodically send branches medially into the DMN to form terminal arbors (Lewis et al. 1980).

These may represent terminations in different regions within one isofrequency slab.

Fibers originating from the low-frequency sacculus terminate separately just ventral to the DMN. Matesz (1979) recognized this region as a separate nucleus saccularis, but it is more likely a separate part of the DMN neuropil (Will et al. 1985b; Will 1988). Lagena afferents terminate in an adjacent region of the neuropil. Therefore, despite the variation in cell type and neurochemical features within the DMN, the terminations within it suggest a relatively uniform, tonotopic auditory terminal field distributed throughout the nucleus, constructed from the input of four different end organs with different frequency representations.

One complication of this is that the sacculus and lagena also provide input to the vestibular nuclei ventral and medial to the DMN (Will et al. 1985a; Will and Fritzsch 1988). In fact, roughly a quarter of the vestibular neurons there are sensitive to auditory stimulation, including frequencies usually associated with the AP and BP (Bricout-Berthout et al. 1984), which may indicate communication among octavolateralis nuclei as well as direct saccular and lagenar input. Although it would not be appropriate to consider these other areas "auditory" like the DMN, it does indicate that auditory and vestibular functions are not strictly separated in anurans.

Two efferent pathways ascend from the DMN (Fig. 8.2A), similar to the dual ascending auditory pathways from the mammalian cochlear nuclei. One is a predominantly crossed connection to the torus semicircularis of the midbrain (Pettigrew 1981; Wilczynski 1981; Will et al. 1985b; Feng 1986a; Edwards and Kelley 2001; see description later in this chapter). Axons in this pathway provide collaterals to neurons within the lateral lemniscus at midbrain levels, which may represent a nucleus of the lateral lemniscus (Feng 1986a). The second is a bilateral connection to the superior olivary nucleus, formed by collaterals of the ascending fibers to the midbrain as well as axons that terminate there (Will et al. 1985b; Feng 1986a). There are also reciprocal commissural connections between the left and right DMNs (Grofová and Corvaja 1972; Will et al. 1985b; Feng 1986a), resulting in a binaural interaction already on the level of the first auditory nucleus. Descending input arises from three higher auditory centers, the torus semicircularis, superior olivary nucleus, and nucleus of the lateral lemniscus (Feng 1986a; Wilczynski 1988; Feng and Lin 1991; Matesz and Kulik 1996). Descending inputs are mainly ipsilateral (Fig. 8.2B).

Although the general pattern of DMN efferents is similar to that of mammals, there is no indication of any differential sources within the DMN of the major ascending projections, and no clear indication how the different DMN cell types identified by Feng and Lin (1996) might contribute to the projections; in fact remarks in a number of papers suggest that all, except perhaps the smallest, cells in the DMN are efferent neurons (Will et al. 1985b; Feng 1986b). Rather, it seems more the case that there is a single, roughly topographic projection from the one primary auditory nucleus to both the torus and superior olivary nucleus.

2.2 Superior Olivary Nucleus

In anurans a superior olivary nucleus (SON) is embedded in the white matter of the ventral medulla centered at the level of the entrance of the eighth cranial nerve and DMN (Fig. 8.2). It is bounded dorsally and ventrally by the crossing axons of DMN and vestibular nucleus neurons forming arcuate commissural connections, the lateral lemniscus, and DMN fibers to the SON. As for the primary auditory nucleus, the SON is a single nucleus without obvious subdivisions. There are no associated periolivary nuclei or nucleus of the trapezoid body. The SON shares with the DMN the characteristic of having GABA (Simmons and Chapman 2002) and acetylcholinesterase (Hall and Bunker 1994) containing neurons and terminals throughout the nucleus. Noteworthy also is a considerable dopaminergic input to the SON whereas noradrenergic fibers are absent (González and Smeets 1991, 1993). To date, however, there has been no systematic study of the morphology of SON cells and hence no known cytoarchitectonic or histochemical basis for regional differentiation of the SON.

The auditory inputs to the SON from the DMN seem to argue against any hidden subdivisions in the SON. The SON receives bilateral projections from the DMN, with heavier input from the contralateral side; as there is no nucleus of the trapezoid body, all connections are direct (Will et al. 1985b; Feng 1986b; Wilczynski 1988). The projection is a single, topographically organized, field representing a tonotopic organization (Fig. 8.4B), with higher frequencies ventrally and lower frequencies dorsally in the nucleus (Feng 1986b; Fuzessery 1988). These extend in the rostrocaudal dimension as isofrequency slabs (Feng 1986b). There is a small, topographic commissural connection between the SONs (Rubinson and Skiles 1975; Feng 1986b). Like the DMN, the SON receives input from auditory centers above it, the torus and nucleus of the lateral lemniscus (Feng 1986b).

The major output of the SON is a bilateral, but predominantly ipsilateral, ascending connection to the torus semicircularis (Rubinson and Skiles 1975; Wilczynski 1981, Feng 1986b; Edwards and Kelley 2001; Fig. 8.2A). As for other parts of the amphibian auditory system, the connection largely preserves topography, and hence tonotopy. Collaterals of these ascending fibers terminate in the nucleus of the lateral lemniscus (Feng 1986b). A small fiber bundle reaches the caudal thalamus (Rubinson and Skiles 1975; Feng et al. 1986b).

2.3 Lateral Lemniscus and Its Associated Nucleus

Ascending fibers from the DMN and SON gather into a lateral lemniscus that occupies a ventrolateral position forward of the SON. It gradually moves dorsally, maintaining its lateral position until entering the torus at its lateral edge.

Cells lateral to the isthmal and midbrain tegmentum, lying within the lateral lemniscus itself, were proposed to be homologous to the mammalian nuclei of the lateral lemniscus (NLL) by Röthig (1927) and Larsell (1934) (Figs. 8.1G,H). This area was given the more neutral name "superficial reticular nucleus" by

Opdam et al. (1976). It does receive collaterals of ascending DMN and SON fibers and connects reciprocally to them (Feng 1986a,b), projects to the torus (Wilczynski 1981), and its neurons are responsive to acoustic stimuli (Rose and Wilczynski 1984). Like the lower brainstem auditory centers, this amphibian NLL is a single nucleus without obvious subdivision. Also like those lower centers, it contains GABA (Simmons and Chapman 2002) and acetylcholinesterase (Hall and Bunker 1994) positive cells.

3. The Auditory Midbrain: The Torus Semicircularis

The torus semicircularis (Figs. 8.1F,G) is a major integrative structure in the anuran brain in which ascending auditory projections from all lower brainstem areas (Fig. 8.2A) and descending input from the forebrain (Fig. 8.2B) converge (Wilczynski 1981; Endepols and Walkowiak 2001). Efferents from the torus are extensive, with ascending projections reaching multiple thalamic nuclei as well as the parts of the subpallial telencephalon and descending projections to lower auditory and motor areas (Fig. 8.2). It is also rich in terminations containing neuromodulatory peptides and neurotransmitters, and it contains steroid hormone binding sites.

Unlike the lower amphibian auditory levels, the torus is composed of several nuclei. However, the nuclei are organized differently than one sees in its mammalian homologue, the inferior colliculus. All nuclei share inputs from a common terminal field formed by the ascending auditory fibers. Their outputs are more widely distributed than are those of the mammalian or avian colliculi and more overlapping in their targets. There is physiological evidence of tonotopy in the torus, but there are no simple anatomical correlates of this, and physiological indications that in many places frequency information is combined. Descriptions from the literature (reviewed in Fuzessery 1988) indicate that there is a preponderance of high-frequency activity rostrally and low-frequency activity caudally, although all frequencies appear to be represented at all levels (Fig. 8.4C). Rostrally, low-frequency activity is most apparent in the central core of the torus, expanding in area caudally so that high frequencies in the caudal torus are apparent mainly in the ventromedial region. Curiously, there is no evidence for a topographic projection from any toral subdivision to any thalamic nucleus. This suggests a very different functional neuroanatomy within the auditory system as it progresses beyond the midbrain.

3.1 Nuclear Organization

The anuran torus semicircularis comprises three main subnuclei: the laminar, principal, and magnocellular nuclei (Potter 1965). The laminar nucleus is located directly beneath the tectal ventricle and covers nearly the entire dorsal and rostral surface of the torus. It consists of parallel sheets of cells alternating with fiber layers. The cell layers are thicker and more numerous (four to six layers up to four cells thick) in the lateral and caudal part of the nucleus, whereas there are only one or two thin layers in the medial and rostral part. Laterally, the laminar nucleus is continuous with layers 1 to 5 of the optic tectum, and it touches the anterior tegmentum rostrally.

The principal nucleus lies ventral and caudal to the laminar nucleus. Its cells have been described as arranged in clusters (Potter 1965) or circular layers (Feng 1983), and are continuous with the thick cell lamina 6 of the optic tectum (Wilczynski 1988). Cell density is highest in the rostral and dorsal part of the nucleus (Feng 1983; Wilczynski and Capranica 1984). In ranid frogs, a cell-sparse area ("ventral toral zone") between the principal nucleus and the tegmentum can be visible, which may be treated as a separate toral area rather than part of the principal nucleus (Wilczynski 1988; Hoke et al. 2004). In other species, for example, in hylid frogs, the neurons of the principal nucleus are distributed more homogeneously; a ventral cell-sparse zone is not present. The most lateral portion of the ventral principal nucleus/ventral toral zone was named by Adli et al. (1999), according to its suggested mammalian counterpart, the reticular cuneiform nucleus, although this nomenclature has not been widely adopted.

The magnocellular nucleus is located in the caudal part of the ventral torus and consists of large scattered cells. It is often illustrated as being lateral in the torus, and this is true for the rostral portion of the magnocellular nucleus. More caudally, however, the large cells extend into the medial torus, almost to the midline. Additionally, two other subnuclei, the commissural and subependymal nucleus, are described (Potter 1965). The subependymal nucleus lies in a midline position between the left and right laminar nuclei; its cells are small and densely packed. The commissural nucleus is located around the midline as well, between the left and right principal nuclei. In ranid frogs, the commissural nucleus is described as a cell-sparse region (Potter 1965; Feng 1983), but can contain numerous cell bodies in other frog groups. The physiological properties of these areas remain obscure, although Hoke et al. (2004) noted immediate early gene activation in this area after acoustic stimulation.

Although some authors describe a similar nuclear arrangement in *Xenopus* (Nikundiwe and Nieuwenhuys 1983; Lowe 1986), other studies propose that the organization of toral nuclei is somewhat different compared to other frog species (Kelley 1981; Edwards and Kelley 2001). According to these latter studies, the laminar nucleus lies caudal to the tectal ventricle and ventral to the caudal part of the optic tectum; its straight cell layers extend dorsoventrally. The principal nucleus is located caudal to the laminar nucleus, and occupies the entire dorsal mesencephalon there. The laminar and principal nuclei are separated from their contralateral counterparts by a cell-sparse zone that may be comparable to the commissural nucleus. The large cells of the magnocellular nucleus can be found only in the lateral torus, ventral to the laminar and principal nuclei. In advanced anurans such as ranids and hylids, the torus comes to occupy its position ventral to the optic tectum as it rotates beneath it from a position caudal to the tectum (the position of the homologous mammalian inferior colliculus) as it enlarges.

The different orientation in *Xenopus* may simply reflect less expansion and/or ventrorostral rotation.

3.2 Cell Types, Immunohistochemistry, Neurotransmitters, and Receptors

Unipolar, bipolar, and multipolar neurons with spherical, ovoidal, or triangular-shaped cell bodies have been described throughout the torus semicircularis (Feng 1983). In the laminar nucleus, cell bodies are mostly medium sized (8 × 10 μm to 9 × 15 μm; Luksch and Walkowiak 1998). Dendrites can be smooth or spiny; they run either parallel or perpendicular to the laminae reaching toward the center of the torus. Frequently, dendrites extend into the principal nucleus, and sometimes they enter the tegmentum rostrally. The complex chemoarchitecture of the torus is summarized in Figure 8.5. Laminar nucleus neurons are immunoreactive

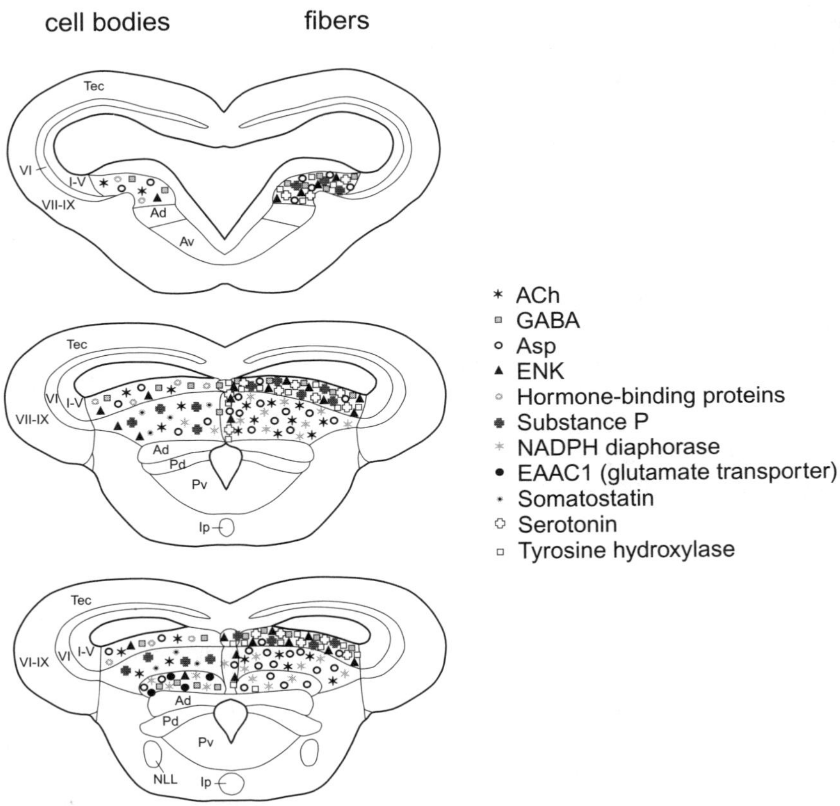

FIGURE 8.5. Chemoarchitecture of the torus semicircularis showing the distributions of various cell types.

for the transmitters acetylcholine, GABA, and aspartate (Hall and Bunker 1994; Endepols et al. 2000; Simmons and Chapman 2002), and possess receptors for GABA-A and kainic acid (Dechesne et al. 1990; Aller et al. 1997). Additionally, the laminar nucleus is rich in neuropeptides: enkephalin, neuromedin U, neuropeptide Y, and galanin have been found in laminar nucleus neurons (Merchenthaler et al. 1989; Lázár et al. 1991, 1993; Maderdrut et al. 1996; Adli et al. 1999, Endepols et al. 2000). An exceptional feature of laminar nucleus neurons is that they concentrate sex steroids such as estradiol and dihydrotestosterone (Morell et al. 1975; Kelley 1980, 1981; di Meglio et al. 1987), and estrogen-binding protein is present in the cell nuclei (Endepols et al. 2000). After estrogen priming, laminar nucleus neurons develop cytosolic progestin receptors (Roy et al. 1986).

In the principal nucleus, cell bodies are medium sized ($9 \times 10\,\mu m$) or small ($6 \times 9\,\mu m$; Luksch and Walkowiak 1998). Medium-sized neurons form clusters in the medial part of the principal nucleus. Their spiny dendrites mainly run perpendicular to the layers of the laminar nucleus and cover large areas within the entire torus. The small neurons are more numerous than the medium-sized ones and are distributed throughout the principal nucleus. Their mostly smooth dendrites are restricted to a small area (Luksch and Walkowiak 1998) and can have radiating patterns without obvious orientational preference (Feng 1983). In *Xenopus*, dendrites of principal nucleus neurons can extend into the laminar nucleus (Edwards and Kelley 2001), which has not been described in other species. Principal nucleus neurons are immunoreactive for acetylcholine, aspartate (mainly in the ventral part), and substance P (Inagaki et al. 1981; Hall and Bunker 1994; Adli et al. 1999; Endepols et al. 2000). Nitric oxide may be used as a retrograde messenger there (Brüning and Mayer 1996; Lázár and Losonczy 1999). Neuropeptides found in principal nucleus neurons include pituitary adenylate cyclase-activating polypeptide (Yon et al. 1992), somatostatin (Tostivint et al. 1996), proneuropeptide Y (Lázár et al. 1993), and galanin (Lázár et al. 1991).

In the magnocellular nucleus, two types of neurons with large ($14 \times 18\,\mu m$) and very large ($16 \times 20\,\mu m$) somata have been described (Luksch and Walkowiak 1998). The first cell type is located in the medial portions of the magnocellular nucleus, and its spiny dendrites cover large areas of the caudal magnocellular and principal nuclei. The second cell type is more frequent in the lateral magnocellular nucleus, and the large, often spiny dendrites are mainly found in the ventral torus. Occasionally, they can extend into the tegmentum. Magnocellular nucleus neurons use GABA, aspartate (only in the lateral part), and acetylcholine as neurotransmitters (Hall and Bunker 1994; Endepols et al. 2000), and they possess GABA-A receptors (Aller et al. 1997). NADPH-diaphorase activity indicates that cells may use nitric oxide as a retrograde messenger (Brüning and Mayer 1996; Lázár and Losonczy 1999). Neuropeptides have not yet been described in the magnocellular nucleus.

3.3 Afferents and Their Organization

Inputs from auditory, other sensory, midbrain, and forebrain sources converge onto the toral nuclei (Figs. 8.2A, 8.6). Auditory input arises mainly in the contralateral primary auditory nuclei (DMN), as well as in the ipsilateral superior olive and nucleus of the lateral lemniscus (Wilczynski 1981). The principal nucleus is the main target of ascending auditory fibers (Feng and Lin 1991; Kulik et al. 1994; Walkowiak and Luksch 1994; Matesz and Kulik 1996), whereas the laminar and magnocellular nuclei receive a weaker input. Where a cell-sparse ventral zone is apparent, the bulk of ascending lemniscal fibers terminate there (Wilczynski 1981, 1988). Because laminar and magnocellular nucleus neurons extend their dendrites into the principal nucleus, they most likely receive direct auditory input as well. In *Xenopus*, the laminar nucleus is strongly innervated by axons of the contralateral DMN and the ipsilateral SON (Edwards and Kelley 2001).

The torus semicircularis also receives ascending afferents from other sensory systems (Fig. 8.6). Somatosensory input arises in the contralateral dorsal column nucleus which projects mainly to the lateral part of the laminar nucleus, but also to the principal and magnocellular nuclei, and (to a lesser extent) to the commissural nuclei via the medial lemniscus (Muñoz et al. 1994, 1995). Furthermore, laminar and magnocellular toral nuclei are innervated by fibers arising in the ventral part of the contralateral dorsal horn at cervical spinal cord levels, running through the ventral and ventrolateral funiculi (Muñoz et al. 1997). A weak projection from the thoracic and lumbar levels has been found as well. Projections from the vestibular system to the principal, magnocellular, and laminar toral nuclei arise in the contralateral ventral and caudal octaval nuclei (Wilczynski 1981; Will et al. 1985b; ten Donkelaar 1998). In *Xenopus*, where the lateral line system is preserved, the contralateral lateral line nucleus projects to lateral parts of the principal and magnocellular nuclei (Will et al. 1985b; Edwards and Kelley 2001).

Toral afferents from the midbrain arise in the ipsilateral nucleus isthmi, which is comparable to the parabigeminal nucleus of mammals (Kulik and Matesz 1997), and all tegmental areas (Feng and Lin 1991). Layer 7 of the optic tectum projects to the principal nucleus (Matesz and Kulik 1996). In addition, all toral nuclei are reciprocally connected to their contralateral counterparts and to other ipsilateral toral subdivisions (Feng and Lin 1991; Matesz and Kulik 1996).

Descending input reaches the torus from several prosencephalic areas, terminating mainly in the laminar and principal nuclei (Figs. 8.2B, 8.6). The central, posterior, posterolateral, and ventromedial thalamic nuclei (Figs. 8.1C,D,E) project to the principal and laminar nuclei. Strong inputs arise from the anterior entopeduncular nucleus and the dorsal/magnocellular part of the suprachiasmatic nucleus (Wilczynski 1981; Feng and Lin 1991; Matesz and Kulik 1996; Edwards and Kelley 2001). Connections from the hypothalamus (*Rana*: Wilczynski 1981; Neary 1988) and anterior preoptic area (*Xenopus*: Edwards and Kelley 2001) terminate in the laminar nucleus. Some fibers also descend from the lateral septal

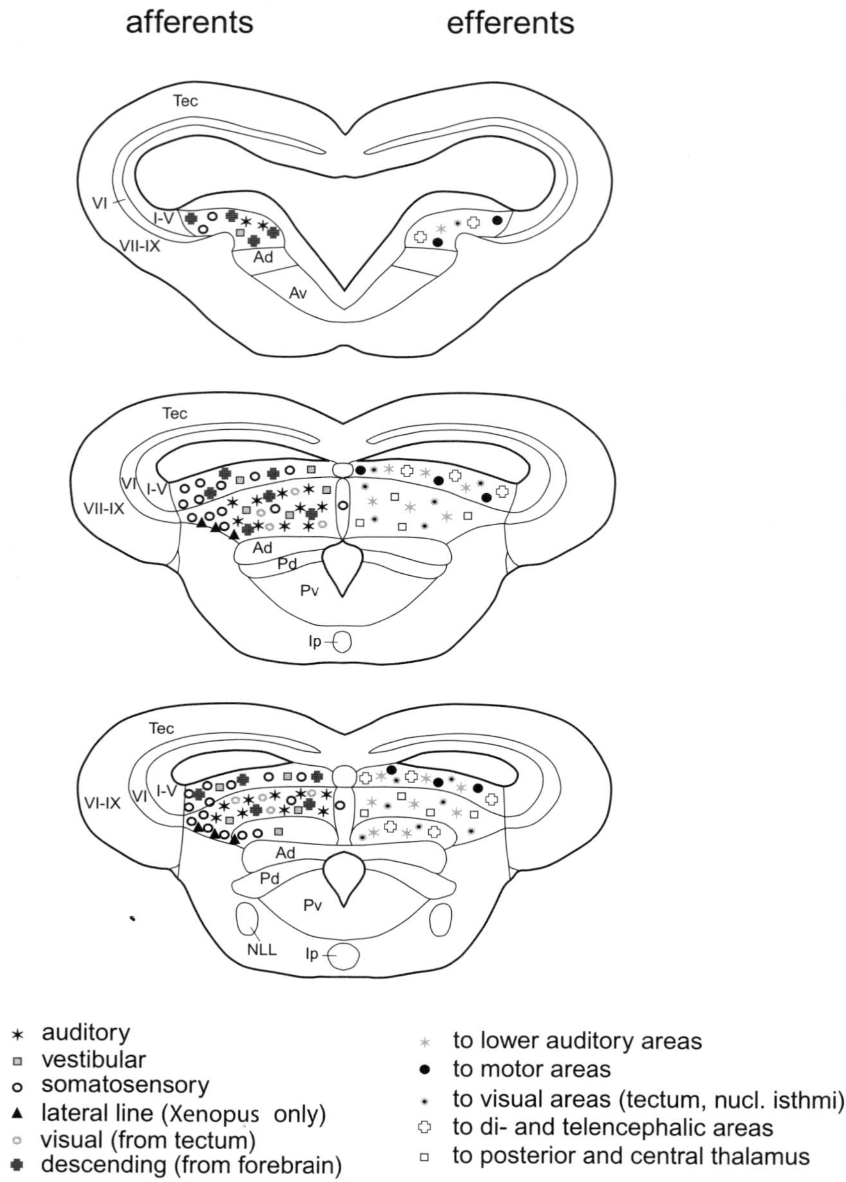

FIGURE 8.6. Summary of connectional organization of the torus semicircularis. Locations of terminals from various sources are marked on the left of each section, locations of efferent neurons targeting different areas are shown on the right.

complex and from the caudal striatum/dorsal pallidum to the laminar nucleus (Marín et al. 1997; Endepols et al. 2005). Axon terminals innervating the laminar nucleus contain numerous neuromodulators and neuropeptides, including dopamine, bombesin, galanin, histamine, neuromedin U, proneuropeptide Y, somatostatin, serotonin, substance P, and enkephalin (for review see Endepols et al. 2000). The magnocellular nucleus seems to receive less descending input.

3.4 Efferents and Their Sources from Among the Toral Nuclei

Within the mesencephalon, all toral nuclei project bilaterally to the optic tectum and tegmental nuclei, and to the ipsilateral nucleus isthmi and secondary isthmal nucleus (Neary 1988; Feng and Lin 1991; Luksch and Walkowiak 1998; Kulik and Matesz 1997). Ascending auditory efferents from the principal nucleus can be traced only up to the central and posterior thalamic nuclei (Feng and Lin 1991; Matesz and Kulik 1996; Luksch and Walkowiak 1998; Endepols and Walkowiak 2001), whereas the laminar and magnocellular nucleus neurons project much farther to di- and telencephalic targets (Figs. 8.2A, 8.6).

Descending projections from all toral nuclei terminate in the lower auditory regions (Figs. 8.2C, 8.6), the ipsilateral NLL and SON, and bilaterally in the DMN (Feng and Lin 1991; Matesz and Kulik 1996; Luksch and Walkowiak 1998; Endepols and Walkowiak 2001; Edwards and Kelley 2001). Laminar nucleus neurons also send projections to motor regions of the medulla (Figs. 8.2D, 8.6), for example, the pretrigeminal nucleus and branchial motor nuclei (Feng and Lin 1991; Strake et al. 1994; Luksch and Walkowiak 1998; Endepols and Walkowiak 2001). Other fibers follow the lateral or medial funiculus and descend deep into the spinal cord, reaching as far as to the upper lumbar segments (Luksch and Walkowiak 1998; Sánchez-Camacho et al. 2001).

4. Forebrain Auditory Pathways

Starting from the midbrain, ascending auditory pathways spread widely throughout the di- and telencephalon (Fig. 8.2A) rather than being focused on specific nuclei as in amniotes. Because virtually all forebrain areas have some auditory input, covering details of cell types, immunohistochemistry, and internal organization would require descriptions of the entire forebrain and is therefore beyond the scope of this chapter. Here we restrict our descriptions to basic connections.

4.1 Thalamic Nuclei

Much of the thalamus receives some toral connections, but the major nuclei receiving toral input are the central, posterior, anterior, and ventromedial thalamic nuclei (Figs. 8.1C,D,E; nomenclature from Neary and Northcutt 1983). New prosomeric maps of the diencephalon (Puelles et al. 1996; Milán and Puelles 2000; Puelles 2001) show that the anterior and central thalamic nuclei as well as the anterior and posteroventral lateral thalamic nuclei (Neary and Northcutt 1983) are derived from the alar region of prosomere 2 (posterior parencephalon), whereas the posterior and posterodorsal thalamic nuclei are found in the more caudal prosomere 1 (synencephalon) and are therefore part of the pretectum. Although the ventromedial/ventrolateral thalamic nuclei appear ventral to the anterior and central thalamic nuclei in transverse sections in adults, they arise rostral to them from the alar region of prosomere 3 (anterior parencephalon).

Toral fibers ascend via a periventricular and a ventrolateral pathway (Neary 1988; Endepols and Walkowiak 2001). The periventricular pathway runs in the periependymal cell-free zone between the dorsal and the ventral thalamus and can be traced up to suprachiasmatic levels. The ventrolateral pathway ascends through the ventrolateral mesencephalon and diencephalon, and eventually joins the lateral forebrain bundle caudal to the optic chiasm. The central and posterior thalamic nuclei receive auditory input from the three main auditory toral nuclei as well as a small input from the superior olivary nucleus (Rubinson and Skiles 1975; Feng 1986b; Hall and Feng 1987; Neary 1988; Feng and Lin 1991; Matesz and Kulik 1996; Luksch and Walkowiak 1998; Endepols and Walkowiak 2001). The ventromedial and anterior thalamic nuclei are the target of fibers arising in the laminar nucleus of the torus semicircularis, as are the anterior lateral and ventrolateral thalamic nuclei, and the suprachiasmatic nucleus (Luksch and Walkowiak 1998; Endepols and Walkowiak 2001).

To date there is no anatomical or physiological evidence that toral auditory inputs to the diencephalic nuclei are topographically organized to form tonotopic maps. Furthermore, although the torus semicircularis is at least predominated by auditory activity (although there are vestibular, somatosensory, and lateral line inputs), the thalamic nuclei receive afferents of roughly equal strength from visual, auditory, somatosensory, and vestibular systems (Hall and Feng 1987; Muñoz et al. 1994, 1995; Roth et al. 2003; Westhoff et al. 2004). Because of this, none of the thalamic nuclei receiving ascending auditory input are comparable to a classic specific thalamic auditory relay nucleus. The anterior and central thalamic nuclei have been argued to be more comparable to the mammalian midline and intralaminar thalamic nuclei. They play a role in the transmission of multimodal associative and limbic information rather than segregated, specific sensory information (Endepols et al. 2003; Roth et al. 2003; Westhoff et al. 2004). We can therefore assume that brain areas that are located rostral to the torus semicircularis cannot be assigned strictly to the auditory system, in spite of their auditory inputs. A region that comes closest to the definition of a "specific auditory forebrain region" is the ventromedial thalamic nucleus, the posterior part of which is thought to be homologous to the mammalian zona incerta (Puelles et al. 1996). Although also multimodal, this area receives auditory information with shorter latencies than the dorsal thalamic nuclei (Roden 2002) and, in contrast to the dorsal thalamic nuclei, does not display habituation (Megela and Capranica 1983). However, it too has many features not seen in specific sensory relay nuclei. It projects to the telencephalon (e.g., to the striatum, septum, medial pallium, amygdalar regions), but most of its efferents are descending (Roden, unpublished observations). Among its main targets are the pretectal and pretoral gray, tegmentum, and torus semicircularis. Similar to the laminar nucleus of the torus semicircularis, the cells of the ventromedial thalamic nucleus concentrate sex steroids (Morell et al. 1975; Kelley et al. 1975; Kelley 1981) and can develop progestin receptors (Roy et al. 1986). Furthermore, ventromedial thalamic neurons contain numerous neuropeptides, such as neuropeptide Y, neuromedin B, cholecystokinin-8, and calcitonin gene-related peptide (Panzanelli et al. 1991; Petkó and Sánta 1992; Lázár et al. 1993; Tuinhof et al. 1994; Petkó and Kovacs 1996).

4.2 Output Pathways from Thalamic Nuclei

There are widespread connections out of the thalamic nuclei (Fig. 8.2A). Although they are not completely independent of each other, for convenience we treat them as three separate categories: thalamotelencephalic connections that target basal ganglia and limbic areas; hypothalamic connections; and descending connections out of the diencephalon. The forebrain areas discussed are illustrated in Figures 8.1A to E.

4.2.1 Telencephalic Targets of Auditory Pathways

There is a considerable ipsilateral projection from the laminar nucleus of the torus semicircularis to the ventral part of the caudal striatum/dorsal pallidum via the ventrolateral pathway, and some neurons project to the diagonal band of Broca and the lateral septal complex (Neary 1988; Endepols and Walkowiak 2001). However, the main ascending input of the striatum/dorsal pallidum complex arises in the lateral anterior and central thalamic nuclei (Wilczynski and Northcutt 1983a; Neary 1988; Marín et al. 1997; Endepols et al. 2004). Given the target (basal ganglia rather than pallial/cortical) and that the same area receives visual and somatosensory input (Wilczynski and Northcutt 1983a), we do not consider this a specific sensory pathway analogous to mammalian geniculocortical connections, but rather a motivational/associative pathway targeting a telencephalic area that modulates motor output (see Walkowiak et al. 1999). The anterior thalamic nucleus projects mainly to the septal complex and the medial pallium (Neary 1984; Northcutt and Ronan 1992; Roden et al. 2005); given the targets, this can be classified as a limbic pathway, although its function remains obscure. Auditory information also reaches the striatum, medial pallium, and septum via the ventromedial thalamic nucleus.

The caudal striatum/dorsal pallidum projects back to the central and ventromedial thalamic nuclei, and to the laminar nucleus of the torus semicircularis and the tegmentum (Wilczynski and Northcutt 1983b; Marín et al. 1997). Neurons of the medial pallium project to the anterior, central, posterior, and ventral thalamic nuclei (Westhoff and Roth 2002).

Through these thalamic efferents, ascending auditory information thus reaches both basal ganglia and limbic regions of the telencephalon. All these regions also receive other sensory input as well. There is no evidence that different sensory modalities are kept separate in the telencephalic areas, and furthermore the auditory input is not topographically organized in any obvious way. This means that just as for the thalamic nuclei, none of the telencephalic areas can be considered specifically auditory centers. Furthermore, there is no part of the anuran telencephalon that can be considered anatomically or functionally equivalent to an auditory cortex, that is, a unimodal, tonotopically organized, pallially derived area devoted to auditory processing.

4.2.2 Hypothalamic Targets of Auditory Pathways

In addition to their ascending connections to limbic regions, thalamic nuclei send substantial efferents to the hypothalamus areas with which these limbic regions are interconnected (Fig. 8.2C). Thalamohypothalamic pathways arise from the central and anterior thalamic nuclei. The central thalamic nucleus connects predominately to the ipsilateral caudal two-thirds of the ventral hypothalamus (Neary and Wilczynski 1986; Hall and Feng 1987; Neary 1988; Allison and Wilczynski 1991). The central thalamic cells of origin lie mainly in the ventral and medial half of the nucleus (Neary and Wilczynski 1986; Allison and Wilczynski 1991), and their terminals occupy a thick band immediately lateral to the hypothalamic cell-dense zone adjacent to the ventricle (Neary and Wilczynski 1986). A small number of central nucleus cells also project to the anterior preoptic area. The major thalamic input to the preoptic area comes, however, from cells scattered throughout the ipsilateral anterior thalamic nucleus (Allison and Wilczynski 1991). Anterior nucleus cells send a smaller projection to the ventral hypothalamus as well. A third strong input to the hypothalamus arises in the septal complex and the medial pallium (Neary 1995; Endepols et al. 2005), providing a link between the limbic and endocrine auditory streams (see below).

Thalamic auditory input to both the ventral hypothalamus and anterior preoptic area is supplemented by significant, primarily ipsilateral, input from the secondary isthmal nucleus (Neary and Wilczynski 1986; Neary 1988; Allison and Wilczynski 1991). This structure lies immediately lateral to the nucleus isthmi in the isthmal tegmentum and its terminals largely overlap those of the thalamic nuclei. It receives a strong input from the torus (Neary and Wilczynski 1986; Neary 1988), is adjacent to the ascending lateral lemniscal fibers and hence may receive an input from them, and its auditory sensitivity has been confirmed physiologically (Bibikov 2003). There is no agreed-upon mammalian equivalent of the secondary isthmal nucleus, but its position and its efferents to hypothalamic and limbic areas suggest similarity with the mammalian parabrachial region.

Thalamic auditory input to the hypothalamus is not unique to amphibians, occurring also from regions adjacent to the medial geniculate and its homologues in mammals (LeDoux et al. 1985) and birds (Cheng and Zuo 1994). They are exceptionally prominent in anurans, however, with thalamic and midbrain connections to the ventral hypothalamus and anterior preoptic area as strong as their connections to telencephalic centers.

4.2.3 Descending Pathways from Thalamic Nuclei

Dorsal as well as ventral thalamic nuclei are the source of descending projections to the lower auditory nuclei (Fig. 8.2B). The most prominent projections arise in the ventromedial and posterior thalamic nuclei and terminate in the laminar and principal nuclei of the torus semicircularis (Wilczynski 1981; Feng and Lin 1991). In addition, the torus receives descending input from the central thalamic nucleus, and there is even a weak projection from the posterior thalamic nucleus to the ipsilateral superior olive (Matesz and Kulik 1996). Apart from their auditory connections, the posterior and ventromedial thalamic nuclei also send descending projections to motor areas in the medulla and the spinal cord (Dicke et al. 1998; Sánchez-Camacho et al. 2001).

5. Discussion and Summary

The basic organization of amphibian brainstem auditory pathways reflects that of a conserved tetrapod auditory system. The ascending pathway has three distinct lower brainstem levels, the DMN, SON, and NLL, prior to reaching a large region of the midbrain roof, the torus semicircularis, where efferents of all the lower levels converge. Within each level, however, tetrapods seem to have adopted idiosyncratic organizations. In fact, as Will et al. (1985a) noted for the primary octavolateralis nuclei, the specific organizations of the auditory processing levels in each vertebrate group might best be considered independently derived features as different tetrapods evolved and differentiated a terrestrial auditory system within a basic tetrapod framework.

In anurans, this consists of a single nucleus at each of the three lower levels even though the input from the ear derives from multiple end-organs. Starting in the midbrain, however, the central anatomy of the auditory system begins to diverge from the familiar mammalian pattern. There, the torus semicircularis represents a key nodal point in the central auditory system and a transition from the familiar auditory features of lower brainstem nuclei to the more integrative and less clearly "sensory" nature of the forebrain. It has some features typical of equivalent midbrain areas in other vertebrates such as the inferior colliculus. It is composed of multiple nuclei, although these cannot be strictly compared to the component nuclei of mammals or birds. Output from all lower centers converges on it, as do some other sensory input and a variety of forebrain connections. Unlike other tetrapods, the input is not clearly segregated into a central, purely auditory component and other more multimodal subnuclei, and in fact the range of forebrain input and neuromodulators associated with it is extraordinary in anurans and target the toral areas most responsible for relaying auditory input to the diencephalon and to motor regions in the medulla and brainstem.

The functional anatomy of auditory connections further indicates growing differences from the standard mammalian model as one proceeds to higher levels. Tonotopy is apparent in lower brainstem centers much as one would expect for an auditory system. By the midbrain, however, tonotopy remains but is supplemented by the construction of more complex feature detectors that combine frequency channels. There is no evidence that ascending toral efferents are topographically organized, suggesting that tonotopy is not present in auditory centers higher than the midbrain.

Ascending auditory input to the diencephalon and telencephalon is extraordinarily widespread. Nearly all thalamic nuclei receive ascending auditory information, and these nuclei in turn provide extensive input to nearly all of the telencephalon except some olfactory areas, and to hypothalamic areas as well. However, despite being widespread, there are no forebrain areas that can be considered truly and exclusively auditory in nature. There appear to be no unimodal sensory centers that are primarily concerned with auditory processing as with the mammalian medial geniculate nucleus or primary auditory cortex. In fact, a clear feature of the forebrain in general is that all sensory input is widespread, with different sensory modalities overlapping in their terminations, so that there is no evidence that sensory streams are segregated from each other. There may, of course, be regions of the frog telencephalon that are homologous to mammalian isocortex and the pallial sensory areas of reptiles and birds (Northcutt 1981; Bruce and Neary 1995); but whatever those may be, in extant amphibians they do not have a differentiated structure marked by unimodal sensory representations we understand to be important for sensory representation and analysis. This remains a puzzling feature of forebrain auditory processing (and of telencephalic sensory processing in amphibians in general) as there is little in the anatomy to suggest the type of auditory sensory representation and analysis familiar to mammalian systems. Rather, it may be that purely sensory processes such as stimulus recognition and localization are mostly complete at the level of the auditory midbrain, and that ascending connections from there are more concerned with linking this analysis with effector systems generating responses to acoustic stimuli, notably conspecific calls.

Conceptualized in this way, the forebrain auditory targets can be thought of as representing three processing streams (Fig. 8.7). One is an audiomotor interface, centered on the posterior and ventromedial thalamic nuclei and striatum (with the central thalamic nucleus providing input to the striatum). Descending connections from the striatum modulate auditory processing in the torus (Endepols and Walkowiak 1999, 2001) and influence acoustically triggered behavior such as phonotaxis and vocalization. Notably, this same efferent pathway modulates visually guided, tectally mediated orientation (Ewert 1997; Patton and Grobstein 1998; Buxbaum-Conradi and Ewert 1999).

A second stream, involving central and anterior thalamic nuclei (plus the secondary isthmal nucleus) input to the preoptic area, hypothalamus, and septal regions, represents an audioendocrine interface. A third to a half of preoptic and hypothalamic nucleus cells are acoustically sensitive (Allison 1992) and

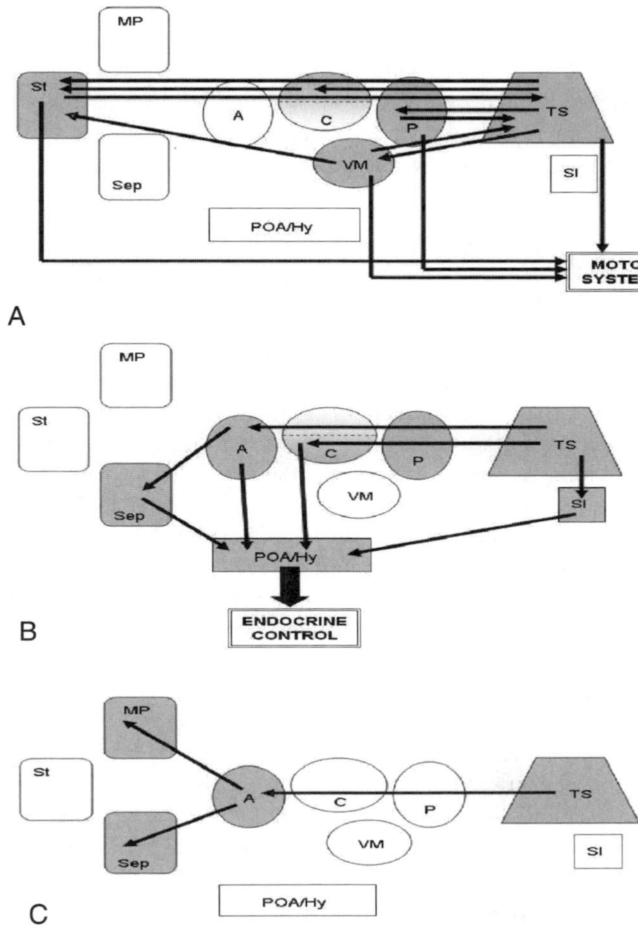

FIGURE 8.7. Three auditory processing streams in the anuran forebrain. A. The "audiomotor interface" connecting the auditory system with areas related to motor control. B. The "audioendocrine interface" connecting the auditory system with endocrine control centers of the preoptic area and hypothalamus. C. The "audiolimbic interface" denoting auditory connections to pallial and subpallial limbic areas; the functional significance of this component is unknown. Not all connections are diagrammed, and there are extensive and complex interconnections among all the illustrated nuclei.

exposure to mating calls elevates gonadal steroids (Burmeister and Wilczynski 2000) and increases GnRH immunoreactivity in septo-preoptic neurons (Burmeister and Wilczynski 2005). The audioendocrine interface can be thought of as modulating the endocrine aspects of social behavior just as the audiomotor interface modulates the overt behavioral responses.

A third stream, mainly from the anterior thalamic nucleus to the medial pallium and septal nuclei targets limbic structures. To date, there is no clear understanding of its function. One would hypothesize that these areas have motivational and mnemonic functions similar to homologous mammalian regions, and the ascending auditory pathway is one of several sensory systems accessing them. This remains an uninvestigated area in anuran neuroethology.

Anatomical and physiological data suggest that much of the forebrain's modulation of auditory processing and the behavior it triggers takes place in the midbrain. The numerous neuromodulatory transmitters and peptides found in axon terminals, especially those entering the laminar nucleus, which is the source of the most widespread toral outputs, indicate that complex modifications of auditory processing or audiomotor integration can take place there. It is therefore reasonable to assume that forebrain areas such as the dorsal thalamic nuclei or the striatum may be involved in sharpening preference functions or changing the sensitivity to auditory stimuli in lower auditory stations, and through this action modifying behavioral responses to acoustic signals.

For anuran amphibians, acoustic communication forms the foundation for their reproductive social behavior. Coupled with the behavioral responses are endocrine changes mediated through auditory interconnections with hypothalamic and limbic areas of the forebrain. For these vertebrates, hearing is intimately tied to communication behaviorally, and also, from what we know about the central auditory pathways, anatomically. Distinguishing "hearing" from "sound communication" may be the best perspective from which to conceptualize the central auditory system of anuran amphibians. "Hearing"—the representation, identification, and localization of acoustic stimuli defining the sensory portion of the system—is consistent with the brainstem components of the auditory system. "Sound communication"—the broader context in which the outcome of the auditory sensory analysis is linked to motor, endocrine, motivational, and mnemonic processes linked to social interactions—characterizes the anatomical organization of auditory pathways throughout the forebrain. The largest single center of the anuran auditory system, the midbrain torus semicircularis, serves as the central station bringing both functions together.

References

Adli DS, Stuesse SL, Cruce WL (1999) Immunohistochemistry and spinal projections of the reticular formation in the northern leopard frog, *Rana pipiens*. J Comp Neurol 404:387–407.

Aitken PG (1981) Autoradiographic demonstration of VIIIth nerve projections in *Rana pipiens*. Neurosci Lett 24:237–242.

Aller MI, Janusonis S, Fite KV, Fernandez-Lopez A (1997) Distribution of the GABAA receptor complex beta 2/3 subunits in the brain of the frog *Rana pipiens*. Neurosci Lett 225:65–68.

Allison JD (1992) Acoustic modulation of neural activity in the preoptic area and ventral hypothalamus of the green treefrog (*Hyla cinerea*). J Comp Physiol A 171:387–395.

Allison JD, Wilczynski W (1991) Thalamic and midbrain auditory projections to the preoptic area and ventral hypothalamus in the green treefrog (*Hyla cinerea*). Brain Behav Evol 37:322–331.

Bibikov NG (2003) Auditory responses in the isthmal region of the frog. Brain Behav Evol 62:169.

Bricout-Berthout A, Caston J, Reber A (1984) Influence of stimulation of auditory and somatosensory systems of the activity of vestibular nuclear neurons in the frog. Brain Behav Evol 24:21–34.

Bruce LL, Neary TJ (1995) The limbic system of tetrapods: A comparative analysis of cortical and amygdalar populations. Brain Behav Evol 46:224–234.

Brüning G, Mayer B (1996) Localization of nitric oxide synthase in the brain of the frog, *Xenopus laevis*. Brain Res 741:331–343.

Burmeister SS, Wilczynski W (2000) Social signals influence hormones independently of calling behavior in the treefrog (*Hyla cinerea*). Horm Behav 38:201–209.

Burmeister SS, Wilczynski W (2005) Social signals regulate gonadotropin-releasing hormone neurons in the green treefrog. Brain Behav Evol 65:26–32.

Buxbaum-Conradi H, Ewert J-P (1999) Responses of single neurons in the toad's caudal ventral striatum to moving visual stimuli and test of their efferent projection by extracellular antidromic stimulation/recording techniques. Brain Behav Evol 54: 303–356.

Cheng MF, Zuo M (1994) Proposed pathways for vocal self-stimulation: Met-enkephalinergic projections linking the midbrain vocal nucleus, auditory-responsive thalamic regions, and neurosecretory hypothalamus. J Neurobiol 25:361–379.

Dechesne CJ, Oberdorfer MD, Hampson DR, Wheaton KD, Nazarali AJ, Goping G, Wenthold RJ (1990) Distribution of a putative kainic acid receptor in the frog central nervous system determined with monoclonal and polyclonal antibodies: evidence for synaptic and extrasynaptic localization. J Neurosci 10:479–490.

Dicke U, Roth G, Matsushima T (1998) Neural substrate for motor control of feeding in amphibians. Acta Anat 163:127–143.

Di Meglio M, Morrell JI, Pfaff DW (1987) Localization of steroid-concentrating cells in the central nervous system of the frog *Rana esculenta*. Gen Comp Endocrinol 67: 149–154.

Edwards CJ, Kelley DB (2001) Auditory and lateral line inputs to the midbrain of an aquatic anuran; neuroanatomical studies in *Xenopus laevis*. J Comp Neurol 438:148–162.

Endepols H, Walkowiak W (1999) Influence of descending forebrain projections on processing of acoustic signals and audiomotor integration in the anuran midbrain. Eur J Morphol 37:182–184.

Endepols H, Walkowiak W (2001) Integration of ascending and descending inputs in the auditory midbrain of anurans. J Comp Physiol A 186:1119–1133.

Endepols H, Feng AS, Gerhardt HC, Schul J, Walkowiak W (2003) Roles of the auditory midbrain and thalamus in selective phonotaxis in female gray treefrogs (*Hyla versicolor*). Behav Brain Res 145:63–77.

Endepols H, Roden K, Luksch H, Dicke U, Walkowiak W (2004) Dorsal striatopallidal system in anurans. J Comp Neurol 468:299–310.

Endepols H, Roden K, Walkowiak W (2005) Hodological characterization of the septum in anuran amphibians: II. Efferent connections. J Comp Neurol 483:437–457.

Endepols H, Walkowiak W, Luksch H (2000) Chemoarchitecture of the anuran auditory midbrain. Brain Res Rev 33:179–198.

Ewert J-P (1997) Neural correlates of key stimulus and releasing mechanisms: a case study and two concepts. Trends Neurosci 20:332–339.

Feng AS (1983) Morphology of neurons in the torus semicircularis of the northern leopard frog *Rana pipiens pipiens*. J Morphol 175:253–269.

Feng AS (1986a) Afferent and efferent innervation patterns of the cochlear nucleus (dorsal medullary nucleus) of the leopard frog. Brain Res 367:183–191.

Feng AS (1986b) Afferent and efferent innervation patterns of the superior olivary nucleus of the leopard frog. Brain Res 364:167–171.

Feng AS, Lin W (1991) Differential innervation patterns of three divisions of the frog auditory midbrain (torus semicricularis). J Comp Neurol 306:613–630.

Feng AS, Lin W (1996) Neuronal architecture of the dorsal nucleus (cochlear nucleus) of the frog, *Rana pipiens*. J Comp Neurol 366:320–334.

Fritzsch B, Nikundiwe AM, Will U (1984) Projection patterns of lateral-line afferents in anurans: a comparative HRP study. J Comp Neurol 229:451–469.

Fuzessery ZM (1988) Frequency tuning in the anuran central auditory system. In: Fritzsch B, Ryan MJ, Wilczynski W, Hetherington TE, and Walkowiak W (eds) The Evolution of the Amphibian Auditory System. New York: Wiley, pp. 253–273.

Fuzessery ZM, Feng AS (1981) Frequency representation in the dorsal medullary nucleus of the leopard frog *Rana p. pipiens*. J Comp Physiol 143:339–347.

González A, Smeets WJAJ (1991) Comparative analysis of dopamine and tyrosine hydroxylase immunoreactivites in the brain of two amphibians, the anuran *Rana ridibunda* and the urodele *Pleurodeles waltlii*. J Comp Neurol 303:457–477.

González A, Smeets WJAJ (1993) Noradrenaline in the brain of the South African clawed frog *Xenopus laevis*: A study with antibodies against noradrenaline and dopamine-β-hydroxylase. J Comp Neurol 331:363–374.

Gregory KM (1972) Central projections of the eighth nerve in frogs. Brain Behav Evol 5:70–88.

Gregory KM (1974) The stato-acoustic nuclear complex and the nucleus cerebelli of the frog. Brain Behav Evol 10:146–156.

Grofová I, Corvaja N (1972) Commissural projection from the nuclei of termination of the VIIIth cranial nerve in the toad. Brain Res 42:189–195.

Hall JC, Bunker MC (1994) Acetylcholinerase staining in the auditory brainstem nuclei of the leopard frog, *Rana pipiens*. Neurosci Lett 182:222–226.

Hall JC, Feng AS (1987) Evidence for parallel processing in the frog's auditory thalamus. J Comp Neurol 258:407–419.

Hoke KL, Burmeister SS, Fernald RD, Rand AS, Ryan MJ, Wilczynski W (2004) Functional mapping of the auditory midbrain during mate call reception. J Neurosci 24: 11264–11272.

Inagaki S, Senba E, Shiosaka S, Takagi H, Kawai Y, Takatsuki K, Sakanaka M, Matsuzaki T, Tohyama M (1981) Regional distribution of substance P-like immunoreactivity in the frog brain and spinal cord: immunohistochemical analysis. J Comp Neurol 201:243–254.

Kelley DB (1980) Auditory and vocal nuclei in the frog brain concentrate sex hormones. Science 207:553–555.

Kelley DB (1981) Locations of androgen-concentrating cells in the brain of *Xenopus laevis*: autoradiography with 3H-dihydrotestosterone. J Comp Neurol 199:221–231.

Kelley DB, Morrell JI, Pfaff DW (1975) Autoradiographic localization of hormone-concentrating cells in the brain of an amphibian, *Xenopus laevis*. I. Testosterone. J Comp Neurol 164:47–62.

Kulik A, Matesz K (1997) Projections from the nucleus isthmi to the visual and auditory centres in the frog, *Rana esculenta*. J Brain Res 38:299–307.

Kulik A, Matesz K, Székely G (1994) Mesencephalic projections of the cochlear nucleus in the frog, *Rana esculenta*. Acta Biol Hung 45:323–335.

Larsell O (1934) The differentiation of the peripheral and central acoustic apparatus in the frog. J Comp Neurol 60:473–527.

Lázár GY, Losonczy A (1999) NADPH-diaphorase-positive neurons and pathways in the brain of the frog *Rana esculenta*. Anat Embryol (Berl) 199:185–198.

Lázár GY, Liposits ZS, Tóth P, Trasti SL, Maderdrut JL, Merchenthaler I (1991) Distribution of galanin-like immunoreactivity in the brain of *Rana esculenta* and *Xenopus laevis*. J Comp Neurol 310:45–67.

Lázár GY, Maderdrut JL, Trasti SL, Liposits ZS, Tóth P, Kozicz T, Merchenthaler I (1993) Distribution of proneuropedtide Y-derived peptides in the brain of *Rana esculenta* and *Xenopus laevis*. J Comp Neurol 327:551–571.

LeDoux JE, Ruggiero DA, Reis DJ (1985) Projections to the subcortical forebrain from anatomically defined regions of the medial geniculate body in the rat. J Comp Neurol 242:182–213.

Lewis ER, Leverenz EL, Koyama H (1980) Mapping functionally identified auditory afferents from the peripheral origins to their central terminations. Brain Res 197:223–229.

Lowe DA (1986) Organisation of lateral line and auditory areas in the midbrain of *Xenopus laevis*. J Comp Neurol 245:498–513.

Luksch H, Walkowiak W (1998) Morphology and axonal projection patterns of auditory neurons in the midbrain of the painted frog, *Discoclossus pictus*. Hearing Res 122:1–17.

Maderdrut JL, Lázár GY, Kozicz T, Merchenthaler I (1996) Distribution of neuromedin U-like immunoreactivity in the central nervous system of *Rana esculenta*. J Comp Neurol 369:438–450.

Marín O, González A, Smeets WJAJ (1997) Basal ganglia organization in amphibians: Efferent connections of the striatum and the nucleus accumbens. J Comp Neurol 380: 23–50.

Matesz C (1979) Central projections of the VIIIth cranial nerve in the frog. Neuroscience 4:2061–2071.

Matesz C, Kulik A (1996) Connections of the torus semicircularis and oliva superior in the frog, *Rana esculenta*. Acta Biol Hung 47:287–301.

Megela AL, Capranica RR (1983) A neural and behavioral study of auditory habituation in the bullfrog, *Rana catesbeiana*. J Comp Physiol A 151:423–434.

Merchenthaler I, Lázár G, Maderdrtu JL (1989) Distribution of proenkephalin-derived peptides in the brain of *Rana esculantaesculenta*. J Comp Neurol 281:23–39.

Milán FJ, Puelles L (2000) Patterns of calretinin, calbindin, and tyrosine-hydroxylase expression are consistent with the prosomeric map of the frog diencephalon. J Comp Neurol 419:96–121.

Morrell JI, Kelley DB, Pfaff DW (1975) Autoradiographic localization of hormone-concentrating cells in the brain of an amphibian, *Xenopus laevis*. II. Estradiol. J Comp Neurol 164:63–78.

Muñoz A, Muñoz M, González A, ten Donkelaar HJ (1994) The dorsal column-medial lemniscal projection of anuran amphibians. Eur J Morphol 32:283–287.

Muñoz A, Muñoz M, González A, ten Donkelaar HJ (1995) The anuran dorsal column nucleus: Organization, immunohistochemical characterization and fiber connections in *Rana perezi* and *Xenopus laevis*. J Comp Neurol 363:197–220.
Muñoz A, Muñoz M, González A, ten Donkelaar HJ (1997) Spinal ascending pathways in amphibians: Cells of origin and main targets. J Comp Neurol 378:205–228.
Muñoz A, Muñoz M, Marin O, Alonso JR, Arévalo R, Porteros A, González A (1996) Topographical distribution of NADPH-diaphorase activity in the central nervous system of the frog, *Rana perezi*. J Comp Neurol 367:54–69.
Neary TJ (1984) Anterior thalamic nucleus projections to the dorsal pallium in ranid frogs. Neurosci Lett 51:213–2188.
Neary TJ (1988) Forebrain auditory pathways in ranid frogs. In: Fritzsch B, Ryan MJ, Wilczynski W, Hetherington TE, and Walkowiak W (eds) The Evolution of the Amphibian Auditory System. New York: Wiley, pp. 233–252.
Neary TJ (1995) Afferent projections to the hypothalamus in ranid frogs. Brain Behav Evol 46:1–13.
Neary TJ, Northcutt RG (1983) Nuclear organization of the bullfrog diencephalon. J Comp Neurol 213:262–278.
Neary TJ, Wilczynski W (1986) Auditory pathways to the hypothalamus in ranid frogs. Neurosci Lett 71:142–146.
Nikundiwe AM, Nieuwenhuys R (1983) The cell masses in the brainstem of the South African clawed frog. J Comp Neurol 213:199–219.
Northcutt RG (1981) Evolution of the telencephalon in nonmammals. Ann Rev Neurosci 4:301–350.
Northcutt RG, Ronan MC (1992) Afferent and efferent connections of the bullfrog medial pallium. Brain Behav Evol 40:1–16.
Opdam P, Kemali M, Nieuwenhuys R (1976) Topological analysis of the brainstem of the frogs *Rana esculenta* and *Rana catesbeiana*. J Comp Neurol 165:307–332.
Panzanelli P, Mulatero B, Lazarus LH, Fasolo A (1991) Neuromedin B-like immunoreactivity in the brain of the green frog (*Rana esculenta* L.). Eur J Basic Appl Histochem 35:359–370.
Patton P, Grobstein P (1998) The effects of telencephalic lesions on the visually mediated prey orienting behavior in the leopard frog (*Rana pipiens*). II. The effects of limited lesions to the telencephalon. Brain Behav Evol 51:144–161.
Petkó M, Kovacs T (1996) Distribution of cholecystokinin-8-like immunoreactivity in the frog brain and spinal cord. Acta Biol Hung 47:355–369.
Petkó M, Sánta A (1992) Distribution of calcitonin gene-related peptide immunoreactivity in the central nervous system of the frog, *Rana esculenta*. Cell Tissue Res 269:525–534.
Pettigrew AG (1981) Brainstem afferents to the torus semicircularis of the Queensland cane toad (*Bufo marinus*). J Comp Neurol 202:59–68.
Potter HD (1965) Mesencephalic auditory regions of the bullfrog. J Neurophysiol 28:1132–1154.
Puelles L (2001) Brain segmentation and forebrain development in amniotes. Brain Res Bull 55:695–710.
Puelles L, Milán FJ, Martínez-de-la-Torre M (1996) A segmental map of architectonic subdivisions in the diencephalon of the frog *Rana perezi*: Acetylcholinesterase-histochemical observations. Brain Behav Evol 47:279–310.
Roden K (2002) Bedeutung des Vorderhirns für die audiomotorische Integration bei Froschlurchen (*Anura*). PhD thesis, University of Cologne, Germany.

Roden K, Endepols H, Walkowiak W (2005) Hodological characterization of the septum in anuran amphibians: I. Afferent connections. J Comp Neurol 483:415–436.

Rose GJ, Wilczynski W (1984) The anuran superficial reticular nucleus: Evidence for homology with nuclei of the lateral lemniscus. Brain Res 304:170–172.

Roth G, Grunwald W, Dicke U (2003) Morphology, axonal projection pattern, and responses to optic nerve stimulation of thalamic neurons in the fire-bellied toad *Bombina orientalis*. J Comp Neurol 461:91–110.

Röthig P (1927) Beiträge zum Studium des Zentralnervensystems der Wirbeltiere. XI. Über die Faserzüge im Mittelhirn, Kleinhirn und der Medulla oblongata der Urodelen und Anuren. Z Mikrosk Anat Forsch 10:381–472.

Roy EJ, Wilson MA, Kelley DB (1986) Estrogen-induced progestin receptors in the brain and pituitary of the South African clawed frog *Xenopus laevis*. Neuroendocrinology 42:51–56.

Rubinson K, Skiles MP (1975) Efferent projections of the superior olivary nucleus in the frog, *Rana catesbeiana*. Brain Behav Evol 12:151–160.

Sánchez-Camacho, Marín O, ten Donkelaar HJ, González A (2001) Descending supraspinal pathways in amphibians. I. A dextran amine tracing study of their cells of origin. J Comp Neurol 434:186–208.

Simmons AM, Chapman JA (2002) Metamorphic changes in GABA immunoreactivity in the brainstem of the bullfrog, *Rana catesbeiana*. Brain Behav Evol 60:189–206.

Strake J, Luksch H, Walkowiak W (1994) Audio-motor interface in anurans. Eur J Morphol 32:122–126.

ten Donkelaar HJ (1998) Anurans. Octavolateral area and connections. In: Nieuwenhuys R, ten Donkelaar HJ, Nicholson C (eds) The Central Nervous System of Vertebrates, Vol 2. Berlin, Heidelberg: Springer-Verlag, pp. 1216–1228.

Tostivint H, Lihrmann I, Bucharles C, Vieau D, Coulouarn Y, Fournier A, Conlon JM, Vaudry H (1996) Occurrence of two somatostatin variants in the frog brain: Characterization of the cDNAs, distribution of the mRNAs, and receptor-binding affinities of the peptides. Proc Natl Acad Sci USA 93:12605–12610.

Tuinhof R, Gonzalez A, Smeets WJ, Roubos EW (1994) Neuropeptide Y in the developing and adult brain of the South African clawed toad *Xenopus laevis*. J Chem Neuroanat 7:271–283.

Walkowiak W, Luksch H (1994) Sensory motor interfacing in acoustic behavior of anurans. Am Zool 34:685–695.

Walkowiak W, Berlinger M, Schul J, Gerhardt HC (1999) Significance of forebrain structures in acoustically guided behavior in anurans. Eur J Morphol 37:177–181.

Westhoff G, Roth G (2002) Morphology and projection pattern of medial and dorsal pallial neurons in the frog *Discoglossus pictus* and the salamander *Plethodon jordani*. J Comp Neurol 445:97–121.

Westhoff G, Roth G, Straka H (2004) Topographic representation of vestibular and somatosensory signals in the anuran thalamus. Neuroscience 124:669–683.

Wilczynski W (1981) Afferents to the midbrain auditory center in the bullfrog, *Rana catesbeiana*. J Comp Neurol 198:421–433.

Wilczynski W (1988) Brainstem auditory pathways in anuran amphibians. In: Fritzsch B, Ryan MJ, Wilczynski W, Hetherington TE, and Walkowiak W (eds) The Evolution of the Amphibian Auditory System. New York: Wiley, pp. 209–232.

Wilczynski W, Capranica RR (1984) The auditory system of anuran amphibians. Prog Neurobiol 22:1–38.

Wilczynski W, Northcutt RG (1983a) Connections of the bullfrog striatum: Afferent organization. J Comp Neurol 214:321–332.

Wilczynski W, Northcutt RG (1983b) Connections of the bullfrog striatum: Efferent projections. J Comp Neurol 214:333–343.

Will U (1988) Organization and projections of the area octavolateralis in amphibians. In: Fritzsch B, Ryan MJ, Wilczynski W, Hetherington TE, and Walkowiak W (eds) The Evolution of the Amphibian Auditory System. New York: Wiley, pp. 185–208.

Will U, Fritzsch B (1988) The eighth nerve of amphibians: Peripheral and central distribution. In: Fritzsch B, Ryan MJ, Wilczynski W, Hetherington TE, and Walkowiak W (eds) The Evolution of the Amphibian Auditory System. New York: Wiley, pp. 159–184.

Will U, Luhede G, Görner P (1985a) The area octavo-lateralis in *Xenopus laevis* I. The primary afferent projections. Cell Tiss Res 239:147–161.

Will U, Luhede G, Görner P (1985b) The area octavo-lateralis in *Xenopus laevis* II. Second order projections and cytoarchitecture. Cell Tiss Res 239:163–175.

Yon L, Feuilloley M, Chartrel N, Arimura A, Colon JM, Fournier A, Vaudry H (1992) Immunocytochemical distribution and biological activity of pituitary adenylate cyclase-activating polypeptide (PACAP) in the central nervous system of the frog *Rana ridibunda*. J Comp Neurol 342:485–499.

List of Abbreviations

A	anterior thalamic nucleus
AP	amphibian papilla
AVC	anterior vertical canal
BP	basilar papilla
C	central thalamic nucleus
CN	caudal nucleus
DMN	dorsal medullary nucleus
G VIII	vestibulocochlear nerve ganglion
HC	horizontal canal
Hyp	hypothalamus
L	lagena
MOT	motor nuclei
N VII	facial nerve
N VIII	vestibulocochlear nerve
NLL	nucleus of the lateral lemniscus
P	posterior thalamic nucleus
Pm	medial pallium
PoA	anterior preoptic nucleus
PrV	pretrigeminal nucleus
PVC	posterior vertical canal
S	sacculus
SC	suprachiasmatic nucleus
Sep	septal complex
SON	superior olivary nucleus

Str	striato-pallidal complex
Tl	laminar nucleus of the torus semicircularis
Tm	magnocellular nucleus of the torus semicircularis
Tp	principal nucleus of the torus semicircularis
VM	ventromedial thalamic nucleus

9
Function of the Amphibian Central Auditory System

Gary J. Rose and David M. Gooler

1. Introduction

For over three decades, the anuran auditory system has played a prominent role in advancing our understanding of how biologically relevant acoustic information is represented and processed in central nervous systems. Much of this progress can be attributed to the neuroethological nature of this work. Like other classical neuroethological organisms (Heiligenberg 1991), anurans were chosen for study because of their specialized behavior(s). Acoustic communication in particular plays a fundamental role in the reproductive behavior and fitness of anurans (see Wells and Schwartz, Chapter 3), and is, therefore, robust and amenable to experimental analysis. Behavioral experiments have identified, and continue to elucidate, the discriminative capacities of anuran auditory systems, and the types of computations that underlie these capacities. In essence, behavioral experiments have guided neurophysiological investigations by formulating testable hypotheses concerning the function of the auditory system. This interplay between analyses at the behavioral and neural levels is one of the defining characteristics of the neuroethological approach.

Anuran neuroethology has its roots in Capranica's (1965, 1966) evoked calling studies with bullfrogs, supporting the notion of a neural AND logical operation that detects the formantlike simultaneous presence of low- and high-frequency energy peaks in their mating call. Other studies suggest that anurans also have neural specializations for analyzing the temporal structure of acoustic communication signals. These include filters for pulse repetition rate and shape (amplitude modulation, AM; Gerhardt 1988, 2001; Brenowitz and Rose 1994; Rose and Brenowitz 1997), direction of frequency change (frequency modulation, FM; Ryan 1983; Rose et al. 1988), and duration of notes (Narins and Capranica 1978; Penna 1997). In addition, it is important in the context of mate selection and aggressive interactions for anurans to localize sound sources. The small interaural distances for most anurans pose interesting challenges both for the animals, and for the experimenters interested in understanding the multifaceted mechanisms that underlie sound localization.

This chapter summarizes the current understanding of the neural substrates of spectral and temporal processing (Feng et al. 1990), and directionality mechanisms in the auditory systems of anurans.

2. Spectral Processing

The peripheral auditory system in anurans performs a frequency analysis of sounds. Each auditory-nerve fiber is tuned to a particular frequency, determined by its site of innervation within the auditory organs of the inner ear. Anuran amphibians have three distinct auditory end organs, the sacculus, amphibian papilla, and basilar papilla (Lewis and Lombard 1988; Simmons et al. Chapter 7). These organs are responsible for the sensitivity and tuning of the anuran auditory system to very low frequencies, low and mid frequencies, and high frequencies, respectively. Only the low-frequency sensitive fibers that innervate the rostral amphibian papilla show two-tone suppression (see below). The range of frequency tuning across all nerve fibers covers the region(s) of spectral energy in the calls of each species. The spectral structure of communication sounds is represented, therefore, in the relative levels of activity across the array of auditory-nerve fibers. Spectral processing refers to the transformations in how the spectral structure of sounds is represented in the central auditory system. These transformations include sharpening the frequency tuning of neurons, selectivity for the steady-state amplitude and frequency of sound, and logical AND computations. Below, we summarize our current knowledge of spectral processing in central auditory regions.

2.1 Dorsal Medullary Nucleus (DMN)

Primary afferents that innervate each of the auditory end-organs project ipsilaterally and tonotopically to discrete regions of the first-order central auditory area, the dorsal medullary nucleus (DMN), also referred to as the dorsolateral nucleus (DLN; Wilczynski and Endepols, Chapter 8). High frequencies are represented dorsomedially, and mid and low frequencies, encompassing the range of the amphibian papilla, are represented progressively more ventrally. This tonotopy is preserved along the rostrocaudal axis of the DMN, and in the commissural connections between left and right DMNs. Saccular fibers terminate in a separate region ventral to the DMN, hence, the range of spectral sensitivity of units in the DMN parallels that in the auditory nerve.

Like the auditory-nerve fibers that innervate them, DMN neurons have V-shaped frequency tuning curves (Feng and Capranica 1976; Fuzzessery and Feng 1983a). Furthermore, the bandwidths of these frequency tuning functions are similar to those of primary afferents of similar best excitatory frequency (BEF), or characteristic frequency (CF), that is, the frequency at which the neuron has its lowest threshold. Inhibition is restricted to neurons of low CF, where intermediate frequencies have a suppressive effect on the excitation caused by sound at or near the CF; such "tone-on-tone" suppression mirrors that already seen in auditory-nerve fibers (Capranica and Moffat 1980). A recent study, however, revealed that some mid-frequency auditory-nerve fibers (CF $\approx$ 500 Hz) show tone-on-tone suppression for frequencies above or below their CF (Benedix et al. 1994). Further work is needed to determine if these properties are also present in mid-frequency DMN neurons.

2.2 Superior Olivary Nucleus (SON)

The next level of the anuran auditory system, the superior olivary nucleus (SON), is also bilateral, with each SON receiving projections from the ipsilateral and contralateral DMN (Wilczynski and Endepols, Chapter 8). Unlike the superior olivary complex of mammals, the SON in anurans is a single nucleus. As in the DMN, the SON is tonotopically organized, with low frequencies represented dorsolaterally and high frequencies represented ventromedially.

Most ($\approx$81%) neurons in the SON have V-shaped frequency tuning curves that are similar to those for DMN cells (Feng and Capranica 1978; Fuzessary and Feng 1983a; Zakon 1983; Zheng and Hall 2000; Fig. 9.1a). The remaining (19%) of the SON cells, however, have complex frequency tuning functions. These consist primarily of units that have very steep low- and high-frequency flanks to their tuning functions, that is, level-tolerant frequency selectivity (11% of all cells; Fig. 1d), or highly asymmetric tuning curves, for example, "recurved" excitatory tuning curves, wherein frequencies that elicit responses when stimulus

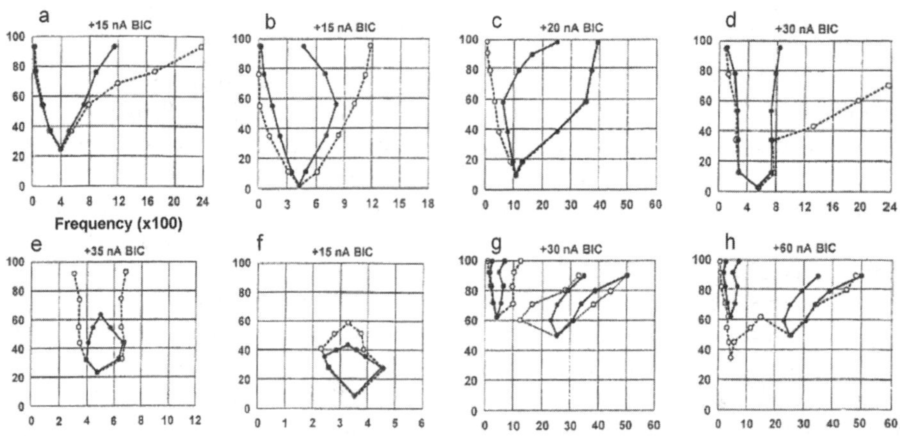

FIGURE 9.1. Frequency tuning functions of seven single units in the superior olivary nucleus before (filled symbols, lines) and after (open symbols, dashed lines) iontophoresis of bicuculline. The magnitude of current used in each case is shown above each plot. g, h: data are from the same cell, tested at two levels of iontophoretic current. Adapted from Zheng and Hall (2000).

amplitude is low fail to do so at higher stimulus amplitudes (5%; Fig. 9.1b,c). In extreme cases, the excitatory response area is limited to a narrow range of frequencies and amplitudes; that is, the tuning curve is "closed" (Figs. 9.1e,f). The latter types constitute only approximately 2 to 3% of the units recorded in the SON. Rarest (1%) are neurons that have W-shaped tuning curves, such that two frequency regions of sensitivity are exhibited (Figs. 9.1g,h).

Possibly the most prominent difference between the frequency tuning properties of neurons in the SON and DMN is that inhibitory flanking regions are seen for all SON cells, and these can be above or below the cells' CF. In the DMN, inhibition appears to be primarily restricted to the high-frequency side of the tuning curves of neurons with low CFs. Thus, it appears that central neural inhibition plays a prominent role in the frequency tuning properties of SON cells. This hypothesis is supported by the results of bicuculline iontophoresis experiments (Zheng and Hall 2000). Blockade of GABAergic inhibition transformed highly asymmetric or level-tolerant tuning curves into the V-shaped functions that are seen peripherally (Figs. 9.1b to d). In most cases, however, bicuculline iontophoresis failed to appreciably alter the structure of closed tuning curves; in the rare exceptions, these response areas were transformed into the level-tolerant type (Fig. 9.1e). Similarly, GABA blockade generally failed to alter the W-shaped tuning functions, or did so only at very high levels of iontophoretic current (compare Figs. 9.1g, h). Either inhibition is not mediated by GABA-A receptors in these cases, or these recordings were from descending fibers from the torus semicircularis (see below).

2.3 *Torus Semicircularis (TS)*

The TS is the primary auditory region in the anuran midbrain and is homologous to the inferior colliculus (Wilczynski and Endepols, Chapter 8). The three primary auditory subdivisions of the TS are the principal nucleus, laminar nucleus, and magnocellular nucleus. Like auditory-nerve fibers, most TS units have a single region of frequency sensitivity; some cells (37%) even show V-shaped excitatory frequency tuning curves. Unlike eighth-nerve fibers, however, TS neurons have well-developed inhibitory receptive fields that are not attributable to peripheral two-tone suppression (Walkowiak 1980; review, Fuzzessery 1988; Hall 1994, 1999). Units with V-shaped tuning functions can be inhibited by frequencies below and/or above their excitatory regions, and bicuculline blocks this inhibition (Figs. 9.2a to f). A major transformation in coding is that approximately 19% of the TS neurons show level-tolerant frequency sensitivity; that is, the frequency band of excitation is restricted to a narrow region regardless of the stimulus intensity. Bicuculline iontophoresis expands the excitatory bandwidth, but does not eliminate the inhibitory flanking regions (Figs. 9.3a,b). This residual inhibition may be due to non-GABA-A transmission and/or to sources in the lower processing centers; the latter hypothesis is consistent with the finding that level-tolerant tuning is also generated in the SON (see previous section). Approximately 16% of TS cells have closed tuning functions; that is, excitatory

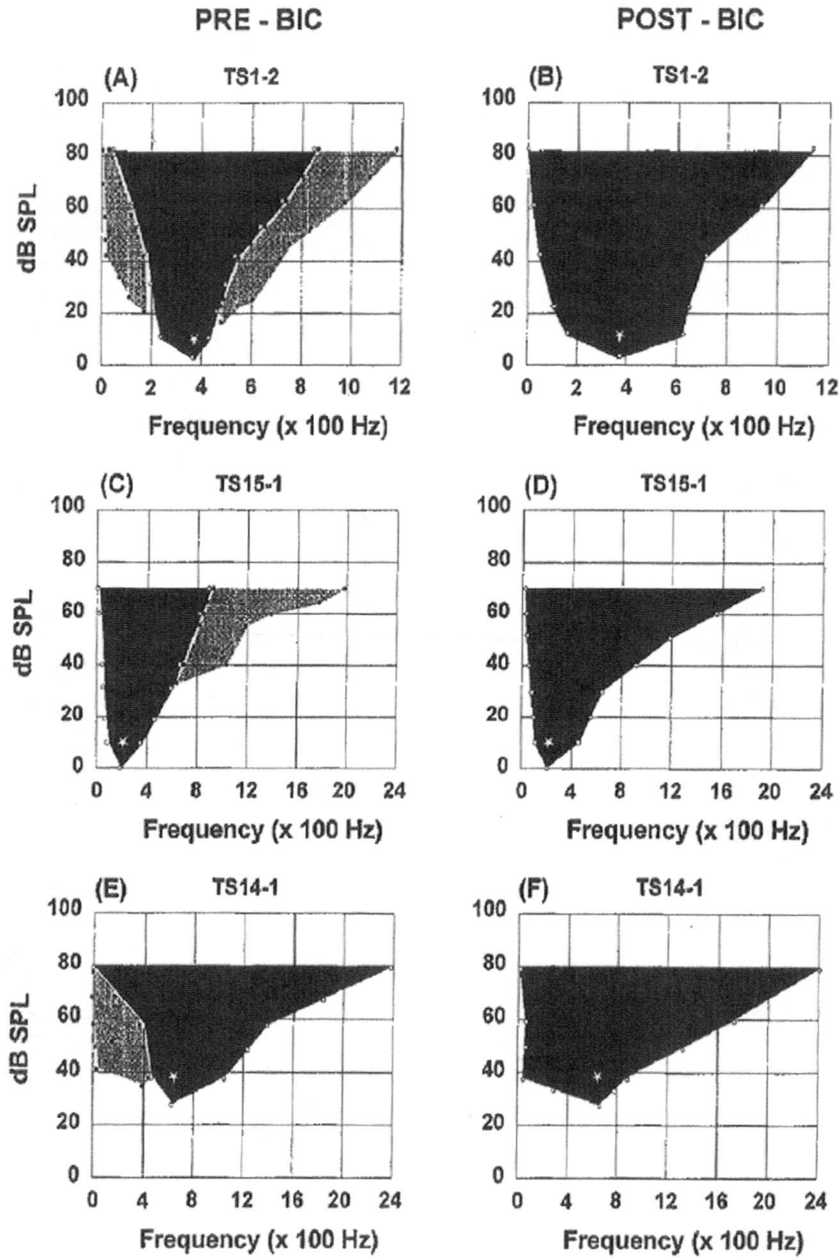

FIGURE 9.2. Frequency tuning functions for three neurons in the torus semicircularis before (a,c,e) and after (b,d,f) iontophoresis of bicuculline. Excitatory (black) and inhibitory (gray) response areas are shown. Inhibitory regions were constructed by measuring the amplitude of tones outside the excitatory region required to inhibit the response to another tone at the best excitatory frequency of the unit (white asterisks). From Hall (1999).

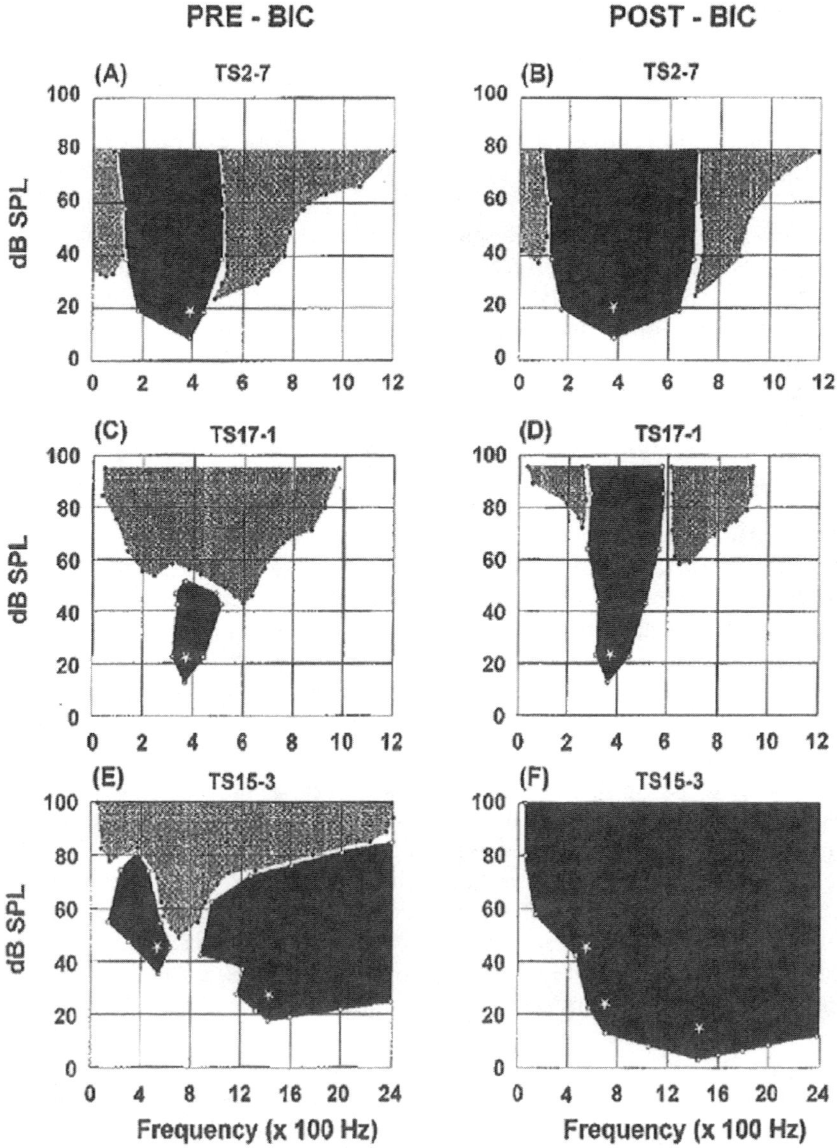

FIGURE 9.3. As in Figure 9.2, but for three neurons in the torus that show more complex excitatory and inhibitory regions. From Hall (1999).

receptive fields consist of a discrete region in frequency-amplitude space. Blocking type-A GABA synapses transforms these tuning functions into level-tolerant types, again never fully eliminating the flanking inhibition (Figs. 9.3c,d). Approximately 9% of the neurons in the TS have two regions of frequency sensitivity (Fig. 9.3e); in cases where these regions are continuous, tuning functions are W-shaped (Hall 1999). Although such neurons are uncommon, like those with closed tuning functions, they are encountered far more frequently in the TS, relative to the SON. Bicuculline iontophoresis results in a single, broad excitatory response area (Fig. 9.3f).

Although most neurons with W-shaped tuning curves respond when energy is present in either excitatory receptive field region, a few show properties characteristic of an AND logical operation; that is, these respond only, or in a facilitated manner, when energy is present in both frequency bands. Units that require energy in both low- and high-frequency bands represent the neural correlate of the AND gate proposed as the filter matched to the spectral structure of the advertisement call (bullfrogs and leopard frogs). This spectral selectivity can be more generally classified as "formant selectivity." Although selectivity of this type is apparently rare in the TS, it is well developed in the posterior thalamic nucleus (Fuzzessery and Feng 1983b).

2.4 Thalamus

Two thalamic regions, the posterior and central nuclei, have been shown to receive projections from the TS and are responsive to acoustic stimuli. The posterior nucleus appears to be innervated primarily by afferents from the laminar nucleus, whereas the principal and magnocellular nuclei project primarily to the central nucleus (Hall and Feng 1987). Single unit recordings indicate that the central and posterior nuclei are specialized for temporal (see below) and spectral processing, respectively. The importance of the thalamus in spectral processing was initially established by evoked potential studies (Mudry et al. 1977), showing that the simultaneous presence of low- and high-frequency energy elicited facilitated responses, that is, evidence for a neural AND logical operation. Recordings from single units in the posterior nucleus have directly shown that many neurons (approximately 33%) in this region respond only when low- and high-frequency energy are simultaneously present (Fuzzessery and Feng 1983b), representing the neural AND gate that was hypothesized from the seminal studies of Capranica (1965). This selectivity is appropriate for processing the spectral properties of the advertisement calls of *Rana pipiens*, for example.

3. Temporal Processing

3.1 Temporal Structure of Communication Signals

Most interspecific calls differ in both spectral and temporal structure, however, intraspecific call types often are spectrally nearly identical and differ solely in their distinctive temporal structure; in addition, the calls of very closely related species, for example, cryptic species, may differ exclusively in their temporal structure. The term "temporal structure" refers to the modulations of signal amplitude (AM), that is, the amplitude envelope, and/or frequency (FM) over time (Figs. 9.4a to c). Temporal "fine structure" refers to the actual time course of the

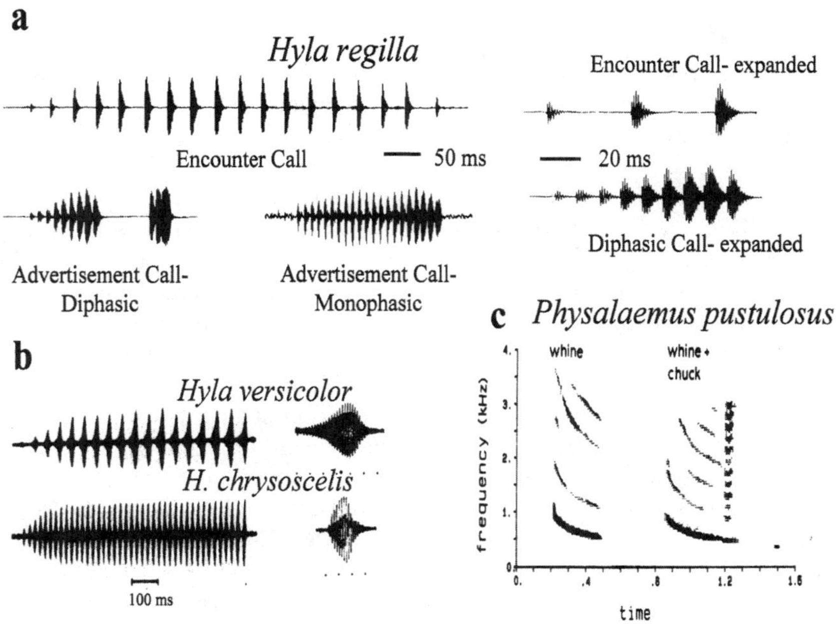

FIGURE 9.4. Representative anuran communication signals. (a) Oscillograms of the encounter and advertisement calls of *H. regilla*; advertisement calls consist of diphasic and monophasic types. (b) Oscillograms of the advertisement calls of *H. chrysoscelis* and *H. versicolor*. (c) Spectrograms of advertisement calls of *Physalaemus pustulosis*; the frequency-modulated (whine) portion can be followed by one or more harmonically rich "chucks".

waveform itself, and is important in cases where substantial energy is concentrated at low frequencies, for example, below ca. 250 Hz.

The most basic temporal features of a vocalization are its duration and rise/fall characteristics; the temporal structure of a tonal call can be described by these parameters. Many communication signals of anurans, however, show distinctive, more complex, patterns of AM (Fig. 9.4a). In some cases, amplitude modulations result from the interactions of harmonically related spectral components of the vocalization (Gerhardt 1988; Gerhardt and Bee, Chapter 5). In most cases, however, AMs result from passive and/or active mechanical processes (Martin 1971; Walkowiak, Chapter 4). The rise and fall characteristics of individual pulses can vary even between the calls of closely related species (Fig. 9.4b), and along with pulse repetition rate (PRR) constitute the primary temporal cues that enable some anurans to discriminate between conspecific and heterospecific calls (Gerhardt 1982, 1988, 2001). In some cases, different intraspecific call types differ virtually exclusively in PRR; individual pulses are highly similar in shape and spectral composition.

3.2 Background

Despite its importance in anuran acoustic communication, the investigation of the representation and central processing of temporal information historically has lagged behind that of spectral processing. For example, at the time of Capranica's 1976 review of anuran auditory systems, the study of temporal processing was in its infancy. In the early 1980s, however, this gap was narrowed with the discovery that a major transformation in the representation of AM occurs between the auditory periphery and the midbrain (Rose and Capranica 1983, 1985; Walkowiak 1988; Rose 1995). With the exception of some units that have very low CFs, auditory-nerve fibers faithfully encode AM or pulse repetition rates up to at least 100 Hz (or pulses/s) in their periodicity of discharges (Rose and Capranica 1985; Walkowiak 1988; Dunia and Narins 1989; Feng et al. 1991). That is, the spikes of a primary afferent tend to occur at a particular phase of the modulation cycle, a property that has been quantified by calculating a "synchronization coefficient." The average response levels (mean spike rate) of auditory-nerve fibers generally increase slightly or are uniform over this range of AM rate or PRR. In the midbrain, however, the response levels of most neurons depend on AM rate. Four classes of selective cells have been recognized: low-pass, high-pass, bandpass, and band-suppression (Fig. 9.5). Only high-pass and some all-pass units show significant synchronization at AM rates above ≈100 Hz. Furthermore, bandpass neurons that show strongest AM selectivity generally exhibit little, if any, synchronization coding. That is, there is a transformation in AM coding, from a periodicity code in the peripheral auditory system to a temporal filter representation in the midbrain. In the following sections, we summarize our current understanding of how temporal information in acoustic signals is represented and processed at the various levels of the central auditory system, with particular attention to the progress that has been made in understanding how this transformation is achieved.

3.3 Dorsal Medullary Nucleus

Responses of DMN neurons to tone bursts show a wide range of adaptation profiles (Hall and Feng 1990). In addition to profiles characteristic of auditory-nerve fibers (Megela and Capranica 1981), which fire throughout the stimulus to varying degrees, many neurons in the DMN show little spontaneous activity and respond in a phasic, or phasic-burst fashion (Hall and Feng 1990, 1991; Feng and Lin 1994). Phasic units generally fire one to two spikes immediately following stimulus onset, whereas phasic-burst units produce a burst of four to ten spikes. These novel response patterns are observed across a wide range of CFs, suggesting their importance in temporal coding. Furthermore, responses of phasic and phasic-burst neurons to tone bursts are strongest when stimulus amplitude rises very quickly; that is, they are rise-time sensitive (Hall and Feng 1988, 1991).

Selectivity for tone burst duration per se appears to be absent in the DMN (Hall and Feng 1991). Over a biologically relevant range of stimulus duration,

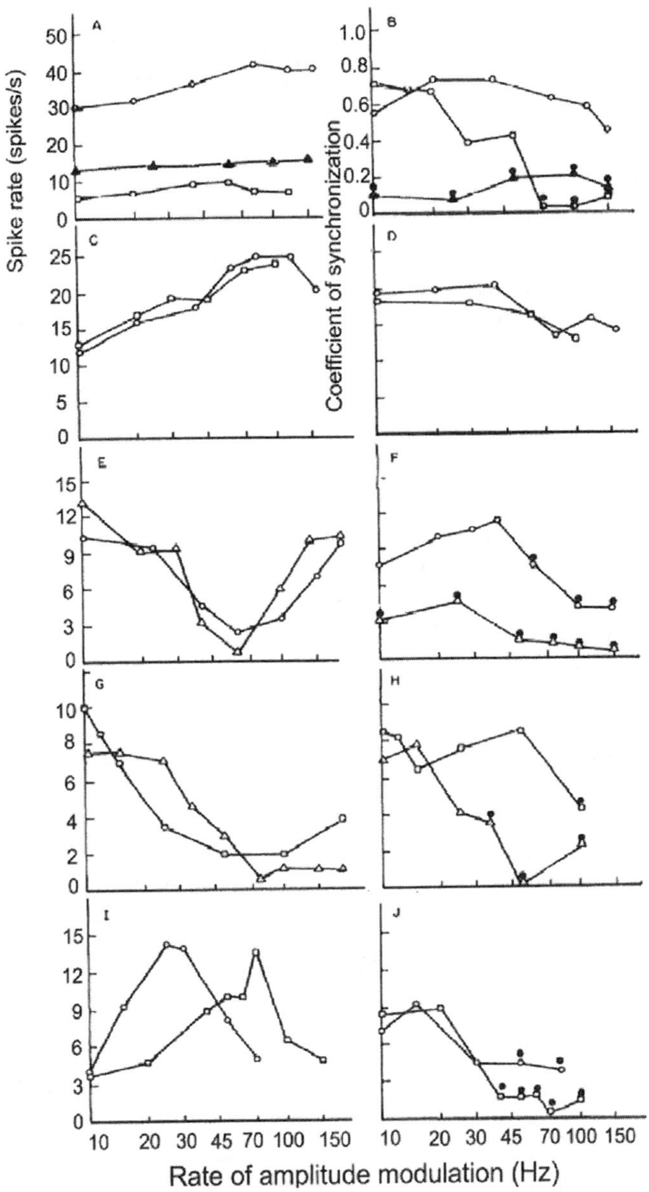

FIGURE 9.5. Response level (spikes/s, left) and synchronization (right) versus the rate of sinusoidal amplitude modulation (AM, noise carrier) for units in the torus semicircularis. Representatives of the five AM-selectivity classes are shown; nonselective (a,b), high-pass (c,d), low-pass (d,e), band-suppression (e,f) and bandpass (g,h). Synchronization values near 1.0 indicate that all spikes occurred at a particular phase of the modulation cycle; nonsignificant synchronization is indicated by filled circles above data points. From Capranica and Rose (1985).

primarylike neurons respond throughout the stimulus; that is, they code stimulus duration by the duration of their response. Phasic and phasic-burst neurons produce a constant response regardless of stimulus duration. As we show, this contrasts markedly with the responses of duration-tuned midbrain neurons.

Neurons in the DMN show synchronization of their discharges for SAM rates up to approximately 150 to 250 Hz, and the cutoffs for these low-pass functions are similar across adaptation classes (Feng and Lin 1994). In this respect, the coding of AM rate via response synchronization (periodicity coding) is highly similar between the auditory nerve (Rose and Capranica 1985; Feng et al. 1991; Dunia and Narins 1989) and the DMN; except for some units that have very low CFs, all auditory-nerve fibers show significant synchronization of their spikes to a particular phase of the modulation cycle for AM rates up to at least 100 Hz. The representation of AM in the latter region differs from that seen in primary afferents, however, when response levels are considered. DMN cells that respond phasically (both phasic and, to a lesser extent, phasic-burst types) prefer fast rise times and respond very weakly to slow AM rates (Fig. 9.6), which are characterized by few modulation cycles per second and slowly rising stimulus amplitude. Consequently, the AM-response functions (generally referred to as modulation transfer functions, $MTFs_{rate}$) of phasic DMN cells generally are high-pass, or bandpass if very fast AM rates are tested. Bandpass selectivity for intermediate rates ($\approx$30 Hz) of AM tones (tonal carriers) in some cases is transformed to high-pass selectivity when a noise source is amplitude modulated (Hall and Feng 1991). Similarly, a few auditory-nerve fibers (ones with sharp frequency tuning) show bandpass MTFs in response to AM tones (Rose 1983). These results suggest that spectral factors contribute to the decline in response level at higher AM rates. The long-term power spectrum of SAM noise is flat, whereas sidebands above and below the carrier frequency are generated when a tone is amplitude modulated. As the AM rate is increased, sidebands occur farther from the carrier frequency (and CF), thereby contributing less energy to the excitation of the cell. Nevertheless, a few units have been recorded in the DMN that show bandpass selectivity to intermediate rates of AM when noise is the carrier (Hall and Feng 1991).

Thus the DMN constitutes an important first stage of processing temporal information, particularly AM. We now turn to the question of how temporal information is represented and processed in the SON, the medullary target of efferents from the dorsal medullary nucleus.

3.4 Superior Olive

Neurons in the SON show temporal response profiles to tones that closely resemble those seen in the DMN, for example, primarylike, phasic-burst, and phasic, along with a small percentage (6%) of "pauser" types (Condon et al. 1995). One difference among these nuclei, however, is that phasic-burst types appear to be more common ($\approx$27% vs. $\approx$7%) in the SON (Condon et al. 1991; 1995).

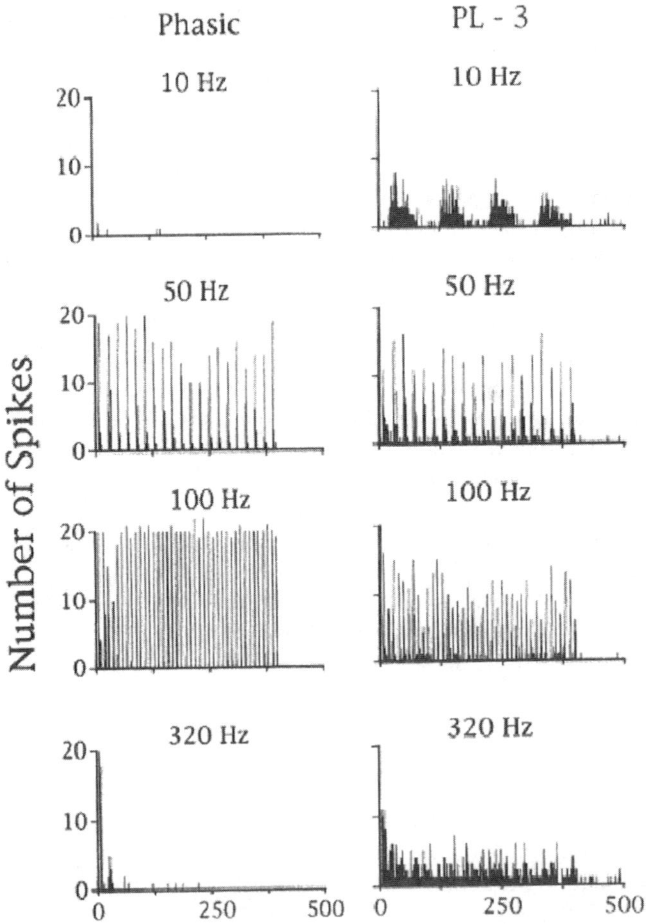

FIGURE 9.6. Peristimulus time histograms of responses of two neurons in the dorsal medullary nucleus to several rates of amplitude modulation. Adapted from Feng and Lin (1994).

As in the DMN, phasic cells in the SON respond best, and in many cases exclusively, to stimuli with rise times ≤15 to 25 ms (Condon et al. 1991). Primarylike neurons, as expected, respond at the same level over a wide range of stimulus rise/fall times. Although most phasic-burst units also show little sensitivity to stimulus rise/fall, approximately 25% exhibit rise/fall sensitivity that is intermediate between phasic and primarylike neurons; these cells respond less strongly for rise/fall times ≥≈50 ms, but this selectivity is less pronounced at amplitudes 20 to 30 dB above threshold.

Similarly, responses to variations in tone duration are correlated with temporal response profiles, as in the DMN. Primarylike neurons respond throughout the stimulus, coding stimulus duration in their duration of response. Phasic units produce a single spike per stimulus presentation, regardless of stimulus duration. The response levels of phasic-burst cells increase with signal duration. Unlike tonic units, however, increases in stimulus duration beyond approximately 80 ms fail to elicit further increases in activity.

Perhaps the most prominent difference in the representation of temporal information between the SON and the DMN is found in the encoding of amplitude modulation. In the SON almost half of the phasic-type neurons show bandpass AM selectivity (Condon et al. 1991), whereas high-pass selectivity predominates in the DMN. Most of the remaining phasic neurons show high-pass AM selectivity. Because phasic neurons can show high-pass selectivity for SAM (i.e., fast recovery times), recovery processes appear to be mechanistically independent of those that underlie the phasic response properties. Phasic SON cells respond weakly at low SAM rates for the reasons described above for DMN neurons, for example, few modulation cycles (pulses) and slow rise/fall times. Unlike in the DMN, however, approximately 27% of all SON cells show low-pass or bandpass response-level MTFs, with cutoffs generally well below 100 Hz. Thus the recovery times (time between successive pulses required to maintain response level) are markedly decreased for these low-pass and bandpass cell types. Finally, rarely, band-suppression units are recorded in the SON; these cells respond strongly to low and high rates of SAM, but poorly to intermediate rates. As we show below, these enigmatic response types are encountered more commonly in the TS.

3.5 Torus Semicircularis

As in the SON, the temporal discharge patterns of TS neurons to tone bursts largely fall into three categories, primarylike (tonic), phasic, and phasic-burst (Gooler and Feng 1992; Penna et al. 1997). Furthermore, the relative proportions of units in these categories (67%, 19%, 14%) also closely resemble those observed in the SON. Similarly, phasic neurons respond best for short rise times, whereas tonic responders are not rise-time selective (Gooler and Feng 1992; Penna et al. 2001).

Tonic (primarylike) cells in the TS reflect tone burst duration in their duration of response. Also similar to that seen in the DMN and SON, the responses of some phasic units are independent of the duration of the tone burst. Unlike cells recorded in the DMN and SON, however, approximately 20% of the neurons in the TS show duration-selective responses (Fig. 9.7) (Narins and Capranica 1980; Gooler and Feng 1992; Penna et al. 1997). For most of these units, maximal responses to tone bursts are seen when the stimulus is of a particular duration, sometimes the shortest duration tested (5 ms). This selectivity, therefore, represents a major transformation that occurs at the midbrain level.

Another important transformation in the torus is seen in the representation of AM rate; this process begins in the DMN and SON, but is particularly well developed in the TS. Early single-unit recordings showed that some neurons respond

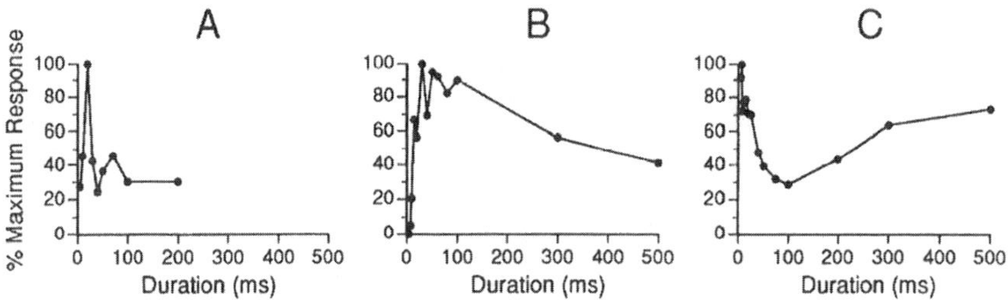

FIGURE 9.7. Normalized response levels versus stimulus duration for three units recorded in the torus semicircularis. From Gooler and Feng (1992).

well to AM tones (Bibikov and Gorodetskaya 1980) or sequences of sound pulses (Walkowiak 1980), but weakly, or not at all, to pure tones. Rose and Capranica (1983, 1985) used sinusoidally amplitude-modulated white noise stimuli to further investigate how AM is represented in the anuran auditory system. With white noise as the carrier in the modulation, long-term spectral properties of the stimulus do not change with AM rate; selective neural responses can, therefore, be attributed to temporal, not spectral, features of the stimulus.

Using this stimulus type, modulation transfer functions (response level vs. AM rate) of TS cells are level-tolerant and can be categorized as low-pass, high-pass, band-suppression, or band-pass (Fig. 9.5). This general conclusion is supported by a large number of studies in which the rate of AM or rate of repetition of sound pulses was varied (Walkowiak 1984; Epping and Eggermont 1986; reviewed in Walkowiak 1988; Eggermont 1990; Gooler and Feng 1992; Diekamp and Gerhardt 1995). The relative proportion of units in each temporal-selectivity class depends on the species studied and the properties of the AM stimuli used (Table 9.1). For example, a higher percentage of units tends to be classified as bandpass when AM tones versus AM noise are used as stimuli (Diekamp and Gerhardt 1995) and when a wide range of AM rates, including very low rates, are tested (Alder and Rose 2000); fewer cells are classified as low-pass or high-pass. A lower percentage of neurons is classified as bandpass when narrower ranges of AM rate are tested (e.g., 10 to 70 Hz; Diekamp and Gerhardt 1995); when tested over a range of approximately 5 to 150 Hz and at approximately 10 dB above each unit's threshold, 40% of the neurons in *Hyla versicolor* were found to be AM bandpass (Rose et al. 1985). This value is similar to the proportions of bandpass neurons seen in other species (Table 9.1).

Interestingly, when AM stimuli were used that matched the natural call characteristics (i.e., natural spectral and AM properties) of gray treefrogs, substantially more units were classified as bandpass (Diekamp and Gerhardt 1995; Table 9.1). Furthermore, approximately 95% of the recorded units were AM selective when tested with "natural AM" versus about 66% for sinusoidal AM. Enhanced selectivity was observed for stimuli with AM characteristics of *H. chrysoscelis* or *H. versicolor*, which have different pulse shape, suggesting that differences in

TABLE 9.1. Proportions of units, recorded extracellularly in the torus semicircularis, assigned across studies to each of the four AM selectivity classes. The species and type of stimulus used in each study is shown above each column. SAM, sinusoidal amplitude modulation; the signal that was amplitude modulated was either a pure tone or noise.

Response type / Stimulus	*Rana pipiens* SAM noise (%)	SAM tones (%)	SAM tones (%)	*Rana temporaria* SAM tones & noise (%)	*Bufo fowleri & americanus* SAM noise (%)	*Hyla versicolor & chrysoscelis* SAM noise (%)	*Hyla versicolor*		
								SAM tones (%)	Nat. AM (%)
Low-pass	17	19	6	6	9	6	36	19	13
High-pass	9	14	21	10	8	7	21	14	42
Bandpass	30	24	56	33	34	40	7	14	29
Band-suppression	9	8	6	21	9	12	4	7	7
	Rose & Capranica 1983, 1985	Gooler et al. 1992	Alder & Rose 2000	Epping & Eggermont 1986	Rose & Capranica 1984	Rose et al. 1985	Diekamp & Gerhardt 1995		

pulse duty cycle (1.0 for SAM, 0.6 for natural AM), not pulse shape, were responsible for these temporal selectivity disparities. Similarly, Rose et al. (1985) found that natural patterns of AM were more effective than SAM in exciting bandpass neurons.

In the two cases studied, the distribution of AM tuning values for TS neurons is species-specific and related to the range of PRRs observed in the calls of each species (Rose and Capranica 1984; Rose et al. 1985). Remarkably, the most selective bandpass cells show little, if any, synchronization of their spikes to a particular phase of the modulation cycle, that is, little periodicity coding of AM rate (Rose and Capranica 1984, 1985; Epping and Eggermont 1986; Eggermont 1990). Across all TS neurons, over one third fail to show significant synchronization at any AM rate (Epping and Eggermont 1986). Of the units that do encode AM rate in their time-locked discharges, few show significant synchronization beyond approximately 50 Hz AM (Rose and Capranica 1985; Eggermont 1990). Thus, as in other vertebrates (Langner 1992), the representation of AM rate is transformed from a periodicity code in auditory-nerve fibers (Rose and Capranica 1985; Dunia and Narins 1989; Feng et al. 1991), to a temporal filter ensemble in the midbrain.

The mechanisms that underlie AM selectivity are only beginning to be understood. Theoretically, bandpass selectivity for SAM might arise from sensitivity to stimulus rise time and duration. However, although some TS neurons do show sensitivity to stimulus rise time and/or duration, these properties generally appear to contribute little to their AM selectivity (Gooler and Feng 1992; Alder and Rose 2000). For example, bandpass selectivity has been observed for square-wave AM, where pulse rise time is constant across AM rates. In most cases, rise-time sensitivity only accounts for the slight differences in the shapes of bandpass functions for square-wave AM versus sinusoidal AM; at slow rates of AM, responses are weaker for sinusoidal AM because cells prefer fast rise times. The phasic response properties of these neurons appear to primarily account for their diminished responses to slow rates of SAM (Hall 1994). Bicuculine injections, presumably reversing their phasic properties, in some cases transform neurons from bandpass to low-pass, and high-pass to all-pass (Hall 1994). Theoretically, the attenuated responses at high AM rates could arise from insensitivity to short-duration pulses. However, bandpass and low-pass neurons, which respond weakly at high AM rates, respond well to short duration pulses that are presented at slow rates (Alder and Rose 2000; Fig. 9.8). At slow PRRs and AM rates, these "recovery neurons" respond phasically to each pulse and, therefore, are low-pass when pulse shape, duration, and number are held constant, and bandpass to SAM. As in the case of AM bandpass neurons in the SON, these findings suggest that recovery processes contribute to the selectivity of these neurons for pulse repetition rate; after the excitation from a stimulus pulse, a recovery period is required before the next pulse can excite the cell.

Do duration and rise time sensitivity account for bandpass selectivity to high SAM rates? Some TS neurons respond best for pulses of fast rise/fall time and short duration, precisely the characteristics of fast SAM rates. Yet, cells that show

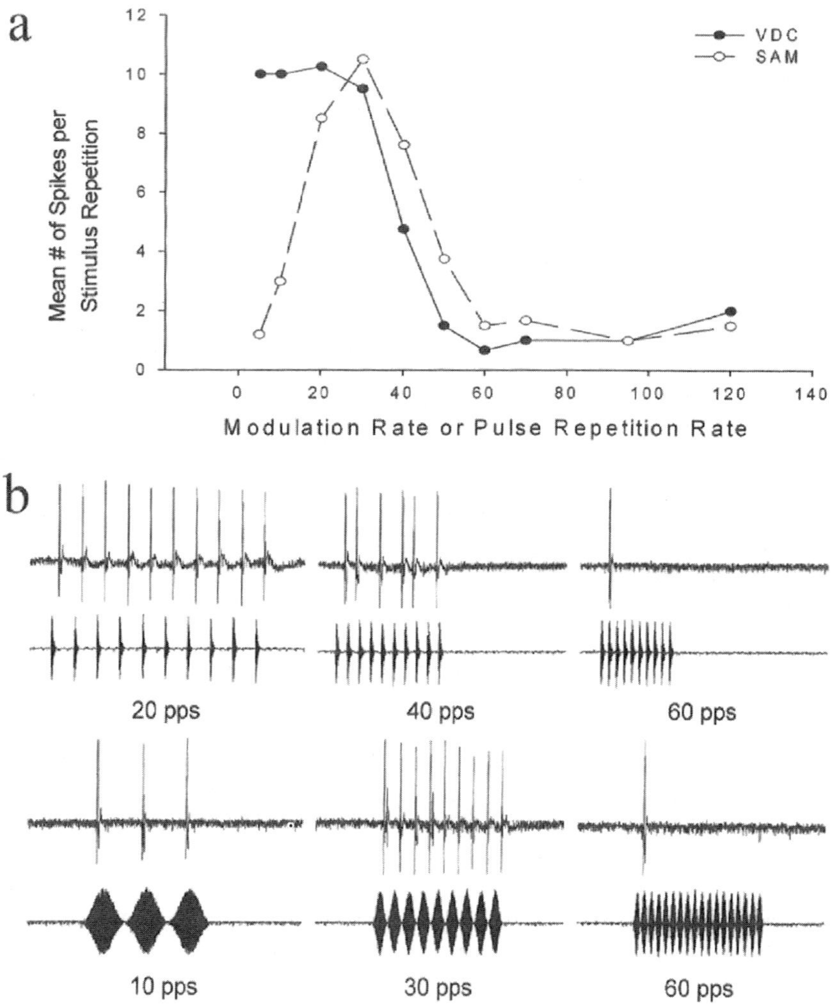

FIGURE 9.8. (a) Response magnitude versus rate of sinusoidal AM or repetition of pulses (10 ms duration) for a "recovery-type" neuron. This unit showed bandpass properties to sinusoidal AM rate, but was low-pass for stimuli in which only pulse repetition rate was varied (pulse number, shape, and duration were constant). (b) Extracellular recordings of responses of this unit to stimuli of these two types.

the strongest AM selectivity fail to respond to such pulses when presented at slow repetition rates. Processes other than just rise time and duration sensitivity must, therefore, underlie the bandpass selectivity of neurons in the anuran TS. This conclusion is, perhaps, not surprising considering that anurans can differentiate between intraspecific calls that differ primarily in PRR, not pulse shape or duration, for example, *Hyla regilla* (Allen 1973; Brenowitz and Rose 1994; Rose and Brenowitz 1997).

Recent work, consistent with behavioral findings (Rose and Brenowitz 2002), has shown that surprisingly long-term integration processes underlie strong selectivity for intermediate or fast PRRs. Neurons of this type only respond after a threshold number of pulses, each separated from adjacent pulses by a cell-specific interpulse interval (time between the onsets of consecutive pulses), have occurred (Alder and Rose 1998, 2000; Fig. 9.9a). The salient temporal feature for eliciting responses is the number of consecutive "correct" intervals not the mean pulse rate (Fig. 9.9b), thereby representing an interval-counting process (Edwards et al. 2002). This integration process appears to account for the

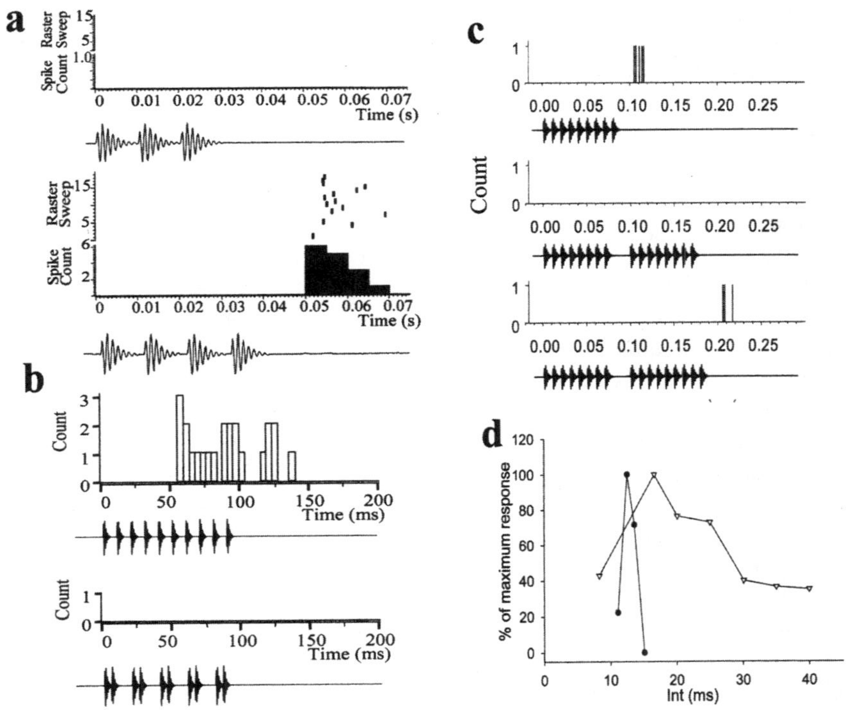

FIGURE 9.9. Properties of integration-type midbrain neurons that respond selectively to fast pulse repetition rates. (a) Raster plots and histograms of responses of a unit to multiple repetitions of stimuli having three (top) or four pulses. (b) Histograms of responses to stimuli consisting of nine consecutive intervals, 10 ms each (top), or alternating intervals of 5 ms and 20 ms (bottom). (c) Effects of a single long (30 ms) interval in resetting the interval-integrating process; this unit had an interval number threshold of 8. (d) Normalized response levels of two units as a function of the duration of a single interpulse interval embedded in a series of optimal intervals.

impressive selectivity of these units for AM or pulse repetition rate. Remarkably, a single interval that falls outside a particular range can reset the integration (Figs. 9.9c,d).

Interestingly, the enigmatic band-suppression neurons, so-called because they respond to slow and fast, but not intermediate, AM rates, are also interval-integrating types (Edwards and Rose 2003). Relative to bandpass neurons, these cells require few (median = 2) intervals for response and have broad interval tolerance. Thus, mechanistically, and perhaps functionally, band-suppression neurons and bandpass interval-integrating neurons appear to belong to a single physiological class.

Most TS neurons that are selective for AM rate derive their selectivity from recovery and/or interval-integrating processes. Thus we suggest that two predominant, physiologically distinct classes of AM selective neurons appear to exist. Neurons that derive their selectivity primarily from recovery properties are bandpass to SAM if phasic, and low-pass if not. Cells that derive their selectivity from interval-integrating properties are strongly bandpass if recovery processes limit responses to high AM rates, or high-pass if recovery times are very short.

The mechanisms that underlie the integration process in the interval-counting neurons and the recovery process of the recovery-type cells are unknown. Theoretical studies have suggested that the interplay between excitation and inhibition might underlie temporal selectivity, particularly interval analysis, in the auditory system (Buonomano 2000; Large and Crawford 2002). In these models, interval analysis stems from differences in the timing, time course, and plasticity (Buonomano 2000) of excitatory and inhibitory inputs. GABAergic inhibition is present in the anuran TS (Hall 1994, 1999), but its role in temporal processing is incompletely understood.

Temporally selective TS neurons are hypothesized to project to the central nucleus of the thalamus (Hall and Feng 1987). Few single-unit studies of temporal processing in thalamic regions have been conducted. Nevertheless the available data indicate that selectivity for temporal features of acoustic signals such as pulse duration and AM rate are particularly well developed in the central nucleus. We now present an overview of these findings.

3.6 Thalamus

As in mammalian auditory systems, the ascending projections from the anuran midbrain (TS or inferior colliculus) terminate in the thalamus. The thalamus of anurans has long been considered to play important roles in mating call recognition (Mudry et al. 1977). Earlier, attention was focused primarily on spectral processing in the thalamus, particularly with regard to the neural substrate of AND logical operations (see previous section). More recently, however, it has become evident that the central nucleus of the thalamus plays a specialized role in temporal processing, representing and enhancing the selectivity for temporal features of sound that are seen in the TS. Although the number of single-unit recordings

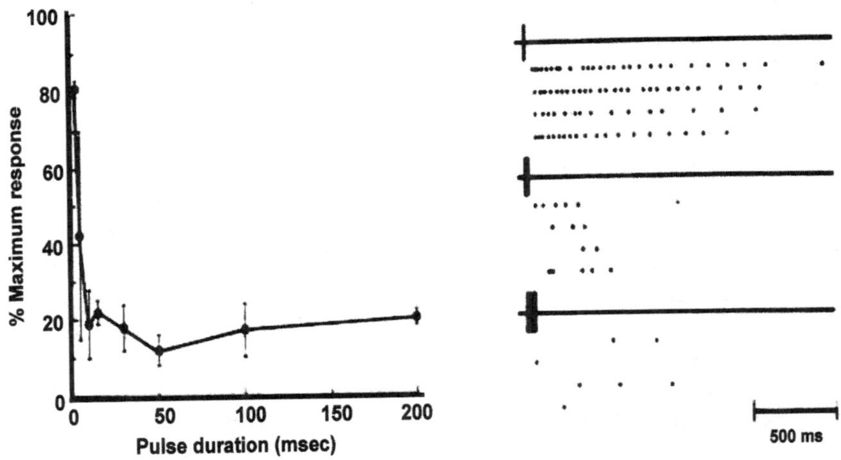

FIGURE 9.10. Normalized response versus tone-burst duration (left) and raster plots of spike timing (right) for a single unit recorded in the thalamus. From Hall and Feng (1986).

in this region is not extensive as of yet, several important findings seem clear at this juncture.

Selectivity for pulse duration, which first appears in the torus, is prominent in the thalamus (Hall and Feng 1986; Fig. 9.10). Approximately 60% of thalamic neurons respond to tone bursts only if they are of very short duration, whereas cells of this type are only rarely (9%) found in the torus. Neurons that respond best when tone bursts have a particular duration are also more commonly found in the thalamus (20 vs. 12%; Hall and Feng 1986).

The transformation in AM coding, from a periodicity code peripherally to a rate-based temporal filter representation centrally, is particularly evident in the thalamus. The responses of neurons in the central nucleus show little coding of AM in the temporal patterns of their discharges, that is, little synchronization (Hall and Feng 1986; Feng et al. 1990). This decline in AM coding by the temporal fluctuations of spike rate is apparently accentuated by the tendency of thalamic cells to produce responses that persist for hundreds of ms after the end of a short tone burst. Based on response levels to various AM rates, thalamic neurons can be described as low-pass (22%), high-pass (26%), bandpass (45%), or band-suppression (7%). In contrast to the TS, AM nonselective response types are not observed in thalamic recordings.

4. Directional Hearing in Anurans

Acoustic communication is critical for mating and territorial behaviors in many anurans. In these behaviors it is important not only to recognize sound patterns that are both spectrally and temporally complex, but to locate the position of a particular calling male (Gerhardt and Bee, Chapter 5). Therefore, the discussion now focuses on neural mechanisms of sound localization in anurans.

4.1 Contributions of the Auditory Periphery to Directional Hearing

Many anurans show good behavioral acuity for localizing sounds despite the apparent limits placed on their use of directional cues (interaural time difference and interaural level difference) by their relatively small heads, lack of pinnae, and restricted range of audible frequencies. However, the tympanic membrane shows frequency-dependent directional vibration patterns and anurans exhibit peripheral specializations that serve to enhance directional hearing. In most anurans the middle ears are acoustically coupled via patent Eustachian-like tubes that open to the mouth cavity. This configuration improves the directional sensitivity of each ear by creating a pressure-gradient receiver (Rheinlander et al. 1979). In a pressure-gradient receiver the motion of the tympanic membranes depends on the pressure difference between the inner and outer surfaces. Directional properties are enhanced because interaural level difference (ILD) and interaural time difference (ITD) cues are greater than they would be in a simple pressure receiver ear (e.g., mammals) where sound impinges on the outer surface only. A limitation of the pressure-difference receiver is that it is highly frequency dependent. Nevertheless, the particular characteristics of acoustical coupling between the ears can yield a directional system that will perform well within a limited range of frequencies.

Pressure-gradient systems work well in anurans because both their communication signals and auditory systems tend to operate within a narrow range of frequencies. Over these frequencies, the resultant sound pressure at the acoustically coupled ears also depends on the multiple routes that sound can travel from different parts of the body to the ear. Besides direct tympanic stimulation, sound can travel to the ears from the body via the nares and mouth (Vlaming et al. 1984), lungs (Narins et al. 1988; Jørgensen 1991; Jørgensen et al. 1991; Hetherington 1992; Ehret et al. 1990, 1994; Christensen-Dalsgaard and Elepfandt 1995), and through proposed extratympanic pathways via osseous/cartilaginous (Lombard and Straughan 1974; Eggermont 1988; Jørgensen and Christensen-Dalsgaard 1997b) and nonosseous/noncartilaginous routes (Narins et al. 1988; Hetherington and Lindquist 1999; Seaman 2002). These sources of acoustic input also demonstrate different frequency and phase-dependent transfer functions creating a complex interaction for directional hearing in a three-dimensional acoustic world. Further enhancements to directional hearing are evident at the level of the auditory nerve and in the central auditory system.

4.2 The Auditory Nerve

Binaural interactions in central auditory neurons are often the focus of studies that investigate neural mechanisms of sound localization. The potential of these interactions for mediating sound localization, however, depends critically on directionality information provided by the peripheral auditory system. The latter also provides a reference from which to evaluate possible functional transformations in auditory processing by central auditory nuclei. Most neurophysiological studies of the auditory nerve in anurans have investigated frequency coding and temporal processing under conditions that decoupled the ears (mouth open), effectively altering the natural pressure relationships at the tympanic membrane. However, a number of studies have investigated auditory-nerve coding of sound field stimuli in frogs with ears normally coupled and positioned in a more natural body orientation. These studies have investigated the influence of sound location on auditory processing and directional properties of auditory-nerve fibers in *Rana pipiens* (Feng 1980; Feng and Shofner 1981; White et al. 1992; Schmitz et al. 1992; Wang et al. 1996; Wang and Narins 1996; Lin and Feng 2001), *Rana temporaria* (Jørgensen and Christensen-Dalsgaard 1997 a,b), and more recently *Hyla cinerea* (Klump et al. 2004).

4.2.1 Directional Sensitivity of Auditory-Nerve Fibers

Recordings from single auditory-nerve fibers in two ranid species under free-field stimulation have demonstrated that the auditory periphery is sensitive to changes in sound direction (Feng 1980; Jørgensen and Christensen-Dalsgaard 1997a). Two directional responses have been established based on the discharge rate of auditory-nerve fibers as a function of azimuth. In order to describe the directional response of the ear the spike rates were converted to dB according to the spike rates associated with the sound levels from the fiber's rate-level function (Feng 1980, 1982). Auditory-nerve fibers with low CFs (<400 Hz) show a figure-eight directivity pattern with poor sensitivity to sounds from 30° in the frontal and 135° in the caudal fields; sources from both ipsilateral and contralateral sides in a range of 60° to 90° of azimuth elicit higher discharge rates indicating high sensitivity to sounds from the lateral fields. Responses of auditory-nerve fibers with high CFs (>500 Hz to >800 Hz) revealed an ovoidal directional pattern emphasizing sensitivity to signals from the ipsilateral azimuths. Consistent with the notion that each ear operates as a pressure-gradient receiver, the directionality of auditory-nerve fibers greatly exceeds that expected simply from measurements of extratympanic sound pressure levels.

The directional response of auditory-nerve fibers expressed in dB constitutes the directional characteristics of the entire acoustic periphery and permits comparison with those of the tympanic membrane alone in the sound field in *Rana temporaria* and *R. esculenta* (Chung et al. 1978, 1981; Pinder and Palmer 1983). The directional responses of high CF auditory-nerve fibers demonstrate the same pattern as the directional characteristics of the eardrum at high frequencies. In contrast, whereas low CF fibers show highly directional (figure-eight) responses, the eardrum shows little difference in direction-dependent vibration pattern for low frequency sounds. These results suggest that the directional characteristics of high CF fibers depends on the directionality of the tympanic membrane, but that the directional response of low CF nerve fibers results from the directional characteristics of the extratympanic pathways. A direct comparison of directional characteristics of tympanic membrane vibration patterns and auditory-nerve responses was carried out in *R. pipiens* (Wang et al. 1996). Auditory-nerve fiber activity was correlated with tympanic membrane velocity in 45% of the fibers (primarily those with CFs between 500 and 1000 Hz) suggesting that the remainder reflect some directional influence of extratympanic pathways.

To gain insight into the influence of the extratympanic and tympanic pathways on the ear's directional characteristics, Feng and Shofner (1981) modified the acoustic characteristics of ear and interaural coupling while recording from the auditory nerve in *R. pipiens*. Acoustic characteristics were modified by three independent manipulations: (1) filling the mouth with moist cotton, (2) loading the contralateral tympanic membrane, and (3) opening the mouth (Fig. 9.11). A comparison of the directional characteristics of the ear under these conditions reveals frequency-dependent effects. Directionality of the ear at middle and high frequencies (>500 Hz) changes from ovoidal to omnidirectional under the first two conditions and to a directional figure-eight pattern with the mouth open. The figure-eight pattern present at low frequencies (<500 Hz) is modified to an ovoidal pattern by filling the mouth, and to a symmetrical and asymmetrical figure-eight pattern by loading the ear and opening the mouth, respectively. These results support the concept that the low-frequency directionality of the ear is determined

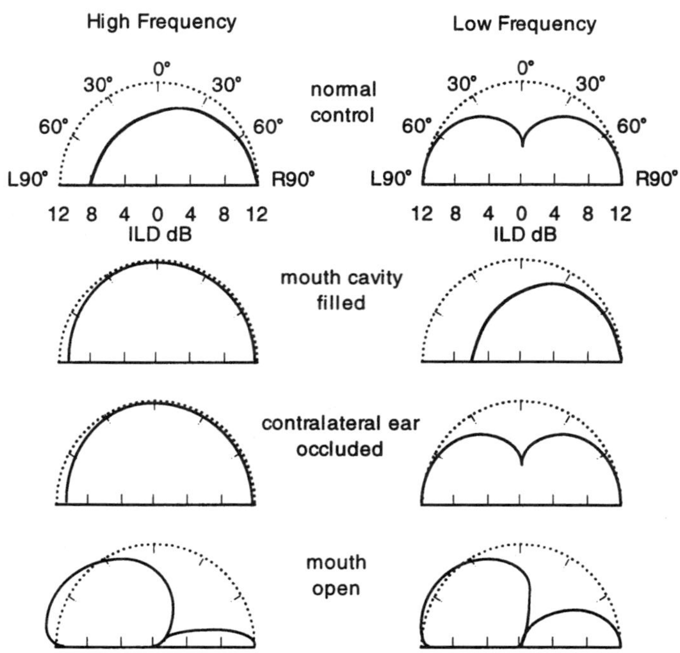

FIGURE 9.11. Directivity patterns of the frog's ear, inferred from recordings of auditory-nerve fibers, under different experimental conditions. After Feng and Shofner (1981).

primarily by extratympanic inputs, whereas directionality at middle to high frequencies results from the acoustical coupling of the middle ear cavities and of the lung to middle ear pathways. As such it indicates that the frog's ear behaves as a combined pressure, pressure-difference receiver.

The results of such frequency-dependent changes in characteristics of the tympanic and extratympanic directionality are evident in the direction-dependent alterations in frequency tuning curves of auditory-nerve fibers (White et al. 1992). Auditory fibers sensitive to mid-frequencies (720 Hz to 1200 Hz) showed the greatest direction-dependent changes in CF and FTC bandwidth with significant increase in CF and decrease in bandwidth at contralateral-posterior azimuths.

Changes in the azimuthal location of a sound source can also be encoded as shifts in the phase- and time-locking of auditory-nerve fiber discharges in anurans. Studies in *R. pipiens* (Schmitz et al. 1992), *R. temporaria* (Jørgensen and Christensen-Dalsgaard 1997b), and *H. cinerea* (Klump et al. 2004) indicate that the magnitude of interaural phase differences and ITDs depends on the direction of sound incidence. Phase-locked responses in the auditory nerves of these species to low-frequency tones showed significant interaural phase shifts with changes in sound direction. As the angle of incidence shifted toward the contralateral side phase advances were evident that resulted in ITDs up 2 ms. This expansion of time differences at the auditory nerve is remarkable in that a typical interaural distance of 2 cm would yield a propagation time difference of only 60 μs. In *R. temporaria* these changes were investigated for stimulus frequencies below 700 Hz and the shifts in ITD were larger for lower-frequency tones that elicited the best phase-locking from auditory fibers (Jørgensen and Christensen-Dalsgaard 1997b). These directional differences likely depend on the changes in extratympanic input with azimuth. In contrast ITD can be encoded by higher-frequency auditory fibers that show poor phase-locking to the tone waveform, but good time-locking to the envelope of amplitude modulated sounds. In *H. cinerea*, auditory-nerve responses to two-tone stimuli composed of 900 Hz and 1200 Hz tones showed time-locking to the stimulus envelope (Klump et al. 2004). The timing of these responses was advanced by as much as 1.3 ms for ipsilateral versus contralateral stimulus presentation. The phase advances for these higher-frequency sounds arriving from ipsilateral azimuths contrasts with the delay seen for lower frequencies. The overall directionality of the ears is likely the cause of phase shifts in the auditory-nerve fiber discharges at higher frequencies; because the ear operates as a pressure-gradient receiver. With greater tympanic vibration for ipsilateral stimulation, threshold amplitude is reached at an earlier point during the rising phase of the amplitude envelope of the stimulus (relative to contralateral presentation).

4.3 Binaural Processing and Directional Sensitivity in the Dorsal Medullary Nucleus

Many DMN neurons are binaural as a result of commissural projections between right and left nuclei (Larsell 1934; Grofova and Corvaja 1972; Feng 1986b; Will

1988; Wilczynski and Endepols, Chapter 8). Anatomical studies of the DMN have shown that the nucleus is comprised of six cytoarchitecturally distinct cell types (Feng and Lin 1996), but it is unclear whether the binaural neurons are associated with a distinct morphology. Feng and Capranica (1976) first studied binaural response properties of neurons in the DMN of *Rana catesbeiana* using closed-field stimulation of each ear (dichotic). In this study, the frog's mouth was kept open, eliminating acoustical interaction between the ears. A little more than half of the 142 cells studied were monaural and 80% were excited by the ipsilateral ear. The 67 binaural cells demonstrated two response types. About 20% of the binaural cells demonstrated an EE response in which stimulation by either ear elicited excitation with similar best excitatory frequencies. The other 80% were EI cells with all but two excited by stimulation of the contralateral ear. In these cells, responses to stimulation by the contralateral ear could be inhibited by stimulation of the ipsilateral ear. Best excitatory and best inhibitory frequencies were similar. The EI cells were sensitive to ITDs of $<150\,\mu s$ which, in a bullfrog, approaches the maximum ITD for a free-field stimulus. DMN neurons were also sensitive to ILDs of 2 to 3 dB, which approximate the maximum ILD attainable in the bullfrog for free-field stimulation. However inhibition was augmented by greater ILDs, up to approximately 30 dB.

Recently, binaural neurons in the DMN of *R. temporaria* have been studied using both closed-field and free-field stimulation (Christensen-Dalsgaard and Kanneworff 2005). Whereas closed-field stimulation permits evaluation of neural interaction independent of acoustical coupling of the ears, free-field stimulation demonstrates binaural interactions typical of the intact auditory periphery. Under free-field stimulation, many DMN neurons demonstrated directional responses similar to those of auditory-nerve fibers. DMN neurons with low-frequency CFs, however, tended to have ovoidal directionality patterns, in contrast to the figure-eight patterns of auditory-nerve fibers. Relative to auditory-nerve fibers, some DMN neurons with high CFs showed increased directionality. These differences probably stem from binaural inhibitory interactions between inputs from the two ears; for example, ipsilateral stimulation can result in earlier and stronger inhibition that reduces excitatory responses from contralateral ear stimulation.

4.4 Binaural Processing in the Superior Olivary Nucleus

Monaural and binaural response types are also seen in the SON. Bilateral projections from each DMN, of which the contralateral pathway is most prominent (Feng, 1986a; Wilczynski 1988; Wilczynski and Endepols, Chapter 8), promote further binaural processing at this stage. Feng and Capranica (1978) studied binaural response properties of neurons in the SON of *H. cinerea* using closed-field stimulation. Fifty-eight percent of the 146 recorded neurons were monaural and were sensitive predominantly to contralateral stimulation. The remaining neurons demonstrated binaural sensitivity; most were EI responses where contralateral stimuli elicited excitation. About 17% of binaural neurons showed EE responses. Neurons in the SON demonstrate ILD and ITD sensitivity, but there appears to be no segregation of neurons sensitive to each cue, as is observed in the barn owl or the mammalian superior olivary complex. ILD sensitivity was studied in a subset of EI neurons using acoustic clicks and the vast majority showed changes in neural discharge over a range of ±10 dB ILD; these responses to stimuli with different ILDs were independent of neuron CF. In contrast, ITD sensitivity was dependent on CF. Neurons with higher CFs (>2600 Hz) showed no change in neural discharge with ITD. Neurons with lower CFs showed a systematic reduction in discharge with increasing lead time between the ipsilateral and contralateral click stimuli over the range of ±500 μs ITD. These neurons showed sensitivity to ILDs and ITDs larger than the maximal physical ILD (<2 dB) and ITD (±45 μs) attainable in *H. cinerea*. Such sensitivity indicates that neurons in the SON can code for enhanced directional cues produced by the auditory periphery under free-field stimulation.

4.5 Binaural Processing and Direction-Dependent Responses in the Torus Semicircularis

4.5.1 Auditory Projections to the TS

The physiology of directional hearing has been studied more completely in the TS than in other central auditory nuclei. The principal nucleus is the primary recipient of projections from the caudal auditory brainstem including bilateral input from the DMN, SON, and the nucleus of the lateral lemniscus (NLL); predominant input is from the contralateral DMN, and the ipsilateral SON and NLL (*R. pipiens*, Feng and Lin 1991; *Bombina orientalis* and *Discoglossus pictus*, Walkowiak and Luksch 1994). The principal nucleus also receives descending auditory input from the posterior thalamic nucleus (Feng and Lin 1991). The magnocellular nucleus receives bilateral input from caudal brainstem auditory nuclei, but primarily from ipsilateral SON and NLL. The laminar nucleus receives ascending inputs predominantly from the ipsilateral SON and also the ipsilateral NLL. However, most of the inputs to the magnocellular and laminar nuclei originate from descending projections of multiple thalamic nuclei (Feng and Lin 1991). The projections to the TS reflect the directionality of the auditory periphery and convergence of input from multiple sites of binaural processing in the caudal brainstem and the thalamus. The extensive bilateral projections to the TS suggest that it is a prominent site for the processing and transformation of auditory signals important for sound localization.

4.5.2 Binaural Processing in the TS

A number of studies have investigated binaural response properties of neurons in the anuran TS yielding somewhat different results depending on whether the stimuli were presented while the ears were coupled or decoupled. Kaulen et al. (1972) showed in an unnamed Chilean frog that about 40% of neurons in the TS responded binaurally with ears coupled; similar numbers of EE and IE types were recorded (almost all showed ipsilateral inhibition). Similar proportions of binaural response types were demonstrated in *R. temporaria* (ears coupled) with the additional feature that EE types were prominent for neurons with high-frequency CFs whereas monaural (EO) and EI response types were typical of neurons with lower-frequency CFs (Hermes et al. 1982; Epping and Eggermont 1985). Epping and Eggermont (1985) also demonstrated that binaural response types may change with sound level and that binaural stimulation could yield facilitation, and partial or complete suppression of neural responses compared to monaural excitation.

An additional complexity in defining binaural response types is that leaving the ears coupled could lead to an overrepresentation of EE neurons (Epping and Eggermont 1985). The ear that is typically inhibitory could appear excitatory because presentation of sound to this ear could inadvertently stimulate the contralateral excitatory ear via crosstalk between the coupled middle ears. To evaluate the effect of decoupling the ears on binaural responses in the TS, Melssen and Epping (1990), also used closed-field stimulation in *R. temporaria*, but with the mouth open. Under these conditions 76% of neurons revealed binaural response types and 88% of these were EI neurons. Another approach to evaluate binaural response properties, but with the frog's mouth in the natural closed position was pursued by Gooler et al. (1996) in *R. pipiens*. To be able to identify EE neurons as those excited by each ear without crosstalk they measured the neural crosstalk due to middle ear coupling in auditory-nerve fibers. Crosstalk was defined as the difference in the linear portion of the rate-level functions for ipsilateral and contralateral stimulation. EE neurons were identified as those where stimulation of each ear produced an excitatory response and the difference between contralateral and ipsilateral thresholds was below the crosstalk at the neuron's CF. Under these conditions 35% of binaural neurons were classified as EE. However, by adjusting the ILD (±10 dB) it was shown that 94% of all binaural neurons showed some degree of binaural inhibition. Changes in ILD revealed large changes in neural activity in 74% of binaural neurons. As the relative sound level at the ipsilateral ear increased, neural discharges fell by ≥50% of that to contralateral monaural stimulation, indicating strong ipsilateral inhibition. These significant changes in inhibition with ILD are relevant to encoding of directional cues as well as the processing of frequency and temporal information in the free-field as discussed in Section 4.5.4.

4.5.3 Directional Sensitivity of Single Neurons in the TS

Under free-field stimulation the majority of neurons in the auditory nuclei of the TS demonstrated azimuth-dependent changes in the discharge rate and/or first spike latency (Feng 1981). The rate-level responses were determined for sounds emanating from a loudspeaker that could be rotated 180° in the horizontal plane to any azimuth in the frontal sound field. These responses can be converted to directionality patterns as described above for the auditory nerve and in fact most were similar to those of auditory-nerve fibers and DMN neurons. The majority of neurons (83%) showed an ovoidal pattern demonstrating strong responses to sounds from contralateral azimuths. A smaller portion of neurons (8%) demonstrated figure-eight or inverted figure-eight patterns in which sounds from more lateral azimuths elicited similar responses, but sounds closer to the frontal midline yielded poorer and better responses, respectively. The remaining 9% showed no changes with sound direction or complex directionality patterns. A significant difference from the auditory nerve and DMN is that the directional patterns are not frequency-dependent in the TS. To investigate central contributions to directionality, directional responses were measured in a subset of frogs after one auditory nerve was sectioned. The result was a significant decrease in directionality compared to the intact condition from 10.4 ± 4.9 dB to 5.6 ± 1.6 dB. This difference in directionality between binaural and monaural systems suggested that directional sensitivity evident at the auditory nerve is further sharpened by binaural processing in the brainstem and thus serves to improve sound localization.

The enhanced "effective" ILDs generated by the acoustics of the anuran ears as pressure-gradient receivers also amplify ITDs as a result of time-intensity trading in auditory-nerve fibers. The encoding of ILDs is represented neurally by differences in discharge rate and latency as great as 15 spikes/dB and 1.4 ms/dB, respectively (Feng 1982). As presented above, neural ITDs can be greatly expanded compared to the small ITDs generated by the distance between the ears. Thus, processing of ILD and ITD becomes highly interrelated in binaural neurons at the level of the TS. The expanded ITDs may also amplify differences in binaural inhibition associated with shifts in sound azimuth. In order for the interaural differences to generate a consistent representation of sound direction they must be compared across frequency bands and be relatively intensity independent. In *R. temporaria*, up to about 60% of neurons in the TS demonstrate intensity-invariant responses to either click stimulation (Melssen and Epping 1992) or amplitude-modulated sounds and a similar proportion show frequency-independent ITDs (Melssen et al. 1990). Seventy-four percent of the TS neurons showed asymmetrical ITD functions indicating a clear shift in response with ITD; most of these neurons responded best to leading contralateral stimulation and were inhibited by ipsilateral stimulation (Melssen et al. 1990). About 20% of the neurons were not selective for ITD and 6% demonstrated a symmetrical response to ITD with best responses at ITD = 0 ms.

Studies of ILD and ITD processing in anurans have revealed that changes in interaural parameters can alter the spike rate and latency of responses. As a result, binaural integration of excitatory and inhibitory inputs in binaural neurons is also modified. It is not surprising, although very interesting, to find that neural responses to temporal (Bibikov 1977; Melssen and van Stokkum 1988; van Stokkum and Melssen 1991) and spectral (Melssen and Epping 1990) features of sounds change with ITD and ILD in many TS neurons. Changes in frequency sensitivity, frequency tuning bandwidth, and CF along with selectivity for amplitude modulation frequency have been documented for TS neurons. This indicates that the identity of sounds produced from different sources in the sound-field could be processed by the same central auditory neurons that encode the source location. It has been suggested that ensembles of neurons could uniquely code this information (Melssen and Epping 1990).

4.5.4 Sound Direction Alters Spectral and Temporal Selectivity in the TS

The integration of spectral and temporal encoding of acoustic information with spatial location is a major task for the neural substrate underlying acoustic communication in anurans. Because mating and territorial behaviors rely on acoustic communication in the free-field it is important to evaluate the neural mechanisms underlying those behaviors in free-field conditions. To begin to understand these mechanisms, studies were carried out to investigate the influence of sound direction on frequency-tuning characteristics (Gooler et al. 1993; Xu et al. 1994) and temporal selectivity for amplitude-modulated sounds (Xu et al. 1996) of neurons in the TS of *R. pipiens*. In these studies, sounds were presented from a loudspeaker that could be rotated 180° in the horizontal plane to any azimuth in the frontal sound-field. For most neurons, changing the sound azimuth from the contralateral to ipsilateral side produced an increase in threshold at CF and a narrowing of FTCs that was most prominent at 10 dB and 20 dB above threshold (Gooler et al. 1993).

The same effect was evident at suprathreshold sound levels where the bandwidth of the isointensity frequency response became narrower when tones were presented from ipsilateral compared to contralateral azimuths with strongest effects around the CF (Xu et al. 1994). Similarly, temporal selectivity for AM sounds increased when stimuli were presented from ipsilateral azimuths. In fact it was shown in a subset of TS neurons that the isointensity frequency response narrowed in parallel with improved amplitude-modulation selectivity for ipsilateral sound directions (Xu et al. 1996).

The results of these studies along with those using dichotic closed-field stimulation suggested that changes in binaural interaction that shaped these responses were likely the result of binaural inhibition. Additional support for a central processing mechanism is that the direction-dependent changes in frequency selectivity in TS neurons (Gooler et al. 1993) were significantly greater than those exhibited by auditory-nerve fibers (White et al. 1992). To investigate the relationship between binaural inhibition and direction-dependent changes in frequency threshold tuning for individual TS neurons, closed-field ILDs were measured to indicate the strength of binaural inhibition free-field FTCs (Gooler et al. 1996). TS neurons that exhibited the largest direction-dependent changes in frequency selectivity were typically those that displayed stronger binaural inhibition. In these neurons, occlusion of the ipsilateral ear (typically inhibitory), abolished direction-dependent frequency selectivity. To test the hypothesis that direction-dependent changes in binaural inhibition alter excitatory FTCs, both inhibitory and excitatory FTCs were measured as a function of sound azimuth in TS neurons (Zhang and Feng 1998). This investigation revealed that the bandwidth of inhibitory tuning expanded as the excitatory FTCs narrowed at ipsilateral azimuths.

That these effects on frequency tuning were due at least in part to inhibitory projections to the TS, was investigated directly (Zhang et al. 1999) using iontophoretic techniques to block inhibitory synaptic transmission. More specifically, they investigated whether blocking inhibition mediated by γ-aminobutyric acid (GABA) and/or glycine would alter direction-dependent changes in frequency selectivity. Iontophoresis of the glycine blocker, strychnine, did not alter direction-dependent frequency tuning. Application of the GABA antagonist bicuculline resulted in a direction-dependent broadening of excitatory FTCs; for stimulation of the ipsilateral ear, where frequency bandwidths were typically narrower, bicuculline induced greater broadening of FTCs (Fig. 9.12). Blocking GABAergic inhibition often reduced direction-dependent frequency tuning dramatically, but not necessarily completely. Some neurons that showed strong binaural inhibition also showed little influence of bicuculline on their ILD function or FTCs (Fig. 9.12). In these cases the lack of local effect of bicuculline suggests that a portion of binaural processing that is evident in the response selectivity of TS neurons takes place in lower brainstem nuclei, possibly by GABA-based inhibition (in *R. pipiens*: Hall 1991; Zheng and Hall 2000; in *B. orientalis, D. pictus,*

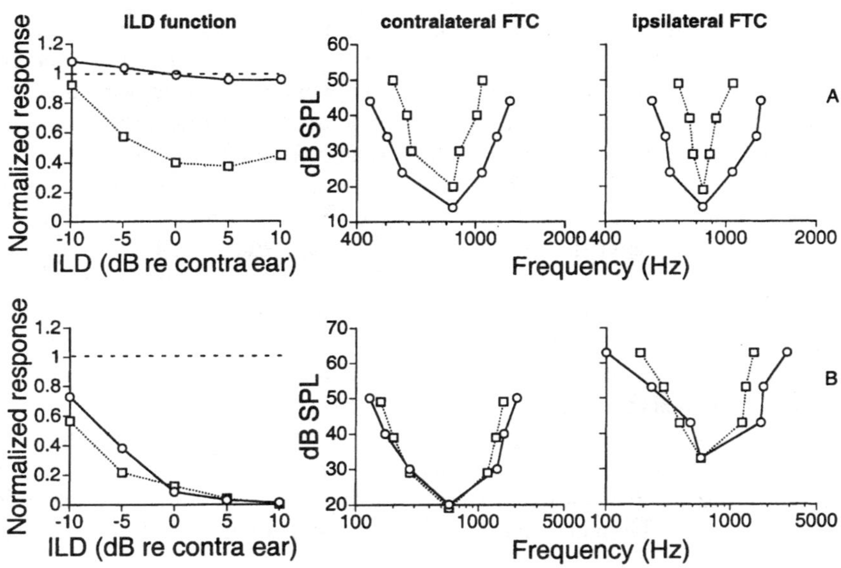

FIGURE 9.12. Examples of ILD functions and closed-field monaural FTCs before and after application of bicuculline. The unit in (A) shows strong binaural inhibition associated with increased levels at the ipsilateral ear. The ipsilateral FTC is narrower than the contralateral FTC. Bicuculline essentially abolishes binaural inhibition causing a broadening of FTCs and a flat ILD function. In (B) the unit shows strong binaural inhibition but little effect of local application of bicuculline on either the ILD function or the FTCs. Dotted and solid lines depict responses in the absence and presence of bicuculline, respectively. From Zhang et al. (1999).

and Xenopus laevis: Endepols et al. 2000), or by more rapid glycine-induced inhibition (Hall 1991).

4.6 Contributions of the Diencephalon and Telencephalon to Sound Localization

The contributions of the diencephalon and telencephalon to sound localization have been evaluated through lesion studies; the effect of focal lesions within these areas on anuran phonotaxis that involves recognition and localization of male mating calls by females was investigated. Schmidt (1988, 1989) studied the results of a series of lesions in the thalamus and telencephalon on phonotaxis. In one set of experiments, females of *Bufo americanus* continued to approach a loudspeaker emitting male mating calls despite extensive lesions including the pretrigeminal nucleus, the telencephalon, dorsal thalamus, and dorsal and medial regions of the TS (Schmidt 1988). However, lesions of the preoptic area and adjacent septal area (Schmidt 1989; in *H. versicolor,* Walkowiak et al. 1999) reduced the frequency of successful phonotactic approaches to the loudspeaker. Because lesions of diencephalic and telencephalic structures reduced the probability of a response, it was suggested that these regions do not alter the motor program for phonotaxis (Schmidt 1988). Instead, the lesions may alter hormonal sensitization to, or acoustic release of, the behavior (Schmidt 1988, 1989). Studies in *H. versicolor* (Walkowiak et al.1999) corroborated the result that phonotaxis was not affected by lesions in the dorsomedial thalamus. However, lesions of the striatum and superficial and deep thalamic structures, but particularly lesions in the TS severely impaired phonotactic performance (Endepols et al. 2003). This study distinguished between effects of lesions on call preference and on phonotaxis, but otherwise it is difficult to know which aspects of phonotactic behavior are altered by particular lesions. It is possible that reduced ability to localize sounds is sufficient to diminish the probability of a phonotactic response or to increase the time to locate the sound source as a result of impaired ability to localize calls.

An understanding of the roles of particular diencephalic and telencephalic nuclei in phonotaxis and directional hearing requires further study. However, the studies described may suggest a role for thalamic structures via the phonotactically critical TS. Because the laminar nucleus of the TS projects to spinal motor neurons (Luksch and Walkowiak 1998) an interruption of descending thalamic projections to the laminar nucleus may alter the communication of sensory-generated motor instructions (Endepols and Walkowiak 2001).

5. Conclusions and Future Directions

After approximately three decades of work, we now have a reasonably good understanding of how some forms of spectral and temporal information are represented and processed at lower levels of the anuran central auditory system. In the brainstem, only the small superficial isthmal reticular nucleus (homologue of the nucleus of the lateral lemniscus) remains to be thoroughly studied; single auditory units that respond to AM stimuli have been recorded (Rose and Wilczynski 1984; Bibikov 2002), however, this region has been comparatively neglected and further studies are needed.

Beyond the brainstem, the function of the auditory system is less clear (Wilczynski and Endepols, Chapter 8). Clearly, thalamic regions show specializations for processing spectral or temporal information, but the study of this area has just begun. Little attention has been given to the roles of auditory inputs to nonthalamic forebrain regions. Consistent with the finding that the hypothalamus receives afferents from thalamic nuclei (Allison and Wilczynski 1991), neurons in the preoptic area and ventral hypothalamus show changes in activity (primarily increases) in response to advertisement call stimuli (Allison 1991); white noise was a relatively ineffective stimulus, suggesting that highly processed auditory information is represented. Future work is needed to identify the spectral and temporal selectivity of these neurons. The roles of auditory inputs to neurons in other forebrain regions (e.g., striatum, septal nuclei, medial pallium) are unknown. These polysensory areas (Wilczynski and Endepols, Chapter 8) provide extensive descending feedback to the midbrain, and may have modulatory actions. In general, however, the functions of these areas and their output are poorly understood.

5.1 Spectral Processing

The frequency tuning functions of units at each of the major auditory nuclei in the brainstem have been well characterized. In addition, iontophoresis and recording studies have clearly demonstrated that GABA-A receptors play a role in mediating the inhibitory processes that underlie the complex tuning functions of many of these cells. Nevertheless, important questions still remain. In several regions of the central auditory system, units are occasionally recorded that have complex tuning functions typical of units at higher levels. Coincidentally in these cases, the inhibition that is important in shaping this tuning cannot be eliminated by bicuculline iontophoresis, or is only eliminated at high currents. These findings raise the question of whether such recordings are from descending afferents, rather than neurons in the particular region of recording. This hypothesis could be tested by recording from single units before, during, and after inactivation of neurons in regions where these descending afferents originate. Experiments of this type could also provide needed information concerning the influences of descending feedback projections on the response properties of neurons at low levels. Second, bicuculline fails to completely block inhibition even in some cases where frequency tuning functions are common for the region being investigated. In particular, neurons in the torus that exhibit level-tolerant tuning (inhibitory regions below and above the units' CF) or closed tuning functions showed inhibition that was only partially attenuated by bicuculline (Fig. 9.3). The "residual" inhibition could be non-GABAergic (type A), or might reflect processes taking place in the caudal auditory brainstem. Intracellular recording is needed to resolve this issue.

5.2 Temporal Processing

Much progress has been made in understanding how temporal information is represented and processed in anuran auditory systems. In particular, a great deal is now known concerning how AM (rate and form) and sound duration are coded; neurons at each level have been classified in terms of their selectivity for these temporal parameters. We are just beginning to gain insight into the mechanisms that underlie temporal selectivity. Clearly integration and recovery processes, along with rise-time and duration sensitivity, play fundamental roles in generating selectivity for AM. Furthermore, it is now clear that "AM band-suppression" cells derive their selectivity from integration processes, like integration-type bandpass cells; thus these neurons belong to a single functional class. The mechanisms underlying these integration (interval-counting) and recovery processes, however, are unknown. Theoretical studies have suggested that the interplay between excitation and inhibition might underlie temporal selectivity, particularly interval analysis, in the auditory system (Buonomano 2000; Large and Crawford 2002). In these models, interval analysis stems from differences in the timing, time course, and plasticity of excitatory and inhibitory inputs. GABAergic inhibition is present in the anuran TS (Hall 1994, 1999), but its role in temporal processing is incompletely understood. Also, because the frequency tuning of inhibition is determined indirectly (via its effects on concurrent excitation), it is likely that the extent of inhibition has been underestimated.

The predominance of inhibition needs to be explained. Although inhibition clearly can alter the frequency tuning of neurons, the possibility that inhibition might, in some cases, function primarily to generate selectivity for temporal features remains to be investigated (Hall 1999).

Finally, only one neurophysiological investigation of FM processing in anurans has been made (Narins 1983). Anurans represent a highly tractable model system in which to study how FM information is represented in the auditory system and the mechanisms that underlie FM selectivity. Little is known, however, about either of these processes.

5.3 Directional Hearing

It is likely that the processing of sound location and pattern recognition merge within the ascending anuran auditory system. It seems that the "what" or sound pattern recognition and the "where" or location of the caller are integrated at least at the level of the TS (Melssen and Epping 1990; van Stokkum and Melssen 1991; Gooler et al. 1993; Xu et al. 1994; Gooler et al. 1996; Xu et al. 1996; Zhang et al. 1999; Lin and Feng 2001, 2003). How are sound identity and location encoded and how is that information coordinated to instruct motor output? In the absence of a spatiotopic map, which still needs a mechanism for translation into motor commands, ensemble encoding (Melssen and Epping 1990) may provide a means of integrating both sets of auditory information. Consider the distributed responses underlying the representation of different calls in the TS of *Physalaemus pustulosus* (Hoke et al. 2004). In this study, immediate early gene (*egr-1*) expression showed different distributions in response to conspecific calls across subdivisions of the TS. The laminar nucleus in particular demonstrated activity best "related to the biological meaning of acoustic signals, that is, their differential ability to trigger natural behavioral responses, rather than simple acoustic properties." Recall that it is the laminar nucleus that has the potential for descending motor control (Luksch and Walkowiak 1998). Further study is needed to determine how sound direction is integrated with call recognition and hormonally mediated gating of behavioral motivation to instruct motor behavioral responses during phonotaxis.

Finally, it is important to remember that anurans in a breeding pond often need to recognize conspecific calls and locate the caller while being challenged by calls from 15 to 25 different species and thousands of individuals. However, they are able to perceptually extract and localize signals in noisy surroundings as described for human listeners by the cocktail-party effect. Behavioral studies in anurans have underscored the role of binaural hearing in improving signal detection in noise in the free-field, but only a few recent studies have investigated the neural mechanisms associated with spatial separation of sound sources and release from masking. These studies are well described within this volume (See Feng and Schul, Chapter 11).

References

Alder TB, Rose GJ (1998) Long-term temporal integration in the anuran auditory system. Nature Neurosci. 1:519–522.

Alder TB, Rose GJ (2000) Integration and recovery processes contribute to the temporal selectivity of neurons in the midbrain of the northern leopard frog, *Rana pipiens*. J Comp Physiol A 186:923–937.

Allen DM (1973) Some relationships of vocalization to behavior in the Pacific treefrog, *Hyla regilla*. Herpetologica 29:366–371.

Allison JD (1991) Acoustic modulation of neural activity in the preoptic area and ventral hypothalamus of the green treefrog (*Hyla cinerea*). J Comp Physiol A 171:387–395.

Allison JD, Wilczynski W (1991) Thalamic and midbrain auditory projections to the preoptic area and ventral hypothalamus in the green treefrog (*Hyla cinerea*). Brain Behav Evol 37:322–331.

Benedix JH, Pedemonte M, Velluti R, Narins P (1994) Temperature dependence of two-tone rate suppression in the northern leopard frog, Rana pipiens pipiens. J Acoustic Soc Am 96:2738–2745.

Bibikov NG (1977) Dependence of the binaural neurons reaction in the frog tours semicircularis on the interaural phase difference. Sechenov Physiol J USSR 63:365–373.

Bibikov NG (2002) Responses of the auditory neurons located in the isthmal region of the lake frog. Sensornye-Sistemy 16(1):23–34.

Bibikov NG, Gorodetskaya ON (1980) Single unit responses in the auditory center of the frog mesencephalon to amplitude-modulated tones. Neirofiziologiya 12:264–271.
Brenowitz EA, Rose GJ (1994) Behavioural plasticity mediates aggression in choruses of the Pacific treefrog. Anim Behav 47:633–641.
Buonomano DV (2000) Decoding temporal information: A model based on short-term synaptic plasticity. J Neurosci 20(3):1129–1141.
Capranica RR (1965) The evoked vocal response of the bullfrog: A study of communication by sound. MIT Res Monogr 33. Cambridge, MA: MIT Press.
Capranica RR (1966) Vocal response of the bullfrog to natural and synthetic mating calls. J Acoust Soc Am 40:1131–1139.
Capranica RR (1976) Morphology and physiology of the auditory system. In: Llinas R, Precht W (eds) Handbook of Frog Neurobiology. Berlin: Springer, pp. 551–575.
Capranica RR, Moffat AJM (1980) Nonlinear properties of the peripheral auditory system of anurans. In: Fay RR, Popper AN (eds) Comparative Studies of Hearing in Vertebrates. Berlin: Springer Verlag, pp. 139–166.
Christensen-Dalsgaard J, Elepfandt A (1995) Biophysics of underwater hearing in the clawed frog, *Xenopus laevis*. J Comp Physiol A 176:317–324.
Christensen-Dalsgaard J, Kanneworff M (2005) Binaural interaction in the frog dorsomedullary nucleus. Brain Res Bull, 66:522–525.
Chung SH, Pettigrew A, Anson M (1978) Dynamics of the amphibian middle ear. Nature 272(5649):142–147.
Chung SH, Pettigrew A, Anson M (1981) Hearing in the frog: Dynamics of the middle ear. Proc Roy Soc Lond B 212:459–485.
Condon CJ, Chang SH, Feng AS (1991) Processing of behaviorally relevant temporal parameters of acoustic stimuli by single neurons in the superior olivary nucleus of the leopard frog. J Comp Physiol 168:709–725.
Condon CJ, Chang S-H, Feng AS (1995) Classification of the temporal discharge patterns of single auditory neurons in the frog superior olivary nucleus. Hearing Res 83:190–202.
Diekamp B, Gerhardt HC (1995) Selective phonotaxis to advertisement calls in the gray treefrog, *Hyla versicolor*: Behavioral experiments and neurophysiological correlates. J Comp Physiol 177:173–190.
Dunia R, Narins PM (1989) Temporal resolution in frog auditory-nerve fibers. J Acoust Soc Am 85:1630–1638.
Edwards CJ, Rose GJ (2003) Interval-integration underlies amplitude modulation band-suppression selectivity in the anuran midbrain. J Comp Physiol A 189:907–914.
Edwards CJ, Alder TB, Rose GJ (2002) Auditory midbrain neurons that count. Nature Neurosci 5:934–936.
Eggermont JJ (1988) Mechanisms of sound localization in anurans. In: Fritzsch B, Ryan MJ, Wilczynski W, Hetherington TE, Walkowiak W (eds) The Evolution of the Amphibian Auditory System. New York: Wiley, pp. 307–336.
Eggermont JJ (1990) Temporal modulation transfer functions for single neurons in the auditory midbrain of the leopard frog intensity and carrier-frequency dependence. Hear Res 43:181–198.
Ehret G, Keilwerth E, Kamada T (1994) The lung-eardrum pathway in three treefrog and four dendrobatid frog species: some properties of sound transmission. J Exp Biol 195:329–343.
Ehret G, Tautz J, Schmitz B, Narins PM (1990) Hearing through the lungs: Lung-eardrum transmission of sound in the frog Eleutherodactylus coqui. Naturwissenschaften. 77: 192–194.

Endepols H, Walkowiak W (2001) Integration of ascending and descending inputs in the auditory midbrain of anurans. J Comp Physiol A 186:1119–1133.
Endepols H, Feng AS, Gerhardt HC, Schul J, Walkowiak W (2003) Roles of the auditory midbrain and thalamus in selective phonotaxis in female gray treefrogs (*Hyla versicolor*). Behav Brain Res 145:63–77.
Endepols H, Walkowiak W, Luksch H (2000) Chemoarchitecture of the anuran auditory midbrain. Brain Res Rev 33:179–198.
Epping WJM, Eggermont JJ (1985) Relation of binaural interaction and spectro-temporal characteristics in the auditory midbrain of the grassfrog. Hearing Res 19:15–28.
Epping WJM, Eggermont JJ (1986) Sensitivity of neurons in the auditory midbrain of the grassfrog to temporal characteristics of sound II. Stimulation with amplitude modulated sound. Hear Res. 24:55–72.
Feng AS (1980) Directional characteristics of the acoustic receiver of the leopard frog (*Rana pipiens*): A study of the eighth nerve auditory responses. J Acoust Soc Am 68:1107–1114.
Feng AS (1981) Directional response characteristics of single neurons in the torus semicircularis of the leopard frog (*Rana pipiens*). J Comp Physiol 144:419–428.
Feng AS (1982) Quantitative analysis of intensity-rate and intensity-latency functions in peripheral auditory nerve fibers of northern leopard frogs (*Rana p.pipiens*). Hear Res 6:241–246.
Feng AS (1986a) Afferent and efferent innervation patterns of the superior olivary nucleus of the leopard frog. Brain Res 364:167–171.
Feng AS (1986b) Afferent and efferent innervation patterns of the cochlear nucleus (dorsal medullary nucleus) of the leopard frog. Brain Res 367:183–191.
Feng AS, Capranica RR (1976) Sound localization in anurans. I. Evidence of binaural interaction in dorsal medullary nucleus of bullfrog (*Rana catesbeiana*). J Neurophysiol 39:871–881.
Feng AS, Capranica RR (1978) Sound localization in anurans II. Binaural interaction in superior olivary nucleus of the green treefrog (*Hyla cinerea*). J Neurophysiol 41: 43–54.
Feng AS, Lin WY (1991) Differential innervation patterns of three divisions of frog auditory midbrain (torus semicircularis). J Comp Neurol 306:613–630.
Feng AS, Lin W-Y (1994) Phase-locked response characteristics of single neurons in the frog "cochlear nucleus" to steady state and sinusoidal amplitude modulated tones. J Neurophysiol 72:2209–2221.
Feng AS, Lin WY (1996) Neuronal architecture of the dorsal nucleus (cochlear nucleus) of the frog (*Rana pipiens pipiens*). J Comp Neurol 366:320–334.
Feng AS, Shofner WP (1981) Peripheral basis of sound localization in anurans. Acoustic properties of the frog's ear. Hear Res. 5:210–216.
Feng AS, Hall JC, Gooler DM (1990) Neural basis of sound pattern recognition in anurans. Prog Neurobiol 34:313–329.
Feng AS, Hall JC, Siddique S (1991) Coding of temporal parameters of complex sounds by frog auditory-nerve fibers. J Neurophysiol 65:424–445.
Fuzessery ZM (1988) Frequency tuning in the anuran central auditory system. In: Fritszch B, Wilczynski W, Ryan MJ, Hetherington TE, Walkowiak W (eds) The Evolution of the Amphibian Auditory System. New York: Wiley, pp. 253–273.
Fuzessary ZM, Feng AS (1983a) Frequency selectivity in the anuran medulla: Excitatory and inhibitory tuning properties of single neurons in the dorsal medullary and superior olivary nuclei. J Comp Physiol 150:107–119.

Fuzessery ZM, Feng AS (1983b) Mating call selectivity in the thalamus and midbrain of the leopard frog (*Rana p. pipiens*): Single and multiunit activity. J Comp Physiol 150:333–344.

Gerhardt HC (1982) Sound pattern recognition in some North American treefrogs (Anura: Hylidae): Implications for mate choice. Am Zool 22:581–595.

Gerhardt HC (1988) Acoustic properties used in call recognition by frogs and toads. In: Fritszch B, Wilczynski W, Ryan MJ, Hetherington TE, Walkowiak W (eds) The Evolution of the Amphibian Auditory System. New York: Wiley, pp. 275–294.

Gerhardt HC (2001) Acoustic communication in two groups of closely related treefrogs. Adv Study Behav 30:99–167.

Gooler DM, Feng AS (1992) Temporal coding in the frog midbrain: The influence of duration and rise-fall time on the processing of complex amplitude-modulated stimuli. J Neurophysiol 67:1–22.

Gooler DM, Condon CJ, Xu J, Feng AS (1993) Sound direction influences the frequency-tuning characteristics of neurons in the frog inferior colliculus. J Neurophysiol 69:1018–1030.

Gooler DM, Xu J, Feng AS (1996) Binaural inhibition is important in shaping the free-field frequency selectivity of single neurons in the inferior colliculus. J Neurophysiol 76:2580–2594.

Grofova I, Corvaja N (1972) Commissural projection from the nuclei of termination of the VIIIth cranial nerve in the toad. Brain Res 42:189–195.

Hall JC (1991) GABA and glycine immunoreactive neurons and terminals in the auditory brainstem and thalamus of the northern leopard frog, *Rana pipiens pipiens*. Soc Neurosci Abstr 17.

Hall JC (1994) Central processing of communication sounds in the anuran auditory system. Amer Zool 34:670–684.

Hall JC (1999) GABAergic inhibition shapes frequency tuning and modifies response properties in the midbrain of the leopard frog. J Comp Physiol A 185:479–491.

Hall JC, Feng AS (1986) Neural analysis of temporally patterned sounds in the frog's thalamus: processing of pulse duration and pulse repetition rate. Neurosci Lett 63: 215–220.

Hall JC, Feng AS (1987) Evidence for parallel processing in the frog's auditory thalamus. J Comp Neurol 258:407–419.

Hall JC, Feng AS (1988) Influence of envelope rise time on neural responses in the auditory system of anurans. Hearing Res 36:261–276.

Hall JC, Feng AS (1990) Classification of the temporal discharge patterns of single auditory neurons in the dorsal medullary nucleus of the northern leopard frog. J Neurophysiol 64:1460–1473.

Hall JC, Feng AS (1991) Temporal processing in the dorsal medullary nucleus of the northern leopard frog (*Rana pipiens pipiens*). J Neurophysiol 66:955–973.

Heiligenberg WF (1991) The neural basis of behavior: A neuroethological view. Ann Rev Neurosci 14:247–267.

Hermes DJ, Eggermont JJ, Aertsen AM, Johannesma PI (1982) Spectro-temporal characteristics of single units in the auditory midbrain of the lightly anaesthetised grass frog (Rana temporaria L.) investigated with tonal stimuli. Hear Res 6:103–126.

Hetherington TE (1992) The effects of body size on functional properties of middle ear systems of anuran amphibians. Brain Behav Evol 39:133–142.

Hetherington TE, Lindquist E (1999) Lung-based hearing in an "earless" anuran amphibian. J Comp Physiol 184:395–401.

Hoke KL, Burmeister SS, Fernald RD, Rand AS, Ryan MJ, Wilczynski W (2004) Functional mapping of the auditory midbrain during mate call reception. J Neurosci 24: 11264–11272.

Jørgensen MB (1991) Comparative studies of the biophysics of directional hearing in anurans. J Comp Physiol A 169:591–598.

Jørgensen MB, Christensen-Dalsgaard J (1997a) Directionality of auditory nerve fiber responses to pure tone stimuli in the grassfrog, *Rana temporaria*. I. Spike rate responses. J Comp Physiol A 180:493–502.

Jørgensen MB, Christensen-Dalsgaard J (1997b) Directionality of auditory nerve fiber responses to pure tone stimuli in the grassfrog, *Rana temporaria*. II. Spike timing. J Comp Physiol A 180:503–511.

Jørgensen MB, Schmitz B, Christensen-Dalsgaard J (1991) Biophysics of directional hearing in the frog *Eleutherodactylus coqui*. J Comp Physiol A 168:223–232.

Kaulen R, Lifschitz W, Palazzi C, Adrian H (1972) Binaural interaction in the inferior colliculus of the frog. Exp Neurol 37:469–480.

Klump GM, Benedix JH Jr, Gerhardt HC, Narins PM (2004) AM representation in green treefrog auditory-nerve fibers: Neuroethological implications for pattern recognition and sound localization. J Comp Physiol A 190:1011–1021.

Langner G (1992) Periodicity coding in the auditory system. Hearing Res 60:115–142.

Large EW, Crawford JD (2002) Auditory temporal computation: Interval selectivity based on post-inhibitory rebound. J Comp Neurosci 13:135–142.

Larsell O (1934) The differentiation of the peripheral and central acoustic apparatus in the frog. J Comp Neurol 60:473–527.

Lewis ER, Lombard RE (1988) The amphibian inner ear. In: Fritzsch B, Ryan MJ, Wilczynski W, Hetherington TE, Walkowiak W (eds) The Evolution of the Amphibian Auditory System. New York: Wiley, pp. 93–123.

Lin WY, Feng AS (2001) Free-field unmasking response characteristics of frog auditory nerve fibers: Comparison with the responses of midbrain auditory neurons. J Comp Physiol A 187:699–712.

Lin WY, Feng AS (2003) GABA is involved in spatial unmasking in the frog auditory midbrain. J Neurosci 23:8143–8151.

Lombard RE, Straughan IR (1974) Functional aspects of anuran middle ear structures. J Exp Biol 61:71–93.

Luksch H, Walkowiak W (1998) Morphology and axonal projection patterns of auditory neurons in the midbrain of the painted frog, *Discoglossus pictus*. Hear Res 122: 1–17.

Martin WF (1971) Mechanics of sound production in toads of the genus *Bufo*: Passive elements. J Exp Zool 176:273–294.

Megela AL, Capranica RR (1981) Response patterns to tone bursts in the peripheral auditory systems of anurans. J Neurophysiol 46:465–478.

Melssen WJ, Epping WJM (1990) A combined sensitivity for frequency and interaural intensity difference in neurons in the auditory midbrain of the grassfrog. Hear Res 44:35–50.

Melssen WJ, Epping WJM (1992) Selectivity for temporal characteristics of sound and interaural time difference of auditory midbrain neurons in the grassfrog: A system theoretical approach. Hear Res 60:178–198.

Melssen WJ, van Stokkum IHM (1988) Sensitivity for interaural time-difference and amplitude-modulation in the auditory midbrain of the grassfrog. In: Duifhuis H, Horst JW, Wit HP (eds) Basic Issues in Hearing. London: Academic, pp. 279–284.

Melssen WJ, Epping WJM, van Stokkum IHM (1990) Sensitivity for interaural time and intensity difference of auditory midbrain neurons in the grassfrog. Hear Res 47:235–256.

Mudry KM, Constantine-Paton M, Capranica RR (1977) Auditory sensitivity of the diencephalon of the leopard frog (*Rana p. pipiens*). J Comp Physiol 114:1–13.

Narins PM (1983) Responses of torus semicircularis cells of the coqui treefrog to FM sinusoids. In: Ewert J-P, Capranica RR, Ingle DJ (eds) Advances in Vertebrate Neuroethology. New York: Plenum, pp. 889–894.

Narins PM, Capranica RR (1978) Communicative significance of the two-note call of the Puerto Rican treefrog, *Eleutherodactylus coqui*. J Comp Physiol 127:1–9.

Narins PM, Capranica RR (1980) Neural adaptations for processing the two-note call of the Puerto Rican treefrog, *Eleutherodactylus coqui*. Brain Behav Evol 17:48–66.

Narins PM, Ehret G, Tautz J (1988) Accessory pathway for sound transfer in a neotropical frog. Proc Natl Acad Sci USA 85:1508–1512.

Penna M (1997) Selectivity of evoked vocal responses in the time domain by frogs of the genus Batrachyla. J Herpetol 31:202–217.

Penna M, Lin WY, Feng AS (1997) Temporal selectivity for complex signals by single neurons in the torus semicircularis of Pleurodema thaul (Amphibia: Leptodactylidae). J Comp Physiol A 180:313–328.

Penna M, Lin WY, Feng AS (2001) Temporal selectivity by single neurons in the torus semicircularis of Batrachyla antartandica (Amphibia: Leptodactylidae). J Comp Physiol A 187:901–912.

Pinder AC, Palmer AR (1983) Mechanical properties of the frog ear: Vibration measurements under free- and closed-field acoustic conditions. Proc R Soc Lond B 219:371–396.

Rheinlander J, Gerhardt HC, Yager DD, Capranica RR (1979) Accuracy of phonotaxis by the green treefrog (*Hyla cinerea*). J Comp Physiol 133:247–255.

Rose G, Capranica RR (1983) Temporal selectivity in the central auditory system of the leopard frog *Rana pipiens*. Science 219:1087–1089.

Rose G, Capranica RR (1985) Sensitivity to amplitude modulated sounds in the anuran auditory nervous system. J Neurophysiol 53:446–465.

Rose GJ (1995) Representation of temporal patterns of amplitude modulation in the anuran auditory system and electrosensory system. In: Covey E, Hawkins HL, Port RF (eds) Neural Representation of Temporal Patterns. New York: Plenum, pp. 1–24.

Rose GJ, Brenowitz EA (1997) Plasticity of aggressive thresholds in *Hyla regilla*: Discrete accommodation to encounter calls. Anim Behav 53:353–361.

Rose GJ, Brenowitz EA (2002) Pacific treefrogs use temporal integration to differentiate advertisement from encounter calls. Anim Behav 63:1183–1190.

Rose GJ, Capranica RR (1984) Processing amplitude-modulated sounds by the auditory midbrain of two species of toads: Matched temporal filters. J Comp Physiol 154:211–219.

Rose GJ, Wilczynski W (1984) The anuran superficial reticular nucleus: Evidence for homology with the nucleus of the lateral lemniscus. Brain Res 304:170–172.

Rose GJ, Brenowitz EA, Capranica RR (1985) Species specificity and temperature and temperature dependency of temporal processing by the auditory midbrain of two species of treefrogs. J Comp Physiol 157:763–769.

Rose GJ, Zellick R, Rand SA (1988) Auditory processing of temporal information in a neotropical frog is independent of signal intensity. Ethology 77:330–336.

Ryan MJ (1983) Frequency modulated calls and species recognition in a neotropical frog. J Comp Physiol 150:217–221.

Schmidt RS (1988) Mating call phonotaxis in female American toads: Lesions of central auditory system. Brain Behav Evol 32:119–128.

Schmidt RS (1989) Mating call phonotaxis in female American toad: Lesions of anterior preoptic nucleus. Horm Behav 23:1–9.

Schmitz B, White TD, Narins PM (1992) Directionality of phase locking in auditory nerve fibers of the leopard frog *Rana pipiens pipiens*. J Comp Physiol A 170:589–604.

Schneider H (1988) Peripheral and central mechanisms of vocalization. In: Fritszch B, Wilczynski W, Ryan MJ, Hetherington TE, Walkowiak W (eds) The Evolution of the Amphibian Auditory System. New York: Wiley, pp. 275–294.

Seaman RL (2002) Non-osseous sound transmission to the inner ear. Hear Res 166: 214–215.

van Stokkum IH, Melssen WJ (1991) Measuring and modeling the response of auditory midbrain neurons in the grassfrog to temporally structured binaural stimuli. Hear Res 52:113–132.

Vlaming MSMG, Aertsen AMHJ, Epping WJM (1984) Directional hearing in the grass-frog (*Rana temporaria* L.). I. Mechanical vibrations of tympanic membrane. Hear Res 14:191–201.

Walkowiak W (1980) The coding of auditory signals in the torus semicircularis of the fire-bellied toad and grass frog: Responses to simple stimuli and to conspecific calls. J Comp Physiol 138:131–148.

Walkowiak W (1984) Neuronal correlates of the recognition of pulsed sound signals in the grass frog. J Comp Physiol A 155:57–66.

Walkowiak W (1988) Central temporal encoding. In Fritszch B, Wilczynski W, Ryan MJ, Walkowiak W (eds) The Evolution of the Amphibian Auditory System. New York: Wiley, pp. 275–294.

Walkowiak W, Luksch H (1994) Sensory motor interfacing in acoustic behavior of anurans. Amer Zool 34:685–695.

Walkowiak W, Berlinger M, Schul J, Gerhardt HC (1999) Significance of forebrain structures in acoustically guided behaviour in anurans. Eur J Morphol 37:177–181.

Wang J, Narins PM (1996) Directional masking of phase locking in the amphibian auditory nerve. J Acoust Soc Am 99:1611–1620.

Wang J, Ludwig TA, Narins PM (1996) Spatial and spectral dependence of the auditory periphery in the northern leopard frog. J Comp Physiol A 178:159–172.

White TD, Schmitz B, Narins PM (1992) Directional dependence of auditory sensitivity and frequency selectivity in the leopard frog. J Acoust Soc Am 92:1953–1961.

Wilczynski W (1988) Brainstem auditory pathways in anurans. In: Fritszch B, Wilczynski W, Ryan MJ, Hetherington TE, Walkowiak W (eds) The Evolution of the Amphibian Auditory System. New York: Wiley, pp. 209–231.

Will U (1988) Organization and projections of the area octavolateralis in amphibians. In: Fritzsch B, Ryan MJ, Wilczynski W, Hetherington TE, Walkowiak W (eds) The Evolution of the Amphibian Auditory System. New York: Wiley, pp. 185–208.

Xu J, Gooler DM, Feng AS (1994) Single neurons in the frog inferior colliculus exhibit direction-dependent frequency selectivity to isointensity tone bursts. J Acoust Soc Am 95:2160–2170.

Xu J, Gooler DM, Feng AS (1996) Effects of sound direction on the processing of amplitude-modulated signals in the frog inferior colliculus. J Comp Physiol A 178: 435–445.

Zakon H (1983) Reorganization of connectivity in amphibian central auditory system following VIIIth nerve regeneration: Time course. J Neurophysiol 49:1410–1427.

Zhang H, Feng AS (1998) Sound direction modifies the inhibitory as well as the excitatory frequency tuning characteristics of single neurons in the frog inferior colliculus. J Comp Physiol A 182:725–735.

Zhang H, Xu J, Feng AS (1999) Effects of GABA-mediated inhibition on direction-dependent frequency tuning in the frog inferior colliculus. J Comp Physiol A 184:85–98.

Zheng W, Hall JC (2000) GABAergic inhibition shapes frequency tuning and modifies response properties in the superior olivary nucleus of the leopard frog. J Comp Physiol 186:661–671.

10
Plasticity in the Auditory System across Metamorphosis

ANDREA MEGELA SIMMONS AND SETH S. HOROWITZ

1. Introduction

Many species of anuran amphibians undergo a developmental process called metamorphosis during which free-living, herbivorous, nonreproductive larvae (tadpoles) transform into partly terrestrial, carnivorous, reproductively active adults. Metamorphosis in anurans is a period of rapid morphological and physiological change affecting all sensory, motor, and vegetative systems. The pattern of larval development and the extent of change during metamorphosis vary (McDiarmid and Altig 1999); some species (e.g., *Eleutherodactylus*) undergo direct development, whereas others (such as the bullfrog, *Rana catesbeiana*) have relatively lengthy tadpole periods. Pipids, such as the African clawed frog, *Xenopus laevis*, remain aquatic as adults, and metamorphic change is not as extensive as that observed in anuran species which become partly terrestrial. Development in anurans thus encompasses diverse patterns of change and is a rich resource for analysis of theories and mechanisms of plasticity, offering a vertebrate system in which neural and anatomical development can be examined in nonfetal animals.

Metamorphosis features regression of structures important only in larval forms, transformation of larval into adult structures, and development of new structures necessary for the adult (Fritzsch et al. 1988). The metamorphic transition involves considerable alteration in external morphology (Gosner 1960; Nieuwkoop and Faber 1994); in behaviors such as respiration, locomotion, and feeding (Etkin 1964; Stehouwer 1988; Burggren and Infantino 1994); and in peripheral and central nervous system structure and functioning (Fritzsch et al. 1988; Lannoo 1999). Metamorphosis also involves changes in neural and hormonal foundations allowing later emergence of reproductive behaviors. All sensory systems undergo some sort of modification or reorganization during the larval period, and at different time courses (Spaeti 1978). In particular, the metamorphic shift from aquatic tadpole to amphibious frog imposes substantial changes in the type of auditory input available to the organism, based on the different acoustic properties of underwater and terrestrial environments (Bass and Clark 2002). Adult anurans rely on vocalizations for mate attraction and territorial defense (Gerhardt and Bee, Chapter 5). The ability to hear species-specific vocalizations develops during the larval and froglet periods, but how this is accomplished is still largely unexplored. Because species such as *R. catesbeiana* have relatively extensive larval periods, during which they are exposed to underwater sound sources, auditory experience may play a strong role in shaping both the neuroanatomy and the physiology of the auditory system across metamorphic development. The peripheral substrates and central mechanisms for processing airborne sounds develop in a sequential fashion throughout metamorphosis (Boatright-Horowitz and Simmons 1997), but the extent to which this developmental sequence is genetically mediated or environmentally modifiable is unknown. During the postmetamorphic period, when the animal is exposed to both underwater and airborne sounds, the peripheral and central auditory systems continue to exhibit anatomical and functional modifications as body and brain size increase (Shofner and Feng 1981, 1984; Boatright-Horowitz and Simmons 1995).

This chapter outlines what is known about auditory system development over metamorphosis in both semi-terrestrial (*Rana*, *Hyla*) and permanently aquatic (*Xenopus*) anurans. Much of the discussion focuses on the two species that have a well-defined tadpole stage and for which most data have been gathered, *Rana catesbeiana* and *Xenopus laevis*. We do not cover developmental literature on endotrophic species, for which little is known about changes in auditory function. We also describe changes occurring during early postmetamorphic development, and highlight areas where intensive study is still needed.

2. Description of Metamorphic Stages

The time course of metamorphosis is typically described by staging tables based primarily on changes in gross cellular configurations (in embryonic stages) and in external body morphology (in postembryonic larval stages). Several different staging systems are used in the literature (Shumway 1940; Taylor and Kollros 1946; Kopsch 1952; Gosner 1960), making comparisons based on different systems difficult. In addition, some older literature describes tadpoles generically, without reference to a particular stage, or only in terms of body size or length (e.g., Larsell 1934; Paterson 1949/50; Sedra and Michael 1959; van Bergeijk 1959). Because body size is influenced by diet, temperature, and rearing condition (Corse and Metter 1980; Nieuwkoop and Faber 1994; McDiarmid and Altig 1999), it can vary widely among individuals even at the same stage of development, making classification based on this parameter unreliable. McDiarmid and Altig (1999) propose that the Gosner (1960) classification, first described for *Rana*, be used as a standard staging system, and they provide a table of approximation between Gosner stages and other staging systems often used. In this chapter, we use the Gosner system to classify amphibious anurans, and *Xenopus* is categorized according to the Nieuwkoop–Faber (NF) system. Other classification systems have been transformed into equivalent Gosner stages using the table in McDiarmid and Altig (1999). Table 10.1 provides rough equivalents between

TABLE 10.1. Comparison of developmental stages in *Rana* and *Xenopus*.

Developmental phase	Gosner stage	*Rana* Criteria	NF stage	*Xenopus* Criteria
Embryonic/ prefeeding	1–19	Rapid cell division, gastrulation, neural plate and tube formation, tail budding, beginning of muscular function	1–34	Rapid cell division, gastrulation, neural plate and tube formation, tail budding, beginning of muscular function
Hatchling	20–25	Transition to feeding and free-swimming tadpole; disappearance of external gills	35–45	Transition to feeding and free-swimming tadpole; lateral line system externally visible
Early larval	26–30	Development of hind limb buds	46–53	Development of hind limb and fore limb buds
Late larval	31–41	Lengthening of hind limbs, differentiation and lengthening of toes, appearance of metatarsal and subarticular tubercles; disappearance of cloacal tail piece	54–61	Appearance of fingers; differentiation of toes; shortening of tentacles
Metamorphic climax	42–46	Emergence and differentiation of fore limbs; resorption of tail; remodeling of head	62–66	Resorption of tail; remodeling of head; appearance of adult skin

Sources: Gosner 1960; Nieuwkoop and Faber 1994; McDiarmid and Altig 1999.

Gosner and NF stages. Photographs of *Rana* and *Xenopus* tadpoles at various stages of development are shown in Figure 10.1.

Gosner (1960) stages begin at fertilization (stage 1), with embryonic stages extending up to stage 19. Stages 20 to 25 are hatchling stages, the time of transition from an immobile embryo to a freely-living tadpole. During hatchling stages, external gills atrophy and the animal begins to swim. The Shumway (1940) system extends from fertilization to approximately the end of the hatchling period. The tadpole (larval) stages (stages 26 to 41) are characterized by the progressive development and emergence of hind limbs and changes in body shape and size. As shown in Figure 10.1, tadpoles in stages 25 to 30, the early larval stages, lack any external limbs, but begin to develop undifferentiated hind limb buds. Stages 31 to 41, the late larval stages, feature a range of hind limb development, from initial differentiation to full development of toes. Metamorphic climax (stages 42 to 46) encompasses a period of rapid morphological change during which the chondrocranium begins to ossify, fore limbs emerge and differentiate, the narrow tadpole oral disc transforms to the wide frog mouth, the

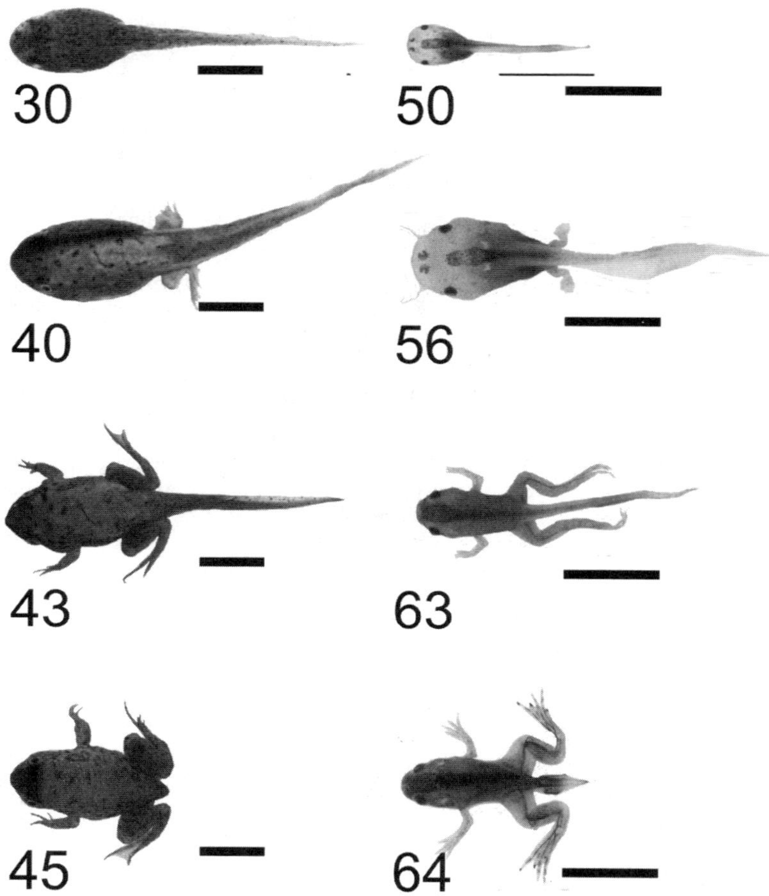

FIGURE 10.1. Comparison of dorsal views of *Rana catesbeiana* (left) and *Xenopus laevis* (right) tadpoles at similar stages of somatic development. Staging of *Rana* is per Gosner (1960) and that of *Xenopus* is per Nieuwkoop and Faber (1994). Scale bars = 1 cm for all images.

tongue doubles in length, the gills and tail resorb, and the eyes migrate from lateral to more dorsomedial positions on the top of the head (Etkin 1964; McDiarmid and Altig 1999). In ranids and other species becoming amphibious after climax, the lateral line system degenerates during this period (Fritzsch et al. 1984, 1988). Metamorphosis is complete at stage 46. Taylor–Kollros (1946) stages I to XX correspond approximately to Gosner stages 26 to 46. After the end of metamorphic climax, ranids are categorized based on snout-vent length (SVL) as froglets (SVL < 5.5 cm), subadults (SVL 5.5 to 10 cm), or adults (SVL > 10 cm; Boatright-Horowitz and Simmons 1995).

The staging system of Nieuwkoop and Faber (1994; Table 10.3) is based on 66 stages from fertilization to completion of metamorphosis. Because *Xenopus* tadpoles are somewhat transparent, both internal and external morphological criteria can be used for staging purposes.

One caveat for any staging system is that changes in external bodily morphology may not correspond neatly to internal development. An example is stage 25 in ranids, during which external body morphology remains constant but there is massive growth of the central nervous system and overall body size (Spaeti 1978). In *Xenopus*, changes in skeletal morphology over metamorphic development correlate poorly with changes in external morphology (Trueb and Hanken 1992). Moreover, there is rapid growth in the ear vesicle between NF stages 46 and 47, during which external morphology changes little (Nieuwkoop and Faber 1994).

3. Anatomical Development

3.1 Comparative Overview

Adult amphibious anurans hear sounds and detect substrate vibrations through two peripheral pathways. One pathway is the tympanic/columellar input to the inner ear, consisting of the external tympanum and the columellar attachments (extrastapes and stapes) to the oval window. The other is the opercularis system, consisting of the operculum located at the caudal end of the oval window and attached via the opercularis muscle to the shoulder girdle (Mason, Chapter 6). Neither of these pathways is functional in tadpoles younger than metamorphic climax stages (Hetherington 1987). Coupled with the lack of a swim bladder, this suggests that if tadpoles relied on either of these pathways for detection of sounds, they would display relatively poor frequency sensitivity and high thresholds similar to those observed in hearing generalist fish (Fay and Simmons 1999). Components of the tympanic and opercularis pathways develop progressively and sequentially throughout the larval period, with species differences appearing primarily in the rate of development of the tympanic/columellar pathway, with less variability in the rate of development of the opercularis pathway (Sedra and Michael 1959; Hetherington 1987). The development of the opercularis system is temporally correlated with the internal development of the fore limbs. Components of the tympanic pathway first develop during the final steps of preparing for terrestrial existence in metamorphic climax, and in some species this transduction route is not mature until sometime after completion of climax (Hetherington 1987). The progressive development of external conduction pathways suggests that hearing sensitivity in tadpoles will vary over the larval period.

The outer ear of adult *Xenopus* differs from that of amphibious ranids. Instead of an external tympanum, these frogs have a tympanic disc, which is suspended in a tympanic annulus and covered with skin. The columella is embedded in this disc, extends through the air-filled middle ear cavity, and is coupled to the inner ear by a stapes. Adult *Xenopus* have a small operculum fused to the posterior wall of the otic capsule (Wever 1985), but whether this functions in a manner similar to the opercularis system of ranids is not clear.

3.2 Otic Capsule

In *R. catesbeiana*, the otic capsule forms early in embryological development (around stage 17) from an invagination of the otic placode. At this point, it is filled only with connective tissue. The otic capsule at stage 24 approximates a pointed spheroid 200 μm mediolaterally and 225 μm rostrocaudally. At stage 25, the otic capsule is partly cartilaginous and has undergone considerable growth, increasing to a mediolateral diameter of 600 μm and a rostrocaudal length of 950 μm (Horowitz et al. 2001; Fig. 10.2A). It has also undergone substantial chondrification, forming the basis of the bony labyrinth (Sokol 1981). The otic capsule is located in the caudolateral region of the chondrocranium and comprises about 1/4 of its entire length, consistent with measurements in *Pelodytes* (Sokol 1981). The shape approximates a flattened sphere, and shows a linear increase in

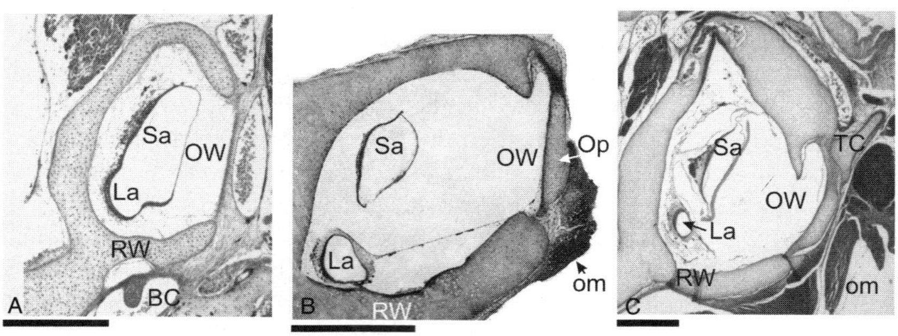

FIGURE 10.2. Trichrome stained horizontal sections (10 μm) through the otic capsule in stage 25 (A), stage 41 (B), and stage 44 (C) *R. catesbeiana* tadpoles. Sections were taken at approximately the same dorsal-ventral plane at the level of the ventral margin of the round window (RW), showing the saccule (Sa) and lagena (La) inner ear organs. The section illustrated in (B) is slightly rotated, accounting for its relatively elongated shape. The bronchial columella (BC) is evident in the stage 25 tadpole (A), whereas the structure has degenerated in the stage 44 animal (C). The oval window (OW) is on the lateral (right) margin. In the stage 25 animal (A), there are no structures overlaying the oval window. In the stage 41 animal (B), the operculum cartilage (Op) is connected to the opercularis muscle (om) and appears to fully overlay the OW. There is no evidence of a tympanic columella (TC) at this stage. By stage 44 (C), the developing *pars interna* of the TC is visible, both overlying and obstructing the oval window. The nerve root leading to the saccule (Sa) is visible as is the lagenar region (La). The round window (RW) has no remaining trace of the bronchial columella at this stage. Rostral is to the top, lateral is to the right. Scale bar = 1 mm.

minor and major diameters between stages 26 and 39. After stage 39 and into metamorphic climax stages, the thickness and complexity of the cartilaginous walls of the otic capsule increase substantially (Fig. 10.2C), although gross external growth of the otic capsule ceases until the postmetamorphic growth period.

In *Xenopus* larvae, the otic placode begins to form between NF stages 23 to 27 (Bever et al. 2003). The otic capsule forms from an invagination of the placode, begins to chondrify around NF stage 47, and is well ossified by NF stage 61 (Trueb and Hanken 1992).

3.3 Inner Ear Organs

In all larval anurans studied to date, the inner ear organs develop prior to the middle ear system. The inner ear develops during embryonic stages from an ectodermal thickening of the otic placode. Invagination of the placode forms the otocyst, which becomes divided with growth into separate chambers that will contain the auditory and otolithic organs (Kil and Collazo 2001). Hertwig (1987; Hertwig and Schneider 1986) described the pattern of formation of the inner ear organs in *R. temporaria* as commencing at early hatchling stages with the initial formation of the semicircular canals and initial differentiation of the otocyst into the utricle and saccule. The BP begins to develop around stage 26, the AP around stages 26 to 32, and the lagena around stage 36. All sensory organs in the inner ear have formed by stage 40, when the hind limbs are fully differentiated but before emergence of the fore limbs (Hertwig 1987). Also in *R. temporaria*, Spaeti (1978) observed the formation of the BP beginning at stage 22 and the AP at stage 26, with stereocilia and the anlage of the tectorial membrane visible at stage 26 in both organs. Between stage 26 and metamorphosis, both organs expand in size and in number of hair cells. In *R. pipiens*, the AP develops before the BP and the lagena, and it begins to differentiate before nerve fibers innervate the sensory epithelium (Larsell 1934). Smirnov (1993) reported that in more advanced frogs such as the hylids, ranids, and microhylids, the AP has achieved its adult form by the beginning of metamorphosis, whereas in more primitive frogs such as the discoglossids and pipids, the AP continues to develop into the postmetamorphic period. Figure 10.2 shows photographs of the otic capsule in the plane of the saccule in *R. catesbeiana* tadpoles at stages 25, 41, and 44. As shown in Figure 10.2A, the saccule takes up a larger proportion of the otic labyrinth at stage 25 than at the later stages. Moreover, the saccule and lagena are not clearly separated, but appear to take up different spatial areas in the same recess. By stage 41 (Fig. 10.2B), the saccule and lagena are clearly separated into their own recesses. Both organs continue to differentiate up until the end of metamorphic climax.

The morphogenesis of hair cells begins with the formation of a kinocilium on a supporting cell, followed by the transformation of microvilli into stereocilia (Lewis and Li 1973). In both stage 26 and adult *R. catesbeiana*, the saccule contains two different types of hair cells. One has short stereocilia and a longer kinocilium with no bulb and is found at the edge of the sensory epithelium. The other has long stereocilia and a kinocilium terminating in a large bulb, and is found near the center of the epithelium. Lewis and Li (1973) suggested that these two types of hair cells represent immature and mature cells, respectively. The number and proportion of juvenile hair cells are much larger in auditory (saccule, AP, BP) than in vestibular (cristae, utricle, and lagena) organs. This suggests that vestibular receptors complete development earlier in life, whereas the auditory receptors continue to mature and increase in number throughout the larval and adult periods (Li and Lewis 1979). In *Bufo marinus* (Corwin 1985), the numbers of saccular hair cells increase about fourfold from metamorphic climax to adulthood, linearly with body size. The functional implications of this continual turnover of hair cells have not been fully described.

Spatial gradations in the mass of the tectorial membrane of the AP change over the postmetamorphic period, suggesting that the frequency sensitivities of auditory nerve fibers innervating this organ also change with growth (Shofner and Feng 1984). There is also a large increase in the volume and mass of the tectorial membrane, implying downward shifts in the resonant frequency of the AP over development.

Fate maps of the *Xenopus* inner ear at the placode (NF stages 23 to 27) and otocyst (NF stages 28 to 31) stages show that every region of these structures gives rise to sensory organs, and that cells from different regions of the inner ear intermingle during these stages (Kil and Collazo 2001). Paterson (1949/50) reported a developmental progression beginning with the initial formation of the semicircular canals, to the appearance of the utricle and saccule, then culminating in the appearance of the AP, the lagena, and the BP. More recent work confirmed this sequence, specifying initial appearance of the saccular and utricular maculae at NF stages 45 to 47, with the AP and the BP developing later (NF stages 47 to 50; Diaz et al. 1995; Kil and Collazo 2001; Bever et al. 2003). By NF stages 50 to 52, the inner ear has expanded approximately twelvefold in size, and its gross morphogenesis appears complete (Paterson 1960; Nieuwkoop and Faber 1994; Bever et al. 2003). The saccule, AP, and BP are smaller in size than in postmetamorphic and adult animals, but have the same morphology (Lopez-Anaya et al. 1997). Diaz et al. (1995) showed that the total number of hair cells in the saccule increases considerably during metamorphosis, with the rate of increase falling off during postmetamorphic development. They also confirmed that, in *Xenopus* as well as in *Bufo* and *Rana*, hair cells with short stereocilia located at the edge of the sensory epithelium are immature.

3.4 Middle Ear

3.4.1 Structural Input to the Otic Capsule

3.4.1.1 Ranid Bronchial Columella

Witschi (1949, 1955) proposed that hearing in ranid tadpoles was mediated by the bronchial columella, which he described as a series of stiff fibers running from the primary bronchus of the lung, piercing the dorsal aorta, and connecting to the round window of the ipsilateral otic capsule. According to his hypothesis, the impedance mismatch between the aquatic environment and the air-filled lungs would induce changes in pressure and/or particle motion in the otic capsule via the round window. The bronchial columella could presumably induce fluid displacement in the inner ear by converting oscillatory movements of the lungs into pulsations of the round window. This system would operate in a manner similar to the swimbladder/Weberian ossicle/otic capsule connection in otophysan fishes, with the difference that the site of transmission to the inner ear would be the round window instead of the oval window.

Witschi's hypothesis has been challenged on several grounds. Witschi (1955) and Hetherington (1987) both noted that the bronchial columella is present in some species of *Rana* but is absent in *Hyla*, *Bufo*, *Scaphiopus*, and *Alytes*; the generality of this mechanism for sound reception in larval anurans is thus limited. Paterson (1960) questioned how the reversal in the roles of the round and oval windows would be accomplished during metamorphosis. Horowitz et al. (2001) showed, in larval *R. catesbeiana*, that the bronchial columella (Fig. 10.2A) is a flexible, fibrous, and unchondrified structure with limited stiffness, providing a tendonlike attachment between the round window and the primary bronchus. It is composed of fibroblasts early in the larval period and becomes more collagenous near metamorphic climax stages. This composition does not provide sufficient rigidity to effectively couple the lungs and round window; on the other hand, a flexible attachment might play a role in pressure equalization as tadpoles move through the water column (Horowitz et al. 2001). Such a function would be biologically useful for the animal, as Eustachian tubes do not develop (at least in toads) until the early postmetamorphic period (Sedra and Michael 1959). Both Witschi (1955) and Horowitz et al. (2001) noted that the bronchial columella degenerates during metamorphic climax stages; thus, even if this structure did play some role in mediating acoustic sensitivity in early larval stages, any hearing sensitivity during metamorphic climax stages must have some other structural basis.

3.4.1.2 Xenopus Bronchial Diverticulum

The bronchial diverticulum of *Xenopus* tadpoles is first apparent by about NF stage 50 (Nieuwkoop and Faber 1994). It is an air-filled sac extending from the primary bronchus of the lung to the ipsilateral round window (Weisz 1945). Witschi (1950) proposed that this connection forms a substrate for sound conduction but provided no functional data in support of this hypothesis. Weisz (1945) and van Bergeijk (1959) suggested, instead, that the diverticula might serve a hydrostatic function, allowing the animal to maintain a stable position in the water. *X. laevis* tadpoles between NF stages 48 to 52 hang head downward in the water column at roughly a 45° angle in a manner that enables them to swim against their own buoyancy (Wassersug and Souza 1990; Simmons et al. 2004). Hydrostatic pressure changes produced by movement through the water column could alter the resonance characteristics of the lungs, which would be sensed by the inner ear organs via the bronchial diverticula (van Bergeijk 1959). The tadpole could then adjust its orientation in the water column by means of this transduced signal. Note that this model requires that the lungs be inflated with air. Wassersug and Souza (1990) observed that tadpoles in microgravity environments detect changes in buoyancy whether or not their lungs are inflated. This argues against a role for the bronchial diverticula in hydrostatic assessment, but it does not provide evidence for or against a role in auditory perception. The bronchial diverticula, like the bronchial columella in ranids, disappear during metamorphic climax (Paterson 1960).

3.4.2 Fenestral Pathway

In *R. catesbeiana* tadpoles younger than stage 37, the oval window is an opening in the ventrocaudal region of the lateral surface of the otic capsule (Fig. 10.2A). The membrane of the oval window is a thin (~40 μm) structure, covered by epidermal and somatic/connective tissues. Its basic shape and structure remain unchanged from stages 26 to 45, although by stage 36, mesenchymal tissue condenses over the opening that will form the operculum cartilage by stage 39. The diameter of the oval window increases gradually for most of the developmental span until the onset of metamorphic climax, when it doubles in size (Horowitz et al. 2001).

Hetherington (1987) proposed that sound stimuli might be directly transmitted to the inner ear in hatchlings and early larval animals directly through the oval window, in a manner similar to that occurring in carcharhine sharks (Corwin 1981). This would provide a low impedance pathway from the external environment directly to the inner ear. Sound waves would stimulate the auditory endorgans through direct kinetic effects on overlying tectorial or otoconial structures (Hetherington 1987).

Functional evidence for the existence of this fenestral pathway comes from experiments in *R. catesbeiana* (Boatright-Horowitz and Simmons 1997) in which the oval window input pathway was artificially blocked by "earmuffs" placed close to the tadpole's head. In stage 32 tadpoles (with an intact bronchial columella but without a mature opercularis system), neural responsiveness from the auditory midbrain (torus semicircularis, TS) to broadband noise stimulation was decreased by this manipulation. In contrast, blocking the oval window input pathway had no effect on neural responsiveness in metamorphic climax tadpoles with a functioning opercularis system. These data suggest that the fenestral pathway could mediate hearing sensitivity in animals without an opercularis system.

3.4.3 Opercularis System

Histological studies of middle ear development in seven anuran species representing three families show that the opercularis system forms earlier in development than the tympanic/columellar system (Sedra and Michael 1957; Hetherington 1987, 1988). In *Hyla crucifer*, the operculum is first visible at about stage 38 as a mesenchymal condensation over the posterior part of the oval window. It gradually chondrifies, extends anteriorly, and either partially or totally fills the oval window by about stage 40. The position of the operculum in the otic capsule of a stage 41 *R. catesbeiana,* obstructing the oval window, is shown in Figure 10.2B. The position of the operculum between stages 38 to 41 has important functional consequences, discussed further in Section 5.2.

As the stapes begins to form, the operculum partially retracts, assuming a slightly curved shape, and by the early postmetamorphic period takes up less area of the oval window than in tadpoles (Hetherington 1987, 1988). The scapula begins to form around stage 27, and the opercularis muscle (which is derived from the levator scapulae superior muscle) begins to form around stages 37 to 38. Fibers of the developing opercularis muscle insert into the lateral surface of the operculum by stage 42, around the time of appearance of fore limbs. In all anuran species studied to date (Sedra and Michael 1959; Hetherington 1987), the opercularis system is completely formed by the end of metamorphic climax. Ranid tadpoles in metamorphic climax stages rest stationary on the substrates of ponds and other water sources. This posture could exert pressure on the shoulder girdle from the fore limbs, providing a stable substrate for activation of the opercularis system by substrate vibrations.

In *Xenopus*, the operculum is visible at about NF stage 59 as a cartilaginous outgrowth from the posterior border of the oval window (Sedra and Michael 1957). It is attached to the stapes by fibers and on its dorsocaudal surface to the opercularis muscle (Wever 1985). At metamorphic climax (NF stage 66), it was described as "nearly filling" the oval window. Trueb and Hanken (1992) were unable to identify the operculum as described by Sedra and Michael (1957) in tadpoles, but did observe this structure in 1- to 2-month-old postmetamorphic animals. Unlike ranids, *Xenopus* tadpoles hang suspended in the water column during early developmental stages and only position themselves on a substrate during and after metamorphic climax. The opercularis system could thus provide a substrate-mediated excitation pathway in these older animals.

3.4.4 Tympanic/Columellar Pathway

There is considerable species diversity in the timing of the formation of the tympanic/columellar pathway (Hetherington 1987), although it is consistently delayed relative to the formation of the opercularis pathway. Moreover, tympanic pathway development in smaller frogs lags behind that observed in larger frogs (Hetherington 1987). In most species studied to date, the stapes and extrastapes (columella and extracolumella) begin to form in metamorphic climax stages, after the formation of the operculum. In *Hyla crucifer*, the footplate (*pars interna*) of the stapes is visible at about stage 40 as a mesenchymal condensation at the anterior edge of the oval window. Chondrification begins at the oval window, and then gradually extends laterally to form the shaft (*pars media*). The *pars interna* remains cartilaginous into adult life whereas the *pars media* becomes ossified after metamorphosis is complete (Sedra and Michael 1959). The extrastapes (*pars externa*) develops in some species from a center of chondrification underneath the skin on the side of the head, while in other species, it develops as an extension of the stapes (Hetherington 1988). In *H. crucifer*, the extrastapes has extended only a short distance toward the lateral side of the head by the end of metamorphic climax. There is no evidence of a middle ear cavity, tympanic annulus, or tympanum at this time, and it is not until approximately 60 days after the completion of metamorphosis that the tympanic pathway appears structurally mature (Hetherington 1987).

In *R. catesbeiana,* the stapes and extrastapes appear to be developed by stage 43. Externally, a thickening of epidermal tissue on the side of the head at stages 42 to 43 marks the first development of a tympanum. By stage 44, the tympanum can be detected as a pale circle underlying the epidermis, and the tympanic annulus is completely closed (Boatright-Horowitz and Simmons 1997), although the middle ear cavity is still filled with fluid at this point. The tympanum does not appear on the side of the animal's head until 24 hours after metamorphic climax is complete. At this point, electrophysiological responses in the TS to closed-field sound sources can be recorded (Boatright-Horowitz and Simmons 1995). During the early postmetamorphic (froglet) period, the width of the tympanum and the intertympanic distance grow linearly with body size and with age (Fig. 10.3). The tympanum growth curves for male and female *R. catesbeiana* are similar until the animals reach a body size of about 10 cm SVL. In this adult period, the tympanum of male animals continues to grow linearly with body size, whereas that of females shows a much slower rate of growth (Boatright-Horowitz and Simmons 1995).

In *Xenopus,* the middle ear forms from its anlage beginning at about NF stage 56 (Nieuwkoop and Faber 1994). The *pars interna* forms at about NF stage 59

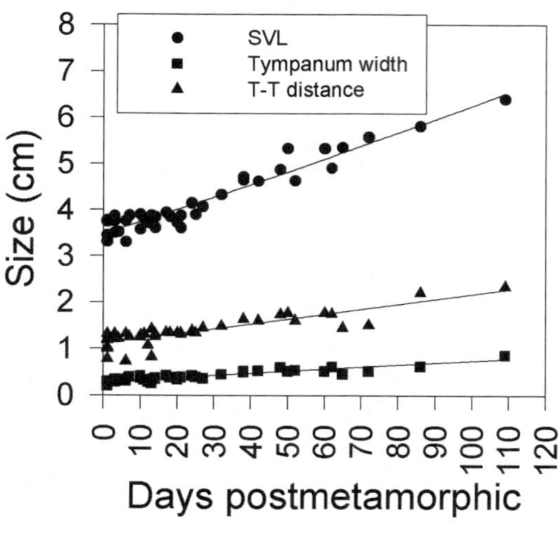

FIGURE 10.3. Relationship between age (days since completion of metamorphosis) and body size (SVL), tympanum width, and intertympanic (T–T) distance in *R. catesbeiana*. For SVL: $r^2 = .93$, $y = .03x + 3.46$. For tympanum width: $r^2 = .84$, $y = .005x + .29$. For T–T distance: $r^2 = .73$, $y = .01x + 1.11$. Data from Boatright-Horowitz (1997).

from a chondrification above the oval window (Sedra and Michael 1957). The *pars media* forms at NF stages 61 to 62 as an extension of the *pars interna* (Sedra and Michael 1957; Trueb and Hanken 1992). At about the same time, two centers of chondrification emerge, one for the *pars externa* and one for the tympanic annulus (Sedra and Michael 1957). Both the *pars externa* and the annulus increase in size, and the *pars media* begins to ossify (Trueb and Hanken 1992). The stapes and extrastapes fuse and the tympanic membrane anlage appears by NF stage 63 (Sedra and Michael 1957; Nieuwkoop and Faber 1994). By the end of metamorphosis (NF stage 66), the tympanic annulus almost completely encloses the distal portion of the extrastapes, the ossification of the *pars media* is complete, and the *pars interna* and the extrastapes remain largely cartilaginous (Vorobyeva and Smirnov 1987; Trueb and Hanken 1992). The extrastapes and the tympanic annulus separate and grow in size during the early postmetamorphic period. In adults, the distal portion of the extrastapes completely fills the area circumscribed by the annulus (Trueb and Hanken 1992).

A summary of the developmental changes in the middle and inner ears is presented in Table 10.2 for *Rana* and in Table 10.3 for *Xenopus*.

TABLE 10.2. Summary of middle and inner ear development in *Rana*.

Gosner stage[1]	Developmental event
16–17	Otic capsule forms in *R. temporaria* (Hertwig 1987)
22–28	Basilar papilla forms in *R. temporaria* (Hertwig 1987; Spaeti 1978)
25	Saccule forms in *R. catesbeiana* (Horowitz et al 2001)
28	Saccule and utricle form in *R. temporaria* (Hertwig 1987)
	Saccule, utricle, lagena, and amphibian papilla form in *R. fusca* (Kopsch 1952)
26–32	Amphibian papilla forms in *R. temporaria* (Hertwig 1987; Spaeti 1978)
32	Basilar papilla forms in *R. fusca* (Kopsch 1952)
36	Lagena forms in *R. temporaria* (Hertwig 1987)
38–41	Operculum develops and transiently blocks oval window in *R. catesbeiana* (Boatright-Horowitz and Simmons 1997)
40	Fusion of rostral triangular and caudal S-shaped patch of amphibian papilla in *R. catesbeiana* (Li and Lewis 1974)
42	Amphibian papilla assumes adultlike morphology in *R. temporaria* (Smirnov 1993)
42–43	Tympanum develops from epidermal tissue on the side of the head in *R. catesbeiana* (Boatright-Horowitz 1997)
43	Tympanic annulus begins to form on posterior section of palatoquadrate in *R. catesbeiana* (Boatright-Horowitz 1997)
44	Bronchial columella lost in ranids (Witschi 1955; Horowitz et al. 2001)
	Peripheral and central lateral-line structures degenerate (Jacoby and Rubinson 1983; Fritzsch et al. 1984)
45	In *R. catesbeiana*: *pars externa* assumes final position and shape; tympanic annulus completely closes; tympanum and overlying epidermis fuse (Boatright-Horowitz 1997)
46	Tympanic-columellar system assumes adultlike morphology in *R. catesbeiana* (Boatright-Horowitz and Simmons 1995)

[1]Gosner (1960).

TABLE 10.3. Summary of middle and inner ear development in *Xenopus laevis*.

Nieuwkoop–Faber stage[1]	Developmental event
23–27	Otic placode forms (Bever et al. 2003)
28–31	Otocyst forms (Bever et al. 2003)
45–47	Saccular macula emerges (Kil and Collazo 2001; Bever et al. 2003)
48–50	Amphibian and basilar papilla form (Bever et al. 2003; Kil and Collazo 2001)
49–52	Lagena forms (Bever et al. 2003)
59	Operculum develops. Columellar footplate forms from chondrification above oval window (Sedra and Michael 1957)
61–62	Formation of columellar shaft as footplate extension (Sedra and Michael 1957; Trueb and Hanken 1992). Emergence of extracolumellar and annular cartilages above muscular process of quadrate (Sedra and Michael 1957)
64	Columella and extracolumella fuse (Sedra and Michael 1957)
66	Annulus encloses distal extracollumella. *Pars media* completely ossified (Trueb and Hanken 1992). Adult number of eighth-nerve axons reached (López-Anaya et al. 1997)
Froglet	Caudal U-shaped patch of amphibian papilla forms (Smirnov 1993)

[1]Nieuwkoop and Faber (1994).

3.5 Auditory Nerve

In *H. regilla* and in *R. pipiens*, the AP is visible and contains hair cells before roots of the dorsal portion of the eighth cranial nerve (nVIII) extend to the organ; however, ventral roots of nVIII do appear to extend to the vestibular organs of the labyrinth at comparable early stages (Larsell 1934). Fritzsch et al. (1988) observed in two species (*Ascaphus* and *Ichthyophis*) that the numbers of nVIII fibers do not significantly increase between larval and adult stages (the specific larval stage of comparison was not given), suggesting that there is more convergence of hair cells onto single afferents in adults than in tadpoles.

Development of nVIII has been studied in *Xenopus* beginning at NF stage 52, when the saccular and papillar branches of the nerve are readily distinguished (Lopez-Anaya et al. 1997). Myelinated axons are visible at this stage, and they increase in number up to the early postmetamorphic period, after which the rate of myelination decelerates. Cross-sectional areas of the axons increase most during postmetamorphic periods. There are no data available on the formation of synaptic connections between the auditory end-organs and nVIII in any species of anuran.

The frequency sensitivity of nVIII has not been directly examined in tadpoles of any species. Shofner and Feng (1981) recorded nVIII responses to pure tones (presented closed-field in air) in *R. catesbeiana* froglets and adults. Their data show that the frequency sensitivities of the AP and BP, as reflected in nVIII tuning curves, change over the postmetamorphic period. In general, froglet nVIII fibers had higher best excitatory frequencies (up to approximately 2500 Hz) than those in adults (up to approximately 1700 Hz), and higher thresholds at high frequencies. Tuning curves were typically V-shaped, as in adults, but broad, high-threshold U-shaped curves were also observed. Shofner and Feng (1981) attributed these developmental differences to the changes in tympanum size (approximately tenfold) between froglets and adults, with corresponding changes in the middle ear transfer function. The actual innervation of nerve fibers to the individual auditory receptor organs was not anatomically confirmed in this study, however, and so it is unclear if nVIII fiber responses reflect input from the saccule, AP, or BP. Because frequency sensitivity changes over development, classifications of fiber innervation based on distributions in adults are not necessarily accurate.

4. Central Auditory Pathways

The neuroanatomy of the auditory system of adult frogs is described by Wilczynski and Endepols (Chapter 8). There are only limited data available on the organization of the central auditory system across metamorphic development, and what data are available are restricted to brain stem nuclei. The brain as a whole changes considerably in size over development, with concomitant changes in nuclear volume and density. Although some nuclei appear to exhibit stable connectivity across larval development, important alterations of connectivity in other nuclei also occur. These alterations reflect the maturation of the auditory periphery, patterns of cell birth and cell death over development, and, in amphibious frogs, the degeneration of the lateral line during metamorphic climax.

4.1 Dorsal Medullary Nucleus

In stage 25 *R. catesbeiana* tadpoles, a tightly packed group of small cells in the medulla located near the entry point of nVIII can be identified as an acoustic nucleus (developing dorsal medullary nucleus, DMN) on the basis of anterograde transport of horseradish peroxidase (HRP) from the auditory end-organs (Jacoby and Rubinson 1983). Between stages 28 to 45, there is a progressive change in the position of this cell mass from this more lateral to a more medial and dorsal location in the medulla (Jacoby and Rubinson 1983; Fritzsch et al. 1984; Kumaresan et al. 1998; Fig. 10.4). By the froglet period, the DMN is located in its adult position (Fig. 10.4D). The change in position of the DMN is accompanied by an increase in nuclear volume and an increase in cell number, reflecting both migration of existing cells (Fritzsch et al. 1984) and birth of new cells (Jacoby and Rubinson 1983; Chapman et al. 2006). The addition of new cells is also reflected in the undifferentiated appearance of most DMN cells in early larval stages. Signs of differentiation are still observed up to late metamorphic climax stages and the early froglet period (Kumaresan et al. 1998). Jacoby and Rubinson (1983)

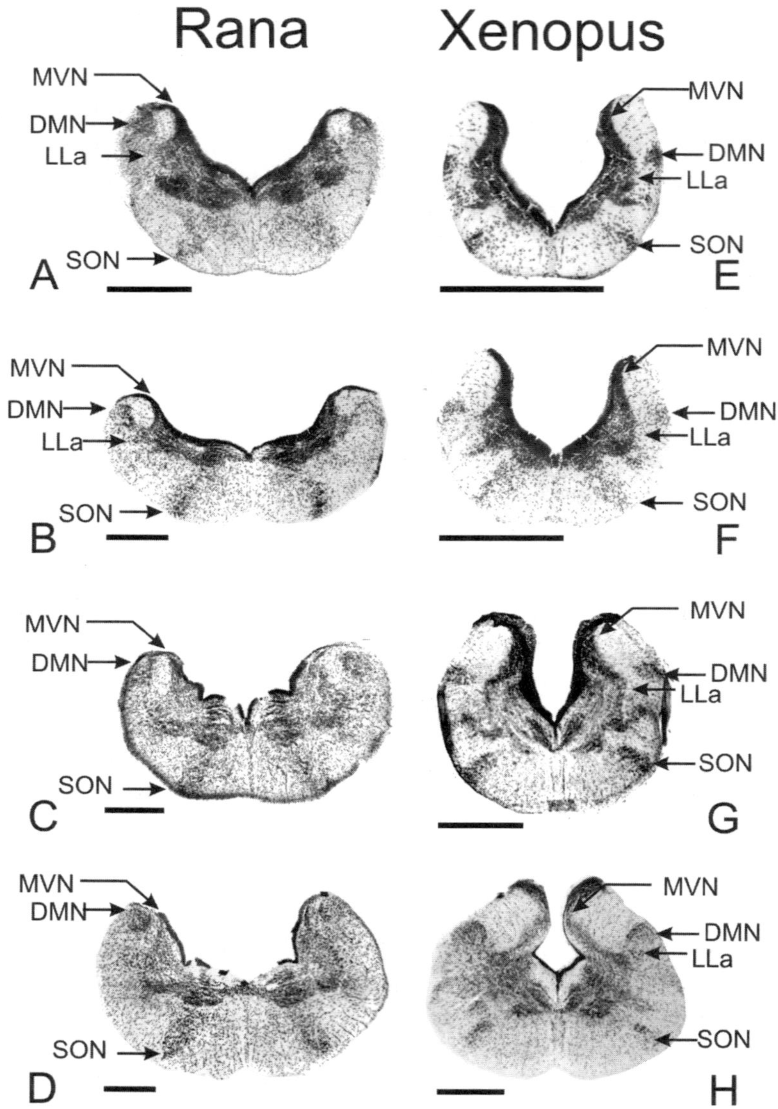

FIGURE 10.4. Cresyl-violet-stained coronal sections (50 μm) through the rostral medulla of *R. catesbeiana* (left) and *X. laevis* (right), at the level of entry of the eighth cranial nerve. *Rana* sections are derived from animals at Gosner stages 30 (A), 36/37 (B), and 43 (C), and a froglet of 4.9 cm SVL (D). *Xenopus* sections are derived from animals at NF stages 51 (E), 53 (F), and 61 (G), and a froglet of 3.1 cm SVL (H). DMN: dorsal medullary nucleus; LLa: anterior lateral line nucleus; MVN: medial vestibular nucleus; SON: superior olivary nucleus. All scale bars 500 μm.

reported that, between stages 28 to 40, afferents from the BP distribute to the dorsal and lateral portions of the DMN, and the afferents from the AP are restricted to the ventral and lateral edges of the nucleus. They argued that the extent of the AP projection zone increases during development, whereas that of the BP decreases. This finding has not been independently replicated. Moreover, terminations of the saccule or lagena in or around the region of the developing DMN have not been traced.

4.2 Lateral Line Projections

In larval *R. catesbeiana* as young as stage 20, afferents from both the posterior and anterior lateral line nerves enter the dorsal medulla and arborize extensively within a distinct lateral line neuropil located lateral to the periventricular area and medial to the DMN (Jacoby and Rubinson 1983). By stage 39, this neuropil region has decreased in size. In *R. temporaria*, the lateral line neuromasts and afferent fibers begin to degenerate between stages 43 and 44, most likely under the influence of thyroid hormone (Fritzsch et al. 1988). Larsell (1934) and Fritzsch et al. (1984) noted signs of degeneration in the anterior lateral line nucleus (LLa) and associated neuropil around stage 42 in a number of species. In *R. catesbeiana*, the LLa can no longer be reliably located in the medulla by stage 43 (Fig. 10.4C).

Based on Golgi-stained medullary sections in hylid tadpoles, Larsell (1934) argued that dendritic processes of a group of cells in the dorsal lateral medulla extend to both the entering nVIII fibers and to entering lateral line fibers. With degeneration of the lateral line afferents, these neurons receive input from nVIII only. Larsell proposed that there is a transformation of function of cells in the dorsal medulla from multisensory (mechanosensitive and auditory) to auditory only across metamorphosis. After the lateral line has degenerated, the brain area previously taken up by the lateral line neuropil would then become populated with neurons sensitive to auditory or vestibular stimulation. Larsell's proposals were not supported by data from anatomical experiments by Jacoby and Rubinson (1983) and by Fritzsch et al. (1984). Results from both studies agree that the lateral line neuropil and the DMN are discrete structures in tadpoles and receive distinct patterns of input with no evidence of converging lateral line/auditory input at any stage of development. It is unclear, however, if intrinsic neurons project between these nuclei in tadpoles.

Lowe and Russell (1982) examined medullary projections of the lateral line nerve in larval (NF stages 56 to 59) and adult *Xenopus*. Although they did not specifically examine nVIII projections, they noted that lateral line projections did not extend to the area of the DMN, suggesting that in *Xenopus*, the DMN and the lateral line nuclei are discrete nuclei in larvae as well as in adults. The DMN retains its lateral position in the dorsal medulla over the course of metamorphic development (Figs. 10.4E to H), consistent with the retention of the lateral line system.

4.3 Vestibular Nuclear Complex

In adult frogs, the vestibular nuclear complex (comprising the medial, lateral, superior, and descending vestibular nuclei) receives input from all otolith organs (Birinyi et al. 2001). The pattern of distribution of afferents from the otolith organs to these nuclei across metamorphosis has not been worked out. Straka et al. (2001) reported that the position of larval rhombomeres in stage 25 *R. catesbeiana* tadpoles is consistent with the location of the vestibular nuclear complex of adult frogs, suggesting that these nuclei do not undergo substantial migration across development. Relative stability or earlier maturation of the vestibular nuclear complex is also implied by results showing that neurons in the lateral vestibular nucleus appear to be fully differentiated by about stage 37, earlier than DMN neurons (Kumaresan et al. 1998). Qualitative examination of the medial vestibular nucleus (MVN) of *R. catesbeiana* shows no obvious change in the location of this nucleus across metamorphosis (Figs. 10.4A to D).

4.4 Superior Olivary Nucleus

There is no evidence in any anuran species of migration of the SON during larval life similar to that of the DMN (see Fig. 10.4). Even in *R. catesbeiana* tadpoles as young as stage 21, the SON can be identified in a similar location as in postmetamorphic froglets and adult frogs. The volume of the SON increases over larval development, and its density decreases. The numbers of both neurons and glia increase during larval and early postmetamorphic life, but at different rates (Templin and Simmons 2005; Chapman et al. 2006).

Studies with various neuroanatomical tract-tracing techniques identify developmental plasticity in projections from the SON to the TS during the larval period (Fig. 10.5). Jacoby and Rubinson (1984) injected HRP into the SON of *R. catesbeiana* tadpoles (stages 35 to 37) and observed retrograde filling of cells, indicating efferent projections, in the ipsilateral medial TS. Injections into the TS itself labeled cells in the ipsilateral SON. This pattern of results is as expected from studies in adult frogs (Wilczynski and Endepols, Chapter 8). Using iontophoresis or pressure injections of HRP, the lipophilic carbocyanine dyes 1,1′-dioctadecyl-3,3,3′,3′- tetramethylindocarbocyanine perchlorate (DiI) and 1,1′-dioctadecyl-5,5′-diphenyl-3,3,3′, 3′-tetramethylindocarbocyanine chloride (DiI-Ph), *Phaseolus vulgaris* lectoagglutinin (Pha-L), and *Vibrio cholera* cholera toxin B subunit (CT-B) into either the SON or the TS, Boatright-Horowitz and Simmons (1997) and Horowitz et al. (2006) observed considerable changes in connectivity over the entire range of metamorphic development. A summary of their results is presented in Figure 10.5. During stages 21 to 38 (Fig. 10.5A), afferent projections to the TS arise bilaterally from the MVN, the LLa, and the DMN, and ipsilaterally from the SON. No afferent projections to the TS from the contralateral SON could be identified. Efferents from the TS were found to project to the MVN and DMN bilaterally and to the ipsilateral SON. Both afferents and

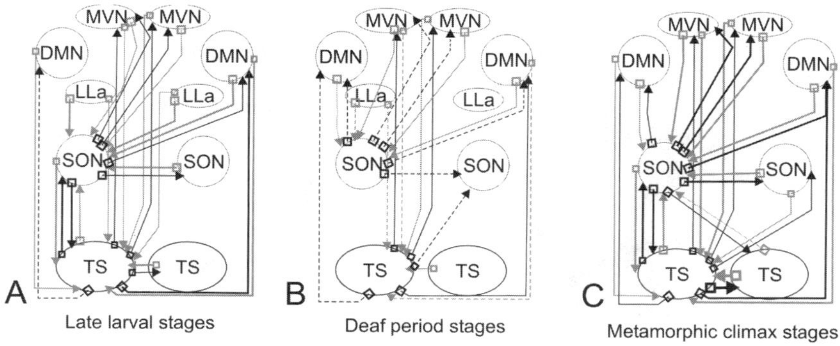

FIGURE 10.5. Schematic of connectivity changes across metamorphic development based on iontophoresis or pressure injection of dye into either the SON or the torus semicircularis (TS). A: Late larval period, stages 31 to 37; B: deaf period, stages 38 to 41; C: metamorphic climax, stages 42 to 46. Connections are based on tracer injection in nuclei on the left side, showing contralateral and ipsilateral connectivity. Substantial rewiring occurs in medullary and midbrain auditory nuclei across development. The most striking changes include the loss of connectivity between the SON and the TS during the deaf period (B). At metamorphic climax (C), SON connectivity is restored, and additional connections between the contralateral SON and TS are formed. Arrows indicate direction of connectivity and line weight (solid vs. dotted) indicates relative number of terminals and/or cell bodies based on anterograde (gray) and retrograde (black) tracer injections. Data from Boatright-Horowitz and Simmons (1997); Horowitz et al (2006).

efferents from the DMN are substantially more numerous on the contralateral side in these stages.

Between stages 38 to 41, dramatic changes in connectivity patterns are seen (Fig. 10.5B). In particular, no or minimal connectivity between the ipsilateral SON and the TS could be observed by any labeling technique. Injections of anterograde tracer into the TS produced label in the contralateral DMN and in the MVN bilaterally, but did not label the ipsilateral SON with sparse label appearing in the contralateral SON at the end of this period. Label of cells in the ipsilateral LLa was still observed, however. Injections of anterograde tracer into the SON produced sparse label in the DMN and MVN bilaterally, the ipsilateral LLa, and the contralateral SON, but no transport rostral of the cerebellar region. Efferent connections between the TS and the SON were also absent or considerably reduced in extent and in intensity. Boatright-Horowitz and Simmons (1997) labeled stages 38 to 41 the "deaf period," partially on the basis of this transient loss of anatomical connectivity between the ipsilateral SON and the TS and partially on the basis of functional changes in the TS (Section 5.2). The period of disconnection between the SON and the TS is synchronous with the transient blockade of the oval window by the formation of the operculum cartilage (Section 3.4.3; Fig. 10.2B).

During metamorphic climax, afferent and efferent connectivity between the TS and ipsilateral SON is restored (Fig. 10.5C). In addition, both afferent and efferent

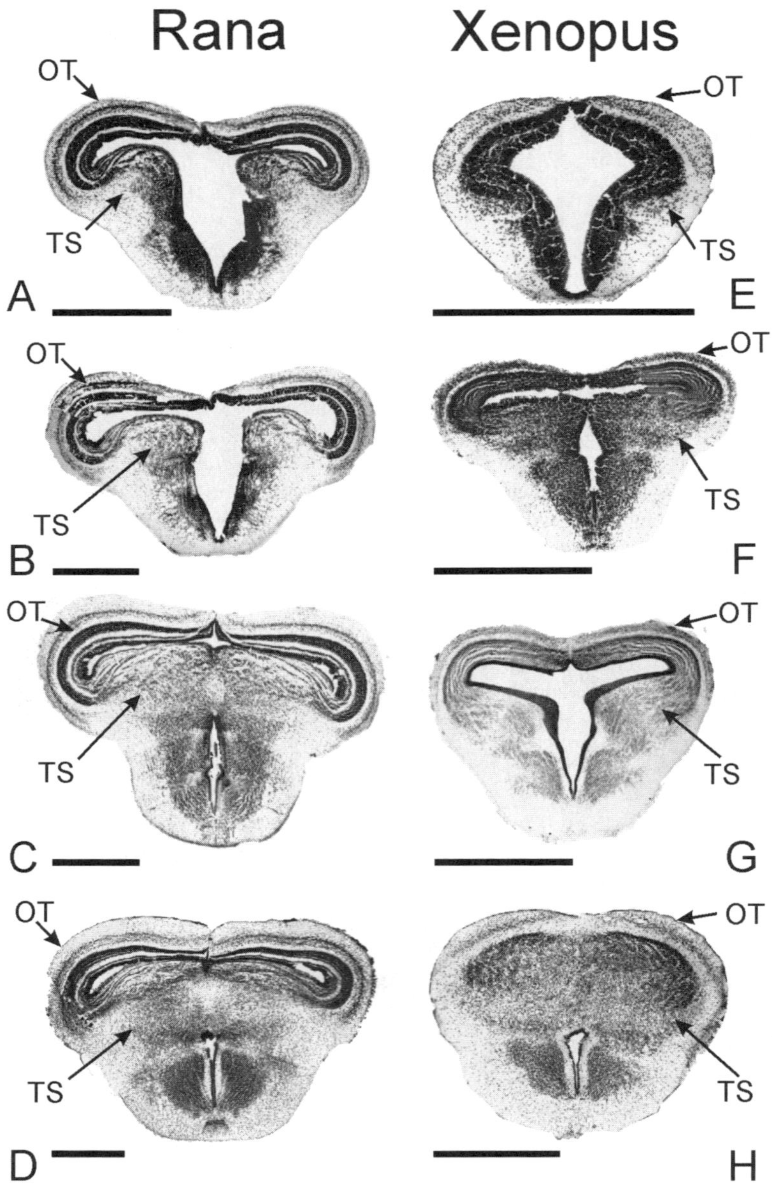

FIGURE 10.6. Cresyl-violet-stained coronal sections (50 μm) through the midbrain (mid to rostral levels) of *R. catesbeiana* (left) and *X. laevis* (right) at different developmental stages. *Rana* sections are derived from animals at Gosner stages 30 (A), 36 (B), and 43 (C), and a froglet of 4.8 cm SVL (D). *Xenopus* sections are derived from animals at NF stages 51 (E), 56 (F), and 61 (G), and a froglet of 2.7 cm SVL (H). OT: optic tectum. All scale bars 1000 μm.

connections between the contralateral SON and the TS are now apparent. There is also an overall increase in the strength of connectivity between most nuclei, aside from the loss of connectivity with the degenerating lateral line nuclei.

4.5 *Torus Semicircularis*

Cresyl violet-stained sections through the TS of *Rana* and *Xenopus* larvae are shown in Figure 10.6. In larval *R. catesbeiana*, the adult organization of the TS cannot be reliably observed until metamorphic climax stages (Kumaresan et al. 1998). Divisions between the subnuclei (particularly the magnocellular and principal nuclei) are difficult to differentiate during the larval period, because of the considerable continuous changes in volume and density of the TS that occur during development. The ventricular laminar region, although distinguishable from the rest of the TS, also undergoes anatomical modification during development. For these reasons, Boatright-Horowitz and Simmons (1997) suggested use of the terms lateral TS, medial TS, and ventricular TS to describe the developing magnocellular, principal, and laminar nuclei, respectively.

In stage 25 to 30 animals (Fig. 10.6A), the medial TS is composed of relatively densely packed round or ovoid cells with no clear laminar organization. The more lateral regions are less cell dense than the medial regions, but have a greater cell density than observed in the adult magnocellular nucleus. This lateral area becomes more cell sparse as development proceeds. The ventricular TS exhibits fewer laminae but layers are thicker. Between stages 25 and 35 to 36, the tectal ventricle is continuous with the third ventricle as far caudal as the level of the nucleus isthmus (Fig. 10.6B). By stages 37 to 38, the TS has increased in size overall by approximately 50% and shows more distinct nuclear organization. The third and fourth ventricles have separated, and a cell mass analogous to the adult commissural nucleus forms. By metamorphic climax, the TS has achieved approximately 75% of its adult size and nuclear boundaries are more discernable (Fig. 10.6C). Lamination of the medial TS is clearly visible.

5. Functional Development

5.1 *Methodological Issues*

A challenge for the functional characterization of hearing across metamorphosis is the technical issues involved in recording auditory responses from small aquatic animals. These include construction of appropriate experimental tanks to minimize standing waves, choice of stimulus variables, and the separation of pressure from particle motion as effective stimuli. These issues are complicated in tadpoles because of the multiple acoustic organs in the anuran inner ear, all of which might change in their sensitivity and selectivity as development proceeds. In adults, the relative functional sensitivities of the AP, BP, and saccule are typically based on distributions of best excitatory frequencies of nVIII fibers, with only a few studies having directly traced these fibers back to their hair cell terminations (Simmons et al., Chapter 7). Using the same frequency distributions in tadpoles to make inferences about inner ear innervation patterns and function implies that frequency sensitivity does not change over development, which is not a valid assumption (Shofner and Feng 1981). Another issue is exemplified in the work of Weiss et al. (1973), who argued that tadpoles have poor hearing sensitivity. Their experiments were based on recordings of evoked potentials in tadpoles kept in air rather than in water, thus using animals that were not only out of their natural habitat but were probably also in poor physiological condition. It is essential to conduct experiments on the functional aspects of hearing in animals kept at least partially submerged, even though this is more difficult for the investigator.

5.2 Developmental Changes in TS Responsiveness

Boatright-Horowitz and Simmons (1997) recorded multiunit activity in response to sounds from the TS of *R. catesbeiana* tadpoles (stages 25 through 46). In these experiments, animals were partially submerged in a small plastic tank and sounds were presented from a loudspeaker positioned above the animal's head, yielding stimulation biased strongly towards pressure changes only. There are considerable differences in the shapes and sensitivities of pure tone audiograms across metamorphic development (Fig. 10.7). Audiograms recorded from the medial TS of tadpoles between stages 27 to 33 exhibit a high-pass shape, with high thresholds to tone frequencies below 1000 Hz and most sensitive frequencies around 2000 Hz. On the other hand, audiograms from tadpoles in metamorphic climax stages are V- or U-shaped, with a broad range of most sensitive frequencies around 1500 to 2000 Hz. Thresholds to low frequencies are lower than at early larval stages, and thresholds to frequencies above 2000 Hz are higher. Between stages 38 to 41, auditory responsiveness is considerably weaker than at early larval or metamorphic climax stages. In some animals in this stage group, no responses at all could be recorded to tones, whereas in others, thresholds are considerably higher across the entire frequency range. This transient lack of responsiveness provides the physiological basis for the definition of the deaf period (Section 4.4).

Neural responsiveness to complex acoustic stimuli also varies considerably over development. In tadpoles younger than about stage 37, the TS shows a broad frequency range of phase-locked activity to amplitude-modulated noise bursts, with significant phase-locking to modulation rates as high as 250 Hz. Between stages 38 to 41, if any neural responses at all are evoked, phase-locking is variable, reflecting the poor auditory responsiveness of the TS during this developmental period. In metamorphic climax stages, strong phase-locking re-emerges, but is now restricted to modulation rates below 100 Hz (Boatright-Horowitz and Simmons 1997; Boatright-Horowitz et al. 1999).

The pattern of neural responsiveness in the TS to airborne sounds reflects the sequential development of peripheral transduction pathways. Neural activity from

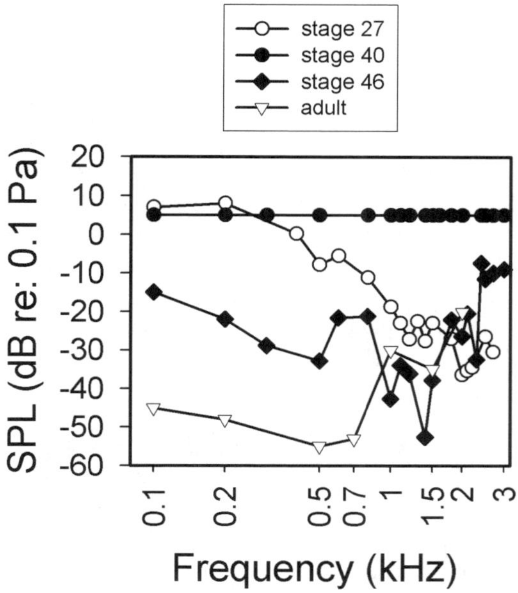

FIGURE 10.7. Multiunit audiograms (threshold, *y* axis; tone frequency, *x* axis) from the TS in three larval and one adult female *R. catesbeiana*. Measurements were taken from animals submerged underwater (just below the water surface for the tadpoles, depth of 0.25 m for the adult female). Sounds were presented to the larval animals from an in-air loudspeaker, and to the adult from an underwater loudspeaker. Recording sites were in the medial TS of the larval animals; recording sites from the adult identified only as TS. The audiogram of the stage 27 tadpole exhibited a high-pass shape, with high thresholds to low frequencies and lowest thresholds at 2000 Hz. In contrast, the audiogram from the stage 46 tadpole was more V-shaped, with lower thresholds to low frequencies and best sensitivity around 1500 Hz. No auditory responses could be evoked from the stage 40 (deaf period) tadpole. The audiogram of the adult showed better sensitivity to low frequencies and weaker sensitivity to higher frequencies. Larval data from Boatright-Horowitz (1997) and Boatright-Horowitz and Simmons (1997); adult data from Lombard et al. (1981).

stages 27 to 37 is due to the operation of the fenestral pathway, whereas activity in metamorphic climax stages reflects the operation of a functional opercularis system. The decline in neural responsiveness during the deaf period (stages 38 to 41) is coincident with the growth of the operculum over the oval window (Fig. 10.2B), and with the transient loss of connectivity between the SON and the TS (Fig. 10.5B). The initial growth and spread of the operculum over the oval window impedes the operation of the fenestral pathway, and, because the opercularis muscle attachment is not yet mature, the opercularis pathway is not yet functional. This results in high threshold audiograms (Fig. 10.7). During metamorphic climax stages, the opercularis pathway is developed, and can mediate auditory function during this time. The tympanic/columellar pathway contributes to neural responsiveness beginning after metamorphic climax, in the postmetamorphic period. TS audiograms to sounds presented in air are sharper and more sensitive in froglets than in tadpoles (Boatright-Horowitz and Simmons 1995). In adult animals, thresholds to low frequency sounds presented underwater are lower than those observed in tadpoles (Fig. 10.7), consistent with the changes in frequency sensitivity of nVIII (Shofner and Feng 1981). Significant phase-locking to amplitude-modulated noise occurs in the same frequency range in froglets as in adults, but in froglets the magnitude of synchronization depends on the route of sound presentation. Open field presentations evoke a broader range of synchronous activity than closed field, consistent with increased contribution of the opercularis and other putative nontympanic pathways (Boatright-Horowitz et al. 1999).

6. Immunohistochemistry of the Developing Auditory System

The pattern of expression of neurotransmitters is important in establishing neuronal homologies and phyletic relationships across species (Baker 1991). During development, changes in expression can serve to identify periods of functional plasticity. The use of histochemical techniques to analyze molecular development in the auditory brain stem across metamorphosis is to date limited. The few studies that have been done show both similarities and differences in maturation of neurotransmitter systems between metamorphosis and brain development in other vertebrates, and between developmental and adult expression in anurans. This is a rich area for further research.

6.1 Acetylcholinesterase

Expression of acetylcholinesterase (AChE), the hydrolytic enzyme of the neurotransmitter acetylcholine, has been used as a marker for periods of plasticity during development, apart from its role in cholinergic transmission in adult brains. Data from *R. catesbeiana* tadpoles (Kumaresan et al. 1998) suggest that metamorphic climax and the early froglet periods are times of considerable plasticity in the auditory brain stem. AChE is expressed beginning in early larval stages. In the DMN, expression peaks during metamorphic climax and the froglet period, coincident with increased differentiation of neurons during these time periods. In contrast, AChE expression in the SON is low and variable until the end of the deaf period, and then increases gradually during metamorphic climax and the froglet period. A dorsal-ventral gradient in expression, with dorsal regions staining more intensely than ventral ones, emerges in the TS during the deaf period and remains throughout adulthood. Expression in the lateral TS peaks during metamorphic climax. AChE label in the lateral vestibular nucleus appears earlier than in the DMN, and does not show any transient increases or decreases in expression throughout development. This suggests that the lateral vestibular nucleus may mature earlier than the DMN, consistent with anatomical findings (Straka et al. 2001).

Expression of choline acetyltransferase (ChAT), the synthetic enzyme for acetylcholine, is a more specific marker of cholinergic transmission. Schlesinger (1981) reported that the brain of larval *Xenopus* shows a "constant" level of ChAT activity over development, but he did not look specifically at auditory nuclei. In contrast, Lopez et al. (2002) reported that ChAT expression follows a caudal to rostral progression, beginning in late embryonic stages (NF stages 38 to 45). No label was observed in the larval *Xenopus* SON or TS over larval development, in contrast to the pattern of AChE staining observed in larval *Rana*. This may indicate the different roles of AChE and ChAT during development.

6.2 GABA

Immunoreactivity for γ-aminobutyric acid (GABA) shows regional differences in the brain of larval *Rana* over development (Simmons and Chapman 2002). Between stages 25 to 30, there is intense staining in the TS and in the lemniscal pathways, whereas the DMN and SON show only diffuse neuropil or puncta label. In the SON, clear label of somata first appears during the deaf period. During metamorphic climax, GABA-positive somata in the medial TS appear organized into discrete layers, indicative of the development of the laminar organization of the principal nucleus seen in adults (Wilczynski and Endepols, Chapter 8). Consistent with data on the expression of AChE, the vestibular nuclear complex shows clear labeled somata earlier than the DMN, again suggesting the earlier maturation of vestibular pathways. This is also consistent with the results of Roberts et al. (1987), who observed strong GABA label in the region of the vestibular nucleus of embryonic *Xenopus*. Changes in label over the rest of *Xenopus* larval development were not examined in that study.

7. Molecular Basis of Metamorphosis

The central molecular tenet of amphibian metamorphosis is that it is driven by thyroid hormone, which changes the expression of genes in target tissues. Thyroid hormone exerts its influence by a wide variety of mechanisms, including interaction with corticosteroids at different times of development (Wright et al. 2003), as well as via direct tissue remodeling action through cell proliferation and apoptosis (Shi 2000). These latter effects are carried out largely by transcriptional regulation mediated by triiodothyroxine receptors. Receptor-mediated remodeling of peripheral tissues has been the primary focus of molecular studies of metamorphosis for decades (Shi 2000), but little is known about its effects on either the peripheral or central auditory system in tadpoles or adults. In mammals, thyroid hormone, acting via specific receptors and gene products, plays an important role in the development of the cochlea (Forrest et al. 2002). It is not known if thyroid hormone influences the development of the inner ear organs in anurans.

Gene expression in the developing auditory system is becoming a rapidly growing field (Keats et al. 2002); however, two issues limit extension of such studies to anuran models. First, most molecular genetic studies in anurans are carried out in *Xenopus laevis* (Heller and Brandi 1999; Serrano et al. 2001), a species whose innate tetraploidy requires careful interpretation of the applicability of gene expression studies to other species. A second and related factor is the limited number of cDNA libraries available for auditory structures in any anuran species. Most genes expressed in auditory precursor or developing cells have only been characterized in an evolutionary or comparative context, and their specific products and role in metamorphic development of hearing are not known. The field of anuran auditory molecular genetics is largely unexplored and is a rich area for future investigations.

8. Behavioral Changes

Tadpole behavior has been studied extensively in an ecological context (Hoff et al. 1999) and in terms of some sensory and motor functions (Stehouwer 1988), but there is little information on metamorphic changes specifically in auditory perception. Tadpoles show such behaviors as schooling (Lum et al. 1982) and rheotaxis (Simmons et al. 2004), both of which are affected by administration of ototoxic drugs. These behaviors reflect in part vestibular function, which may include mediation by the saccule or lagena, but they probably do not reflect auditory function mediated by either the AP or BP. Techniques to study auditory behaviors in tadpoles are critically necessary to examine functional maturation of the auditory periphery and, by extension, the central auditory system.

9. Summary

The time course of metamorphosis is modulated by external factors such as food supply, contaminants, temperature, and light cycle, making it an interesting model for understanding the relationship between genetic and experiental aspects of development. The lengthy tadpole period of some anurans may also allow behavioral plasticity (learning) to occur. The study of metamorphosis as an indicator of either physiological or behavioral plasticity is still in its infancy, however. We have only limited knowledge of functional changes that occur in the central auditory system across development. There is no information on functional changes that may occur during the lengthy period between metamorphosis and adulthood, when males first exhibit the capacity to vocalize. Behavioral measures of hearing in tadpoles are needed to test models based on physiological and anatomical data. Although basic anatomical changes in both the inner ear and the brain stem have been described, detailed functional models of the biophysics of sound transduction in the tadpole period, and how these vary during the metamorphic transition from an aquatic to a terrestrial existence have yet to be formulated. The opercularis system mediates substrate vibrations in adult anurans, but appears to be involved in mediating pressure detection in metamorphic climax *Rana*. Whether and how this system also mediates particle motion or substrate vibrations in tadpoles is not known. Our knowledge of central auditory processing in the tadpole period is limited to a few studies exploring acoustic sensitivity of the TS; to date, nothing is known about the neurophysiology of either the medulla or the telencephalon.

This review focused on developmental changes in two species, one that becomes amphibious after completion of metamorphosis and one that remains aquatic. Because *Xenopus* does not face the demands of switching from hearing in a totally aquatic to hearing in a partly terrestrial environment, the degree of metamorphic change in its auditory system may not be as extensive as in *Rana*. For example, the *Xenopus* DMN, unlike the *Rana* DMN, does not appear to migrate over larval development. Whether this is due to changes in density and distribution of other auditory or nonauditory brainstem nuclei related to differences between aquatic versus terrestrial lifestyles is unknown. We have no data on anatomical development or functional maturation of other nuclei in the *Xenopus* auditory pathway that would allow us to fully test this hypothesis. Moreover, there are little data on metamorphic development and maturation of the auditory pathway in most anuran species. It is not yet clear whether the model of sequential development of peripheral transduction systems developed for *Rana* can be generalized to *Xenopus* or to any other anuran species.

The vertebrate auditory system undergoes extensive growth and reorganization across development (Sanes and Walsh 1998). Species-specific differences in postnatal rates of maturation, parental investment, and environment demand different degrees and rates of neural development for young animals, but there are significant parallels between different developing organisms that can help elucidate common developmental mechanisms. Studies in numerous vertebrate species have demonstrated the need for early sensory experience to establish mature sensory function and have also shown that appropriately patterned stimuli can refine or eliminate sensory responding in the developing animal. Amphibian tadpoles are excellent models for the study of the plasticity of the nervous system due to the extensive neurosensory remodeling that occurs during a lengthy, nonfetal developmental period. Tadpoles, as aquatic organisms whose final form relies, at least in part, on atmospheric hearing show many similarities in their auditory development to that seen in mammalian fetuses. The uterine environment is an aquatic one, sharing many acoustic features with bodies of shallow water, including transmission of external environmental sound with a low-pass filter characteristic, elevated propagation velocity, and relatively high ambient noise due to structurally contiguous sound sources (Bass and Clark 2002). In humans, behavioral responses to externally presented sounds emerge between fetal weeks 19 and 22, with responses to frequencies below 500 Hz emerging before those to frequencies in the range of 1000 to 3000 Hz (Hepper and Shahidullah 1994). Much of the fetal responsiveness to frequency characteristics of sound is driven by a combination of the low-pass characteristics of the uterine environment and maturation of the cochlea. The forces driving auditory system development in tadpoles (a watery environment and sequential development of peripheral transduction pathways) argue that tadpoles deserve consideration as a plausible model for examining the development of hearing in utero.

Acknowledgments. We thank Rebecca Brown, Judith Chapman, Emma Sarro, and Thomas Templin for assistance in preparation of Figures 10.1, 10.4 and 10.6, and Thomas Templin for preparation of the tables. Preparation of this manuscript was supported by NIH/NIDCD grant DC05257 to AMS and SSH.

References

Baker H (1991) Evaluation of species-specific biochemical variation as a means for assessing homology in neuronal populations. Brain Behav Evol 38:255–263.

Bass AS, Clark CC (2002) The physical acoustics of underwater sound communication. In: Simmons AM, Popper AN, Fay RR (eds) Acoustic Communication. New York: Springer-Verlag, pp. 15–64.

Bever MM, Jean YY, Fekete DM (2003) Three-dimensional morphology of inner ear development in *Xenopus laevis*. Develop Dynamics 227:422–430.

Birinyi A, Straka H, Matesz C, Dieringer N (2001) Location of dye-coupled second order and of efferent vestibular neurons labeled from individual semicircular canal or otolith organs in the frog. Brain Res 921:44–59.

Boatright-Horowitz SS (1997) Ontogeny of the bullfrog auditory system across metamorphic development. PhD dissertation, Brown University, Providence RI.

Boatright-Horowitz SS, Simmons AM (1995) Postmetamorphic changes in auditory sensitivity of the bullfrog midbrain. J Comp Physiol A 177:577–590.

Boatright-Horowitz SS, Simmons AM (1997) Transient "deafness" accompanies auditory development during metamorphosis from tadpole to frog. Proc Natl Acad Sci USA 94:14877–14882.

Boatright-Horowitz SS, Garabedian CE, Odabashian KH, Simmons AM (1999) Coding of amplitude modulation in the auditory midbrain of the bullfrog across metamorphosis. J Comp Physiol A 184:219–231.

Burggren WW, Infantino RL (1994) The respiratory transition from water to air breathing during amphibian metamorphosis. Amer Zool 34:238–246.

Chapman JA, Weinstein JL, Simmons AM (2006) Cell proliferation in the *Rana catesbeiana* auditory medulla over metamorphic development. J Neurobiol 66:115–133.

Corse WA, Metter DE (1980) Economics, adult feeding and larval growth of *Rana catesbeiana* on a fish hatchery. J Herpetol 14:231–238.

Corwin JT (1981) Morphology of the macula neglecta in sharks of the genus *Carcharhinus*. J Morph 152:341–362.

Corwin JT (1985) Perpetual production of hair cells and maturational changes in hair cell ultrastructure accompany postembryonic growth in an amphibian ear. Proc Natl Acad Sci USA 82:3911–3915.

Diaz ME, Varela-Ramirez A, Serrano EE (1995) Quantity, bundle types, and distribution of hair cells in the sacculus of *Xenopus laevis* during development. Hear Res 91:33–42.

Etkin W (1964) Metamorphosis. In: Moore JA (ed) Physiology of the Amphibia. New York: Academic, pp. 427–468.

Fay RR, Simmons AM (1999) The sense of hearing of fishes and amphibians. In: Fay RR, Popper AN (eds) Comparative Hearing: Fish and Amphibians. New York: Springer, pp. 269–318.

Forrest D, Reh TA, Rusch A (2002) Neurodevelopmental control by thyroid hormone receptors. Curr Opin Neurobiol 12:49–56.

Fritzsch B, Nikundiwe AM, Will U (1984) Projection patterns of lateral line afferents in anurans: A comparative HRP study. J Comp Neurol 229:451–469.

Fritzsch B, Wahnschaffe U, Bartsch U (1988) Metamorphic changes in the octavolateralis system of amphibians. In: Fritzsch B, Ryan MJ, Wilczynski W, Hetherington TE, Walkowiak W (eds) The Evolution of the Amphibian Auditory System. New York: Wiley, pp. 359–376.

Gosner KL (1960) A simplified table for staging anuran embryos and larvae with notes on identification. Herpetologica 16:183–190.

Heller N, Brändli AW (1999) *Xenopus* Pax-2/5/8 orthologues: Novel insights into Pax gene evolution and identification of Pax-8 as the earliest marker for otic and pronephric cell lineages. Dev Genet 24:208–219.

Hepper PG, Shahidullah BS (1994) Development of fetal hearing. Arch Dis Childhood 71:81-87.

Hertwig I (1987) Morphogenesis of the inner ear of *Rana temporaria* (Amphibia, Anura). Zoomorph 107:103–114.

Hertwig I, Schneider H (1986) Development of the supporting cells and structures derived from them in the inner ear of the grass frog, Rana temporaria (Amphibia, Anura). Zoomorph 106:137–146.

Hetherington TE (1987) Timing of development of the middle ear of Anura. Zoomorph 106:289–300.

Hetherington TE (1988) Metamorphic changes in the middle ear. In: Fritzsch B, Ryan MJ, Wilczynski W, Hetherington TE, Walkowiak W (eds) The Evolution of the Amphibian Auditory System. New York: Wiley, pp. 297–306.

Hoff KV, Blaustein AR, McDiarmid RW, Altig R (1999) Behavior: Interactions and their consequences. In: McDiarmid RW, Altig R (eds) Tadpoles: The Biology of Anuran Larvae. Chicago: University of Chicago Press, pp. 215–239.

Horowitz SS, Chapman JA, Simmons AM (2006) Plasticity of auditory medullary-midbrain connectivity across metamorphic development in the bullfrog, *Rana catesbeiana*. Brain Behav Evol (in press).

Horowitz SS, Chapman JA, Kaya U, Simmons AM (2001) Metamorphic development of the bronchial columella of the larval bullfrog (*Rana catesbeiana*). Hear Res 154: 12–25.

Jacoby J, Rubinson K (1983) The acoustic and lateral line nuclei are distinct in the premetamorphic frog, *Rana catesbeiana*. J Comp Neurol 216:152–161.

Jacoby J, Rubinson K (1984) Efferent projections of the torus semicircularis to the medulla of the tadpole, *Rana catesbeiana*. Brain Res 292:278–381

Keats BJB, Popper AN, Fay RR (2002) Genetics and Auditory Disorders. New York: Springer.

Kil SH, Collazo A (2001) Origins of inner ear sensory organs revealed by fate map and time-lapse analyses. Dev Biol 233:365–379.

Kopsch F (1952) Die Entwicklung des braunen Grasfrosches *Rana fusca* Roesel. Stuttgart: Thieme.

Kumaresan V, Kang C, Simmons AM (1998) Development and differentiation of the anuran auditory brainstem across metamorphosis: An acetylcholinesterase histochemical study. Brain Behav Evol 52:111–125.

Lannoo MJ (1999) Integration: Nervous and sensory systems. In: McDiarmid RW, Altig R (eds) Tadpoles: The Biology of Anuran Larvae. Chicago: University of Chicago Press, pp. 149–169.

Larsell O (1934) The differentiation of the peripheral and central acoustic apparatus in the frog. J Comp Neurol 60:473–525.

Lewis ER, Li CW (1973) Evidence concerning the morphogenesis of saccular receptors in the bullfrog *Rana catesbeiana*. J Morphol 139:351–362.

Li CW, Lewis ER (1974) Morphogenesis of auditory receptor epithelia in the bullfrog. In: Johari O, Corwin I (eds) Scanning Electron Microscopy. Chicago: IIT Res Inst, pp. 791–798.

Li CW, Lewis ER (1979) Structure and development of vestibular hair cells in the larval bullfrog. Annals Otol Rhino Laryn 88:427–437.

Lombard RE, Fay RR, Werner YL (1981) Underwater hearing in the frog, *Rana catesbeiana*. J Exp Biol 91:57–71.

Lopez JM, Smeets WJAJ, Gonzalez A (2002) Choline acetyltransferase immunoreactivity in the developing brain of *Xenopus laevis*. J Comp Neurol 453:418–434.

Lopez-Anaya VL, Lopez-Maldonado D, Serrano EE (1997) Development of the *Xenopus laevis* VIIIth cranial nerve: Increase in number and area of axons of saccular and papillar branches. J Morph 234:263–276.

Lowe DA, Russell IJ (1982) The central projections of lateral line and cutaneous sensory fibres (VII and X) in *Xenopus laevis*. Proc Roy Soc Lon B 216:279–297.

Lum AM, Wassersug RJ, Potel MJ, Lerner SA (1982) Schooling behavior of tadpoles: A potential indicator of ototoxicity. Pharmacol Biochem Behav 17:363–366.

McDiarmid RW, Altig R (1999) Research: Materials and techniques. In McDiarmid RW, Altig R (eds) Tadpoles: The Biology of Anuran Larvae. Chicago: University of Chicago Press, pp. 7–23.

Nieuwkoop PD, Faber J (1994) Normal Table of *Xenopus laevis* (Daudin). New York: Garland.

Paterson NF (1949/50) The development of the inner ear of *Xenopus laevis*. Proc Zool Soc Lond 119:269–291.

Paterson NF (1960) The inner ear of some members of the Pipidae (Amphibia). Proc Zool Soc Lond 134:509–546.

Roberts A, Dale N, Ottersen OP, Storm-Mathisen J (1987) The early development of neurons with GABA immunoreactivity in the CNS of *Xenopus laevis* embryos. J Comp Neurol 261:435–449.

Sanes DH, Walsh EJ (1998) The development of central auditory processing. In: Rubel EW, Popper AN, Fay RR (eds) Development of the Auditory System. New York: Springer-Verlag, pp. 271–314.

Schlesinger C (1981) Ontogenesis of the acetylcholine system in the brain of the South African clawed toad (*Xenopus laevis* Daudin). J Hirnforsch 22:543–553.

Sedra SN, Michael MI (1957) The development of the skull, visceral arches, larynx and visceral muscles of the South African clawed toad, *Xenopus laevis* (Daudin) during the process of metamorphosis (from stage 55 to stage 66). Verh K Nederld Acad Wet Natuurk 51:1–80.

Sedra SN, Michael MI (1959) The ontogenesis of the sound conducting apparatus of the Egyptian toad, *Bufo regularis* Reuss, with a review of this apparatus in Salientia. J Morph 104:359–375.

Serrano EE, Trujillo-Provencio C, Sultemeier DR, Bullock WM, Quick QA (2001) Identification of genes expressed in the *Xenopus* inner ear. Cell Mol Biol (Noisy-le-grand) 47:1229–1239.

Shi YB (2000) Amphibian Metamorphosis: From Morphology to Molecular Biology. New York: Wiley.

Shofner WP, Feng AS (1981) Post-metamorphic development of the frequency selectivities and sensitivities of the peripheral auditory system of the bullfrog, *Rana catesbeiana*. J Exp Biol 93:181–196.

Shofner WP, Feng AS (1984) Quantitative light and scanning electron microscopic study of the developing auditory organs in the bullfrog. J Comp Neurol 224:141–154.

Shumway W (1940) Stages in the normal development of *Rana pipiens*. I. External form. Anat Rec 78:139–144.

Simmons AM, Chapman JA (2002) Metamorphic changes in GABA immunoreactivity in the brainstem of the bullfrog, *Rana catesbeiana*. Brain Behav Evol 60:189–206

Simmons AM, Costa LM, Gerstein HB (2004) Lateral line-mediated rheotactic behavior in tadpoles of the African clawed frog (*Xenopus laevis*). J Comp Physiol A 190:747–758.

Smirnov S (1993) The anuran amphibian papilla development, with comments on its timing, rate, and influence on adult papilla morphology. Zoo Jb Anat 123:273–289.

Sokol OM (1981) The larval chondrocranium of *Pelodytes punctatus*, with a review of tadpole chondrocrania. J Morph 169:161–183.

Spaeti U (1978) Development of sensory systems in the larval and metamorphosing European grass frog. J Hirnforsch 19:543–575.

Stehouwer D (1988) Metamorphosis of behavior in the bullfrog (*Rana catesbeiana*). Dev. Psychobiol. 21:383–395.

Straka H, Baker R, Gilland E (2001) Rhombomeric organization of vestibular pathways in larval frogs. J Comp Neurol 437:42–55.

Taylor AC, Kollros JJ (1946) Stages in the normal development of *Rana pipiens* larvae. Anat Rec 94:7–23.

Templin T, Simmons AM (2005) Cellular and spatial changes in the anuran superior olive across metamorphosis. Hear Res 207:87–98.

Trueb L, Hanken J (1992) Skeletal development in *Xenopus laevis* (Anura: Pipidae). J Morphol 214:1–41.

van Bergeijk WA (1959) Hydrostatic balancing mechanism of *Xenopus* larvae. J Acoust Soc Am 31:1340–1347.

Vorobyeva E, Smirnov S (1987) Characteristic features in the formation of anuran sound-conducting systems. J Morphol 192:1–11.

Wassersug RJ, Souza KA (1990) The bronchial diverticula of *Xenopus laevis*. Are they essential for hydrostatic assessment? Naturwissen 77:443–445.

Weiss BA, Stuart BH, Strother WF (1973) Auditory sensitivity in the *Rana catesbeiana* tadpole. J Herpetol 7:211–214.

Weisz PB (1945) The development and morphology of the larva of the South African clawed toad, *Xenopus laevis*: I. The third-form tadpole. J Morph 77:163–192.

Wever EG (1985) The Amphibian Ear. Princeton, NJ: Princeton University Press.

Witschi E (1949) The larval ear of the frog and its transformation during metamorphosis. Z Naturforsch 4b: 230–242.

Witschi E (1950) The bronchial diverticula of *Xenopus laevis* Daudin. Anat Rec 108:590.
Witschi E (1955) The bronchial columella of the ear of larval Ranidae. J Morphol 96:497–511
Wright ML, Guertin CJ, Duffy JL, Szatkowski MC, Visconti RF, Alves CD (2003) Developmental and diel profiles of plasma corticosteroids in the bullfrog, *Rana catesbeiana*. Comp Biochem Physiol A Mol Integr Physiol 135:585–595.

11 Sound Processing in Real-World Environments

Albert S. Feng and Johannes Schul

1. Introduction

A real-world listening environment is generally noisy and contains multiple auditory objects each producing a distinct sound pattern that overlaps in time and spectrum with sounds from surrounding objects. Frog choruses are excellent examples of complex real-world listening environments. Usually, a large number of frogs participate in a single chorus, making the chorus sound very intense with constantly changing spectral and temporal compositions. A female frog is thus exposed to sounds from many conspecific as well as heterospecific males originating from different distances and directions. Signal identification and source localization are compromised due to the presence of competing sounds. In addition, there are generally acoustic scatterings and reflections from objects in the vicinity of frog choruses; these alter the signal characteristics thereby compounding the listening tasks.

In spite of this, there is evidence that females readily perform mate choice. They sample males' calls, select, and approach a particular male within a chorus, often bypassing other males enroute. In this chapter, the physical characteristics of frog calls and how calls are affected by the frog's natural environment and competing sounds are described first. This is followed by a description of the problems and solutions to hearing in complex acoustic environments, and of the unresolved problems.

Listening in a complex auditory scene in the presence of background noise is challenging, due to the fact that all sounds in the ambiance converge onto the two eardrums. It is therefore up to the brain to sort the information to determine "what" is out there and "where" it is, that is, to perform auditory scene analysis (Bregman 1990). Our current understanding of the mechanism underlying auditory scene analysis is limited. Much work remains before we fully understand how frogs identify and localize signals in the presence of competing sounds, as well as group and segregate auditory streams.

2. Physics of Sound Transmission

Sound transmission is affected by various physical factors in an environment; signals are distorted in both the spectral and temporal domains (see reviews by Wiley and Richards 1978; Michelsen 1978; Forrest 1994; Bradbury and Vehrencamp 1998; Padgham 2004). The effects on spectral characteristics of a sound are complex. In general, higher-frequency signals are more severely attenuated than lower-frequency signals; in certain spatial configurations of sender and receiver, however, the reverse occurs (Wiley and Richards 1978). Because anuran communication signals are usually tonal with sound energy concentrated in 1 to 2 narrow frequency bands over a few octaves, spectral distortions are not as profound as for some other biotic sounds (e.g., the impulsive insect signals).

Sound transmission through an animal's natural habitat affects the temporal characteristics of a sound physically and perceptually. For example, the loss of amplitude during sound propagation leads to a decrease in perceived sound duration, when weaker signal components fall below the detection threshold of the receiver. This effect is robust for frog calls because they typically contain sound pulses having long rise and/or fall times. At the same time, because a receiver often hears sounds arriving via a direct as well as an indirect path, echoes and reverberations from objects along a sound path create extra delays, thereby increasing the duration of sound pulses and obscuring the silent intervals between pulses. Temporal smearing, or degradation of the temporal pattern, of communication signals may be so severe that it becomes the primary factor that limits the communication distance rather than the attenuation of the signal per se (Lang 2000). In general, echoes have a greater influence over higher-frequency sounds because of their shorter wavelengths, the abundance of objects with equivalent sizes in the environment, and the inability of high-frequency sounds to curve around objects along their paths. Therefore, temporal degradation is most pronounced for species whose calls have dominant energy in the high-frequency range.

In addition to echoes, environmental turbulence such as wind gusts induces random amplitude fluctuations, degrading a signal's temporal pattern further. Such amplitude fluctuations are usually of low frequency (below 50 Hz) and independent of the signal's carrier frequency (Wiley and Richards 1978).

In light of the fact that acoustic signals are subject to degradations by the environment, it is likely that some temporal parameter would be better preserved than others during transmission. As described above, the absolute durations of, and the silent intervals between, sound pulses are vulnerable during transmission; the changes are unpredictable as effective pulse duration may shorten due to attenuation, or lengthen due to reverberation. In contrast, the pulse repetition rate that characterizes many anuran calls is minimally affected, and is therefore well suited for transmitting the identity of the signaler.

In birds and insects, acoustic communication signals are often adapted for optimal sound transmission for their specific habitats (Römer and Lewald 1992; Richards and Wiley 1980). The evidence for environmental effects on frog sound communication signals is mixed. There is positive evidence from a study of geographical call variation in two subspecies of cricket frogs (*Acris crepitans crepitans* and *A. c. blanchardii*). One subspecies, *A. c. crepitans*, typically breeds in pine forests, whereas the other subspecies prefers more open habitats. Their advertisement calls differ significantly; the calls of *A. c. crepitans* are well suited for transmission (i.e., do not degrade much) in their forest habitats (Ryan et al. 1990; Ryan and Wilczynski 1991). The calls of *A.c. blanchardii* show large geographical variations; frogs that occupy isolated areas of pine forests tend to produce calls having similar characteristics to the calls of *A. c. crepitans*, suggesting habitat effects. In spite of this, habitat differences alone cannot account for the overall geographical variations (Wilczynski and Ryan 1999).

Systematic studies of transmission of frog calls in the different habitats have provided largely negative results of habitat effects (Kime et al. 2000; Penna and Solis 1998; Bosch and Riva 2004). Indeed, the calls of some species are better suited for transmission outside the species' breeding habitats. The mismatch is presumably due to the fact that detection and localization of frog calls and chorus sound are range dependent, and at long distances acoustic signals only play a minor role (see Section 4). However, other environmental factors, such as noise and the resonance property of the calling site, have been shown to shape their communication signals (e.g., Lardner and bin Lakim 2002; Narins et al. 2004; Penna 2004).

3. Competing Sounds in Natural Environments

During sound transmission through the environment, signals are subject to masking from abiotic and biotic noise (i.e., signals of other noisy animals and vocal signals of conspecifics). For anurans, a major source of abiotic noise is fast-flowing streams or waterfalls that may have amplitudes of above 70 dB SPL in the frequency range between 50 Hz and 4000 Hz with peak energy near 100 Hz (Feng et al. 2002; Narins et al. 2004). Frog species that breed under such conditions tend to produce very high frequency calls that extend into the ultrasonic range (Narins et al. 2004), or involve visual signals to communicate (Hödl and Amézquita 2001; Narins et al. 2003, 2005). Another potential source of abiotic noise is wind; the spectral energy of wind noise is generally limited to frequencies below 100 Hz (Rinberg and Davidowitz 2003) and thus below the frequency range of sound communication for most anurans.

The dominant biotic noise is vocalizations of other sound-producing organisms, for example, insects and sympatric frog species. The amplitude of such noise can reach 80 to 85 dB SPL when measured within a frog chorus (Narins 1982; Schwartz and Gerhardt 1998; Wollerman and Wiley 2002). Frogs employ several strategies to reduce acoustic interference between species (Littlejohn 1977; Garcia and Narins 2001; Chek et al. 2003). One strategy is to call from different geographical locations around a calling site (i.e., from trees or plants in the water) to achieve spatial separation. Another strategy is temporal segregation:

this is done by either restricting calling to different times of an evening or different times of year, or by antiphonal calling (see Wells and Schwartz, Chapter 3; Gerhardt and Bee, Chapter 5). In assemblages of a few males, each male would call with regular timing, either roughly synchronized to a neighbor's call or in alternating fashion (Brush and Narins 1989; Greenfield 1994). In larger choruses, the sequence of calling among the males is difficult to discern and does not appear to adhere to a recognizable temporal pattern (Greenfield and Rand 2000; Wollerman and Wiley 2002).

Another strategy is spectral separation, by placing the dominant call frequency within a band that is not occupied by another species. For example, the community of eight frog species of the genus *Eleutherodactylus* partition the spectral space between 1 and 8 kHz with little overlap between species, providing each species with a "private spectral channel" (Drewry and Rand 1983). Similar situations are found in many temperate and tropical anuran communities. Interestingly, the audiograms of many anurans have regions of enhanced sensitivity (=low thresholds) that correspond to the spectral peaks in male calls (Gerhardt and Schwartz 2001). This "matched filtering" (Capranica and Moffat 1983) serves to reduce the perceived amplitude of heterospecific calls relative to that of conspecific calls, thus improving the signal-to-noise ratio in mixed choruses.

Although spectral separation is commonly observed in multispecies choruses, there are exceptions. Some closely related and sympatric species are known to call with overlapping or even identical spectra, for example, the gray treefrogs *Hyla chrysoscelis* and *H. versicolor* (Gerhardt 2001). Overlaps in call spectra are sometimes asymmetrical, leaving one species with a potentially private frequency channel. For example, two Panamanian treefrogs, *Hyla ebraccata* and *H. microcephala* frequently call in close proximity to each other. The call of *H. microcephala* has a bimodal spectrum with main spectral peaks at 6 kHz and 3 kHz, whereas *H. ebraccata* calls contain a single spectral peak around 3 kHz (Schwartz and Wells 1983). Here, parts of the *H. microcephala* spectrum are not masked by the congener's call, whereas the complete spectrum of *H. ebraccata* is subject to masking by *H. microcephala*.

Another strategy for minimizing interspecific interference is to produce calls having distinct (and thus conspicuous) temporal patterns. Fine-scale temporal parameters such as pulse rate, or pulse duration, are usually quite different between species sharing a chorus, especially between species with similar call spectra; these differences are much larger than the intraspecific variation within a population. For example, for the two gray treefrog species with almost identical call spectra, the pulse rate of *H. chrysoscelis* is twice as fast as that of *H. versicolor* (Gerhardt 2001). Thus, the differences in temporal parameters among the species in a chorus may play an important role in species recognition.

Even when heterospecific callers are absent, the acoustic scene of a frog chorus is still complex, due to the presence of a large number of conspecific male callers. The mechanisms that reduce interference in multispecies choruses (e.g., spectral, temporal, and geographical partitioning) do not apply to unispecies choruses. To minimize interference from their neighbors' calls, callers often engage in vocal interactions, for example, change their call parameters (e.g., duration) or call timing. These are discussed in detail in Wells and Schwartz, Chapter 3 of this volume.

4. Listening in Natural Environments

Given the temporary nature of frogs' breeding sites, how do gravid females localize appropriate breeding sites, and is the chorus sound involved in this task? Gerhardt and Klump (1988a) found in a laboratory study that the barking treefrog (*Hyla gratiosa*), which lives in temporary ponds, responds to chorus sound, whereas the green treefrog (*H. cinerea*), which breeds in permanent ponds does not. The difference in breeding biology of these two species appears to dictate the role of chorus sound in localization of the breeding pond. However, Murphy (2003) found in a recent field study with *H. gratiosa* that chorus sounds have no influence on the number of females arriving at a breeding site, implying that females of this species do not need acoustic cues to localize the breeding site. This discrepancy is perhaps attributed to the particular chorus sound Gerhardt and Klump (1988a) used in their laboratory study. Namely, calls of individual *H. gratiosa* are discernible in their chorus sound (their Fig. 1), making it possible for females to home in on individual calls, albeit at a low signal-to-noise ratio (SNR). Recently, Beckers and Schul (2004) studied the importance of call amplitude for female phonotaxis in gray treefrogs (*Hyla versicolor*) and found that the female's response strength (using the walking speed as a metric) decreases with decreasing call amplitude. Their results show that phonotaxis is less robust with increasing distance, suggesting that in frogs, female phonotaxis operates primarily at short ranges, that is, at moderate to high stimulus amplitudes.

Taken together, these results indicate the probable dichotomy of search strategy; that is, at long ranges females likely rely on nonacoustic cues to localize suitable breeding sites, but at short ranges when the individual calls become conspicuous the acoustic cues are used to detect, identify, and localize reproductively active males. The distance at which the behavioral switch occurs is unclear. Further research is necessary to determine this as well as the threshold for call detection at different SNRs.

5. Acoustic Communication in Noisy Environments

In noisy environments a receiver must first detect signals in the presence of background sound before identifying the signals and discriminating conspecific from heterospecific signals. Additionally, for phonotaxis and territorial interactions, a receiver must localize the perceived signal. Chorus sounds interfere with all three auditory tasks: detection, identification (recognition), and localization. Behavioral studies have addressed such interference and evaluated the performances of the receiver under noisy conditions.

5.1 Signal Detection and Recognition

Whereas the standard approach for determining signal detection involves psychophysical experiments with conditioned animals (review in Klump et al. 1995), this approach is not easily realizable in frogs due to the difficulty conditioning these animals (Elepfandt et al. 2000). Thus, for frogs, acoustic modification of reflexes has instead been used to study signal detection, but mostly under quiet conditions (Strother 1962; Brzoska et al. 1977; Brzoska 1980; Moss and Simmons 1986). Another widely used approach involves studying acoustically guided responses in either gravid females (e.g., phonotaxis), or reproductively active males (e.g., evoked vocal response). Because such responses are specific to particular signals, signal detection is not separable from signal recognition. However, thresholds for signal detection and recognition in noisy conditions appear to be similar. For example, in *Hyla cinerea*, results from female phonotaxis (Ehret and Gerhardt 1980) and reflex modification (Moss and Simmons 1986) yield practically identical masking thresholds even though the latter only measured signal detection without pattern recognition.

When a signal is masked by a background sound, signal detection is compromised. In *Hyla cinerea*, Gerhardt and Klump (1988b) determined the SNR for females to respond to species advertisement calls masked by a conspecific chorus and found that the SNR must be at least 0dB (i.e., when the male call had the same amplitude as the chorus). At SNR of –6dB, females fail to respond. In experiments involving mixed-species chorus sound as masker, females of *H. gratiosa* and *H. ebraccata* require a SNR of at least +3dB for phonotaxis (Wollerman 1999; Murphy and Gerhardt 2002). In these three studies, the SNR was given as the difference between the sound pressure level of the frog call and that of the broadband noise (not the spectrum level), but it is unclear whether the sound pressure level was from a RMS, or a peak, measurement.

Female frogs employ various strategies to improve signal detection; these involve minimizing spectral overlap (Ehret and Gerhardt 1980), and spatial and temporal overlap (Schwartz and Gerhardt 1989, 1995; Wollerman 1999). In signal detection studies in treefrogs described previously, male calls and masking chorus sound were broadcast from a common loudspeaker. When the call and chorus noise were presented from separate loudspeakers that were 90 to 120° apart, call detection thresholds can be lowered by 3 to 6dB (Schwartz and Gerhardt 1989, 1995). The improvement in call detection threshold corresponds to the loss in relative energy of the chorus sound due to the directionality of the frog's hearing system, as determined by laser vibrometry (Jørgensen and Gerhardt 1991).

In addition to chorus sounds, broadband noise has also been used as a masker for quantifying signal detection thresholds. In such studies, the detection thresholds are expressed as critical ratios rather than SNRs (Narins 1982; Moss and Simmons 1986). The critical ratio refers to the ratio between the level of the test tone at detection threshold to the spectrum level of the broadband noise that totally masks the test spectrum level; critical ratios are an indirect measure of the frequency selectivity of the sensory system (Scharf 1970). In *Hyla cinerea*, critical ratios for the two spectral peaks of the species advertisement call at 900 Hz and 3000 Hz are 20 to 25 dB, respectively (Ehret and Gerhardt 1980; Moss and Simmons 1986). For intermediate frequencies and below 900 Hz and above 3000 Hz, the critical ratios are higher, similar to the shape of the pure-tone audiogram (Moss and Simmons 1986). These results support the idea that the anuran auditory system operates as a selective filter which is optimal for detecting species' vocal signals.

As discussed earlier, the pulse rate within a call is the temporal parameter least susceptible to distortion during sound transmission and to masking by chorus sounds. Accordingly, pulse rate has been widely assumed to be the crucial parameter for call recognition (review in Gerhardt and Huber 2002). However, this hypothesis has been formally tested only in a few anuran species, and the results to date have been mixed. For example, females *Hyla chrysoscelis* indeed utilize the pulse rate to recognize conspecific calls. In contrast, however, the sibling species *H. versicolor* evaluates the absolute duration of the pulses and of the silent interval between sound pulses for call recognition, largely independent of pulse rate (Schul and Bush 2002). Similarly, the salient features for call recognition have been found to differ among sibling species of acoustic insects (Schul and Bush 2002; Deily and Schul 2004). For this reason, generalization of behavioral data from one species to another related species must be made with great care.

5.2 How Many Callers Do Frogs Hear in a Chorus?

Frog chorus represents a challenging listening environment that limits the number of calls that each frog can detect. The greater the chorus density, the more severe the listening constraints. Several factors generally determine the hearing limit: (1) the relative amplitudes of individual callers and the background chorus, (2) the distances of nearest neighbors, and (3) the threshold for signal detection in noise. These factors have been measured for some frog choruses and used to estimate the number of callers that a frog can detect in a chorus: these estimates range from 1 to 2 (Brush and Narins 1989; Wollerman 1999) to 3 to 5 (Gerhardt and Klump 1988b).

Perceptual mechanisms potentially limit detection of calls to a smaller number. For example, gain control and attention mechanisms in the sensory pathway (Pollack 1988; Römer and Krusch 2000) may limit the detection to the one to two loudest calls reaching the ears. Such mechanisms have been described for insects (and humans), allowing these organisms to attend selectively to one to two signallers in a dense chorus. For frogs, field observations of females' mate choice or males' evoked vocal response in a chorus provide useful information about the number of callers they can detect in a chorus. Some studies indicate that females predominantly choose the closest male (Grafe 1997; Murphy and Gerhardt 2002). Other studies suggest that females sequentially sample several males (coursing from the calling site of one male to another) before choosing a male (Morris 1991). However, this does not necessarily mean that call detection is limited to one caller, because other reasons (e.g., predation) might influence female sampling behaviors (Grafe 1997). Other anecdotal reports suggest that some frogs use sampling strategies that would require the detection of several males simultaneously. Clearly, further systematic research is necessary to pin down the limit of auditory objects heard and the conditions that shape this limit.

Do frogs perceive the calls of different males within a chorus as separate perceptual objects in the same way humans perceive auditory objects in a complex scene? Namely, can frogs perform auditory grouping and source segregation? This question has not been explicitly addressed, and there seem to be at least two possible alternatives. In one scenario, frogs could perceive individual callers as separate auditory objects thus allowing females to compare and choose between two callers based on their call characteristics. Alternatively, frogs perceive a single image of the entire auditory scene without perception of individual objects. In this case, different callers contribute differentially to the single perceptual object, depending on their amplitude and call characters. Hence, choice between two (or several) callers would be determined by the differential contributions to the auditory scene, and the position of the caller dominating the image would also dominate the direction of the response.

A quantitative analysis of behavioral orienting response has been used to distinguish between perceptions of one or more auditory images in acoustic insects (Wendler 1989). Such experiments indicated that insects (Orthoptera) perceive only a single auditory object (e.g., Wendler 1989; Helversen and Helversen 1995; Helversen et al. 2001). Given that the acoustic behavior in orthopterans is comparable in complexity and selectivity to that in frogs, it seems reasonable to at least consider such a model for the frog auditory system. Both the model with separate perceptual images and the model with the single image can in theory explain the performance of female frogs as found in behavioral experiments. Because of this, there is a need to determine whether frogs perceive calls of individual males in a chorus as separate auditory streams, or whether the complete scene contributes to a single stream. Preliminary studies from Schwartz et al. (2004) indicate that frogs lack the ability to "restore" signal, that is, to fill in signal elements that are missing or inaudible due to the presence of masking noise (see Section 5.2 of Wells and Schwartz, Chapter 3). If their result is validated, the frog's perceptual ability would be in line with that of insects.

5.3 Influence of Chorus on Call Selectivity

Mate choice by female frogs has been studied in great detail in laboratory experiments in terms of call preference (see review in Gerhardt and Huber, 2002). These experiments employ a two-alternative-forced-choice (2AFC) paradigm, with little or no background noise present. In complex and noisy conditions of real-world frog choruses, female selectivity is compromised, as evidenced by the behavior of female *Hyla versicolor*. In 2AFC experiments in a laboratory, females reliably prefer long calls over short calls, and this preference is most pronounced when the two alternative stimuli are shorter than the average male call (Gerhardt et al. 2000). In similar experiments in an artificial chorus in the field (with eight loudspeakers playing male advertisement calls having different call durations), females discriminate against the shortest calls, but show no preferences among calls having average or longer durations (Schwartz et al. 2001). Whereas the results of both experiments show that males that produce below-average call lengths are selected against, a female's preference in the more realistic situation is less pronounced than under quiet conditions in the laboratory.

5.4 Sound Localization in Chorus

Frogs show remarkable sound localization ability in undisturbed sound fields (see Fay and Feng, 1987; Feng and Schellart 1999; Gerhardt and Huber 2002; Gerhardt and Bee, Chapter 5). For example, the spatial acuity of females of *Hyla cinerea* is ~12° when approaching a single sound source (Rheinlaender et al. 1979). This perceptual ability is in part attributed to having a directional receiver. Frogs ears function as a combination pressure/pressure-gradient receiver (Feng and Shofner 1981). Such a system derives its directionality from an interaction of sound waves between the external and internal sides of the tympanums through different pathways. The directionality is a function of the amplitude and phase relationship of the two sound inputs; these cues degrade when sound propagates in frogs' natural habitats, due to echoes reaching the receiver from indirect paths. This degradation is pronounced for the amplitude cues, but the phase relationship remains surprisingly stable (Michelsen and Rohrseitz 1997). Because the directionality of a pressure-gradient receiver depends more on the phase information and is fairly robust to changes in amplitudes, the ears' directionality is likely not going to be seriously affected due to the presence of echoes. This tenet has yet to be validated, however.

The presence of masking noise such as an intense chorus generally has a deleterious effect on the localization performance. Although there is some evidence that frog sound localization is compromised by background noise (e.g., Schwartz and Gerhardt 1995), to our knowledge this issue has not been systematically studied. Also, the precedence effect has never been demonstrated for anurans (see below). These research topics need to be investigated.

In many frog species, vision plays a role in communication; use of optical cues is not limited to diurnal frogs but also occurs in nocturnal species (review in Hödl and Amézquita 2001), because anurans have excellent low-light vision (Larsen and Pedersen 1982; Buchanan 1998). There are two ways vision may affect sound communication. First, the responding animal can use optical "landmarks" to stabilize their phonotactic response, as revealed for acoustic insects (Helversen and Wendler 2000). This is advantageous when the acoustic signals provide directional information intermittently. In this case, the optical cue does not provide any information about the caller itself. Second, optical cues that are synchronous with the acoustic signals, such as movements of the vocal sac, may serve as "signals" that may aid frogs in localizing the caller (Narins et al. 2003, 2005).

The potential role of vision in sound communication has been under-appreciated. Most behavioral studies of phonotaxis have been made under low-light or infrared illuminations; some visual cues remain under these conditions and frogs can potentially use visual cues to identify landmarks during phonotaxis, as suggested by Passmore et al. (1984). In light of this, it is important that future studies consider the possibility of the involvement of vision both in female phonotaxis and in male–male territorial interactions.

6. Mechanisms of Hearing in Complex Environments

A complex sound can be characterized by its distribution of sound energy in different frequency bands, and how this distribution changes with time. In the time domain, a natural sound typically contains a number of discrete components, appropriately ordered in time, each having specific spectral and temporal characteristics. This section describes what is known of the neural basis of hearing in real-world situations. For general anatomy and physiology of the frog auditory system, readers are referred to Wilczynski and Endepols, Chapter 8, Rose and Gooler, Chapter 9, and Fay and Popper (1999). The emphasis of this section is placed on auditory processes relevant to listening in complex acoustic environments.

6.1 Masking and Masking Release

As described previously, the ability to detect, analyze, and localize sound in the presence of competing sounds is vital for frogs' reproductive success. Many studies have investigated the neural basis of masking under controlled laboratory conditions, and these have focused on single auditory-nerve fibers in the periphery (Ehret and Capranica 1980; Megela and Capranica 1982; Zelick and Narins 1985; Narins 1987; Narins and Wagner 1989; Dunia and Narins 1989; Wang and Narins 1996). A review of earlier literature is given in Narins and Zelick (1988). In the presence of a continuous or gated background noise, the rate level function of an auditory-nerve fiber to a tone at the unit's characteristic frequency (CF) undergoes range compression and horizontal shift to higher sound levels (Fig. 11.1). Range compression (or reduction in a unit's dynamic range) is attributed to the presence of background firing; the higher the relative noise level the greater are the range compression and the horizontal shift (Narins 1987). These changes effectively shift the operating point of a fiber, thereby raising its threshold for tone detection.

Lin and Feng (2001) determined quantitatively the magnitude of threshold elevation of auditory-nerve fibers using freefield stimulation with a loudspeaker placed on one side of the frog, ipsilateral to the recorded auditory nerve. Their study shows that, when stimulated with a pure tone at a fiber's CF and a concurrent wideband background noise at 6 dB above its threshold at CF (i.e., a SNR of −6 dB), the presence of noise elevates the fiber's tone detection threshold by

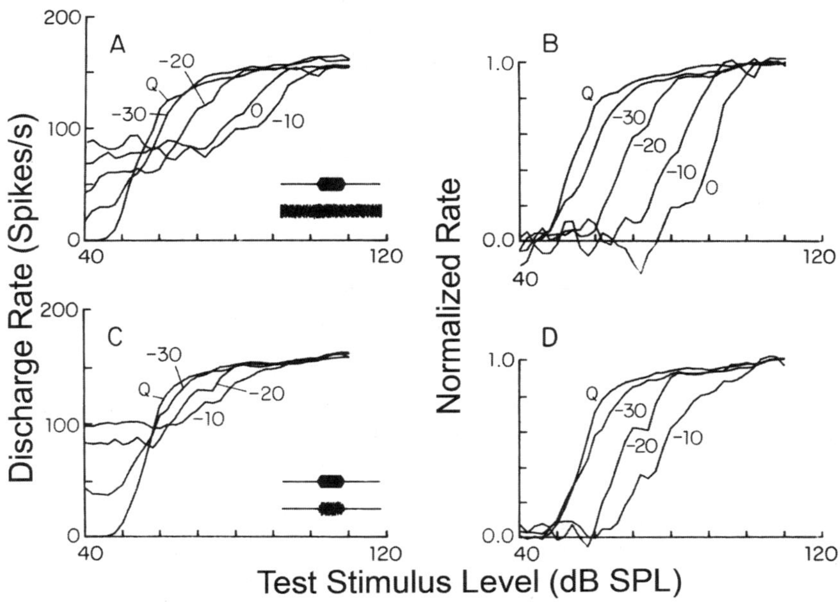

FIGURE 11.1. Tone-evoked rate-level functions for a low-frequency auditory nerve fiber from *Eleutherodactylus coqui* (BF = 260 Hz) obtained in quiet and in various levels of continuous (A) and gated (C) background noise. (B) and (D) are normalized rate-level functions given in (A) and (C), respectively. Reprinted from Narins (1987, Coding of signals in noise by amphibian auditory nerve fibers. Hearing Research 26:145–154), Copyright 1987, with permission from Elsevier.

an average of 4.22 dB. The presence of background noise additionally weakens time-locked discharges of low-frequency auditory-nerve fibers at various test frequencies and lowers a fiber's vector strengths (Freedman et al. 1988; Narins and Wagner 1989). The higher the relative masking noise level, the greater is the deterioration of the vector strength (Fig. 11.2).

Whereas the neural basis of masking is well characterized for the frog's auditory periphery, the literature on masking effects on the central auditory system is deficient. There is no study of masking in brainstem auditory structures other than the torus semicircularis. In the torus, one study has described the masking effect for auditory neurons therein, as a component of investigation on spatially mediated masking release (Lin and Feng 2001). Lin and Feng showed that, similar to auditory nerve fibers, the presence of wideband background noise elevates the toral neuron's tone detection threshold; threshold elevation is frequency independent and more pronounced (with an average elevation of ~11 dB) than for eighth-nerve fibers. The basis for increase in threshold elevation is unclear, however.

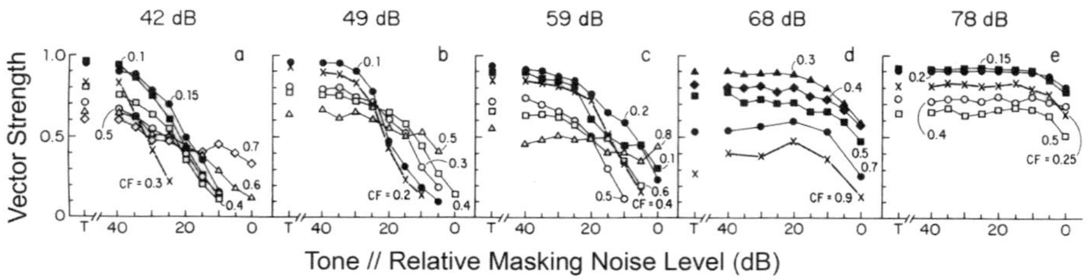

FIGURE 11.2. Effects of masking noise level on time-locking discharge patterns, in terms of vector strength, for five single auditory nerve fibers of *Eleutherodactylus coqui*. The fiber's CF and threshold at CF are shown for each panel. Test tone is presented at 10 dB above the fiber's threshold at the test tone frequency. The frequency of the test tone is indicated for each curve, relative to the fiber's CF. The fiber's response to the tone alone is indicated by the symbols above the "T" on the horizontal axis, whereas the remaining symbols illustrate the response to tones masked by increasing levels of broadband noise. The crosses, closed symbols, and open symbols represent the vector strength for test tone frequencies at, below, and above the fiber's CF, respectively. Reprinted with permission from Peter M. Narins and Ingeborg Wagner, The Journal of the Acoustical Society of America, 85, 1255 (1989). Copyright 1989, Acoustical Society of America.

Physiological studies of masking in the frog auditory system have involved two masking paradigms: transient signal in the presence of a continuous masker, and concurrent or simultaneous masking for which the signal and the masker are presented at the same time. Nothing is known about forward or backward masking, in spite of the fact that one of the calling strategies is antiphonal where a male follows the call of a chorus leader within a precise time window (see Section 3). Clearly, this is an area that needs further research.

As mentioned earlier, spatial separation of signal and masker sources decreases detection thresholds by means of spatially mediated masking release, or spatial unmasking (Schwartz and Gerhardt 1995). Ratnam and Feng (1998) found that, in the auditory midbrain of *Rana pipiens pipiens*, increasing the angular separation between the sources for a broadband noise and an amplitude-modulated (AM) stimulus indeed leads to improvement in signal detection, that is, lowering of the AM detection threshold. Their study confirmed the result of multiunit study of Schwartz and Gerhardt (1995). Ratnam and Feng (1998) further reported that most neurons in the frog torus demonstrating spatially mediated masking release show phasic discharge patterns, suggesting that these units may function as the primary signal detectors under challenging listening conditions.

Lin and Feng (2001) compared the physiological basis of spatial unmasking at the frog auditory periphery and midbrain. These investigators used the frog's natural call as a probe; the probe was positioned at the neuron's best receptive field (i.e., at contralateral 90° for midbrain neurons, or ipsilateral 90° for auditory-nerve fibers). The masker was a concurrent broadband noise presented at a SNR of −6 dB. They found that, at both levels, angular separation of probe and

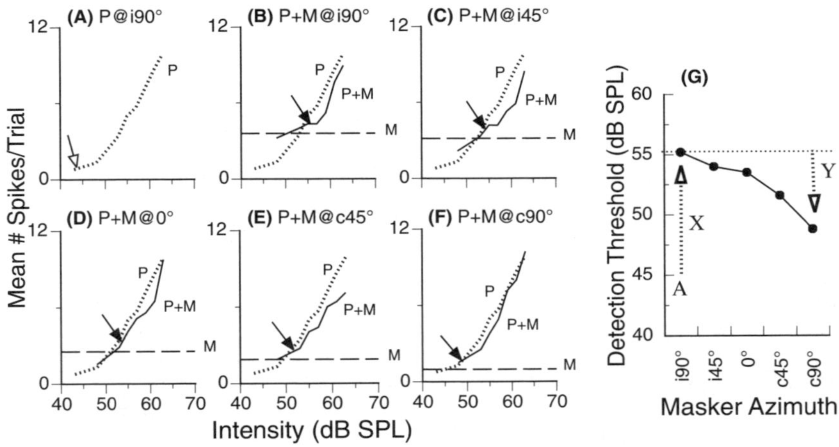

FIGURE 11.3. Spatial separation of probe and masker sources produces a partial masking release for an auditory nerve fiber in *Rana pipiens pipiens* (BF = 1100 Hz). Shown are the fiber's rate level functions to probe alone presented at ipsilateral 90° (dotted line in A to F), and to probe in masker (solid line in B to F). The probe comprises species' mating trill and is presented from a fixed position (ipsilateral 90°), and the masker is a concurrent broadband noise with an initial signal to noise ratio of –6 dB and is presented at different sound azimuths, from ipsilateral 90° to contralateral 90°. The fiber's response to masker alone is shown by dashed line in panels B to F for different masker azimuths. The detection threshold for probe alone is shown by open arrow in A, and for probe in masker by filled arrow in B to F; these are shown as open diamond and filled circles in G, respectively. For panel G, X represents the shift in probe detection threshold due to masking, and Y the fiber's maximum spatially mediated masking release. Data from Lin and Feng (2001).

masker sources produces a masking release, and the greater the angular separation the more pronounced is the masking release (Figs. 11.3, 11.4). For auditory-nerve fibers, the maximum masking release averages 2.9 dB, and that for torus neurons is significantly more pronounced, averaging 9.4 dB (Fig. 11.5).

It is noteworthy that spatially mediated masking release at the auditory periphery is frequency dependent, being more robust at higher than at lower frequencies (Lin and Feng 2001). This suggests that masking release at the auditory periphery is attributed mainly to the physics of the acoustics and the head shadowing effect. The facts are: (1) moving the masker source away from the probe source, and especially toward the opposite side of frog's head, reduces the SNR at the ear proximal to the probe source; and (2) the change in SNR is greater at higher frequencies due to the stronger head shadowing effect.

The greater strength of the spatial unmasking in the auditory midbrain suggests that neural processing taking place in the auditory brainstem is responsible for enhancing the physically based masking release at the auditory periphery. To determine the location and nature of signal processing that is responsible for the

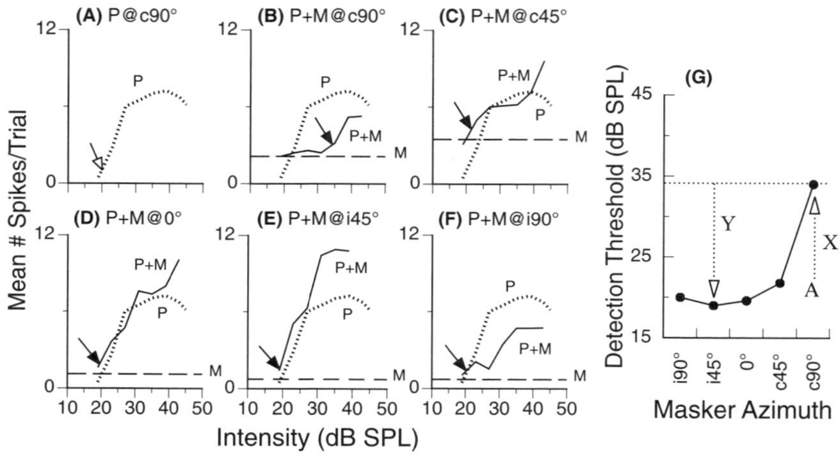

FIGURE 11.4. Spatial separation of probe and masker sources produces a partial masking release for a neuron in the torus semicircularis in *Rana pipiens pipiens* (BF = 1200 Hz). Shown are the unit's rate level functions to probe alone presented at contralateral 90° (dotted line in A to F), and to probe in masker (solid line in B to F). The probe comprises species' mating trill and is presented from a fixed position (contralateral 90°), and the masker is a concurrent broadband noise with an initial signal-to-noise ratio of –6 dB and is presented at different sound azimuths, from ipsilateral 90° to contralateral 90°. The unit's response to masker alone is shown by dashed line in panels B to F for different masker azimuths. The detection threshold for probe alone is shown by open arrow in A, and for probe in masker by filled arrow in B to F; these are shown as open diamond and filled circles in G, respectively. For panel G, X represents the shift in probe detection threshold due to masking, and Y the fiber's maximum spatially mediated masking release. Data from Lin and Feng (2001).

central component of masking release, Lin and Feng (2003) compared the properties of single neurons in the torus in response to spatial unmasking paradigm during: (a) iontophoretic application of bicuculline, an antagonist for GABAa receptor; and (b) two control conditions, that is, before and following withdrawal of drug application. They hypothesized that GABAergic inhibition within the torus, known to give rise to binaural inhibitory interactions and directional sensitivity for single neurons in the torus semicircularis (Zhang et al. 1999), is responsible for the enhancement in spatial unmasking in the midbrain. Results of this iontophoretic study validate the importance of local inhibitory interactions for spatial unmasking: application of bicuculline markedly reduces the strength of spatial unmasking, from an average maximal of ~9 dB under the control conditions to an average of ~4 dB (Fig. 11.6).

In the mammalian auditory midbrain, GABA is known to play a role in numerous auditory processes, from shaping of spatial (or directional) selectivity to enhancing the selectivities to interaural time and intensity differences, and to duration and complex sound. To determine the particular GABAergic process that

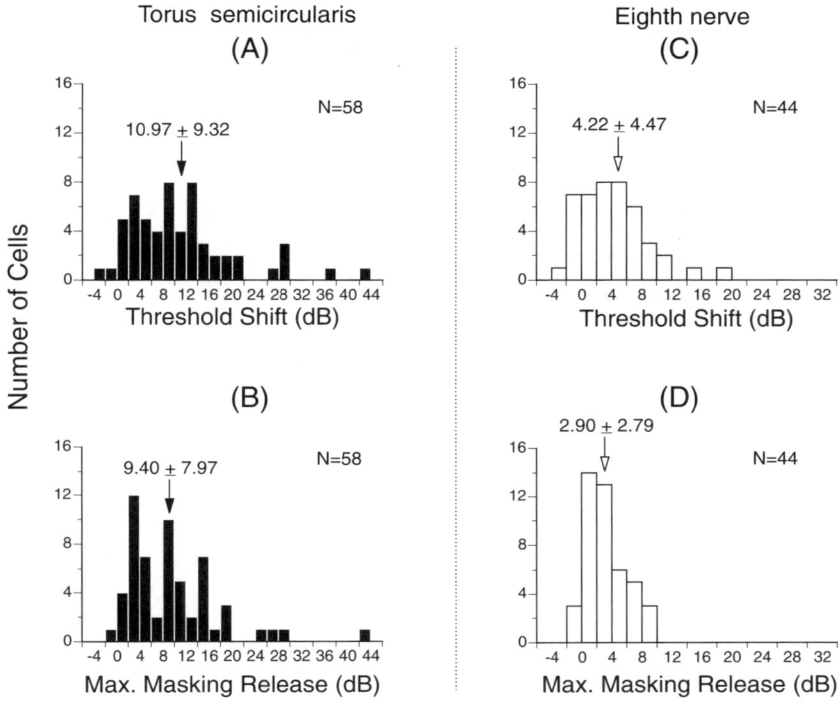

FIGURE 11.5. Distributions of the elevations in probe detection threshold due to masking for single units in the torus semicircularis (A) and auditory nerve (C) of *Rana pipiens pipiens*. Shown also are the distributions of the amounts of maximum release from masking resulting from angular separation of probe and masker sources for torus (B) and auditory nerve (D). The mean (shown by arrow) and standard deviation of each dataset are shown above each graph. Data from Lin and Feng (2001).

underlies spatial unmasking, Lin and Feng (2003) additionally evaluated how the directional sensitivity of torus neurons was affected by drug application. For two-thirds of torus neurons, drug application reduces units' directional sensitivity, concomitant with a reduction in the strength of spatial unmasking. This correlation confirms their working hypothesis that GABAergic interaction primarily increases the unit's directional sensitivity (through an increase in binaural inhibitory interactions) thereby enhancing their spatial unmasking ability.

For one-third of the torus neurons studied, however, application of bicuculline reduces the strength of the units' spatial unmasking without a concomitant reduction in the unit's directional sensitivity. Thus, the reduced strength of spatial unmasking is due to other GABAergic-based auditory processes (e.g., signal selectivity, duration selectivity, and/or attention mechanism). The particular GABAergic process responsible for spatial unmasking, however, has yet to be determined. Also, at this time, the origin of GABAergic input responsible for

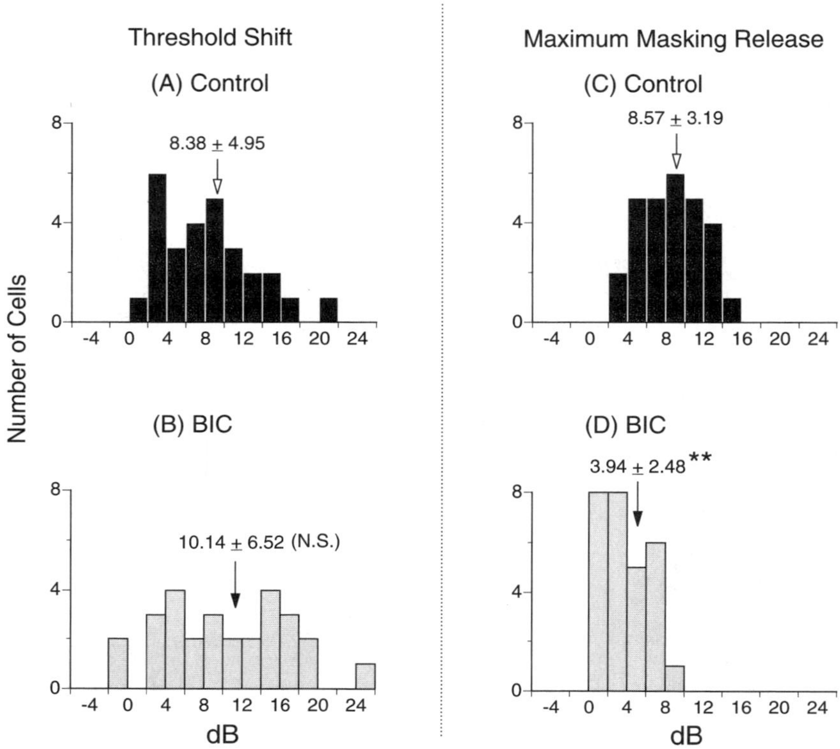

FIGURE 11.6. Distributions of the elevations in probe detection threshold attributable to masking (A, B) and the maximum masking release attributable to spatial separation of probe and masker sources (C, D) for 28 units in the torus semicircularis of *Rana pipiens pipiens* under two different experimental conditions: control condition (A, C), and when bicuculline is applied iontophoretically to the torus (B, D). The results of two-tailed Student's *t* tests from the same dataset from two different experimental conditions are shown in B and D. The mean (shown by arrow) and standard deviation of each dataset are shown above each graph. N.S., Not statistically significant. * Statistically significant difference. Data from Lin and Feng (2003). Copyright 2003 by the Society for Neuroscience.

spatial unmasking is unknown. In particular, it is unclear whether it is due to GABAergic projection from the ascending or the descending pathway.

6.2 Comodulation Masking Release

The above studies on masking release have assumed that spectral cues alone are important. In realistic situations, however, spectrotemporal cues may also be important, as demonstrated behaviorally in various vertebrates through the process of *comodulation masking release* (CMR). CMR, which was first shown

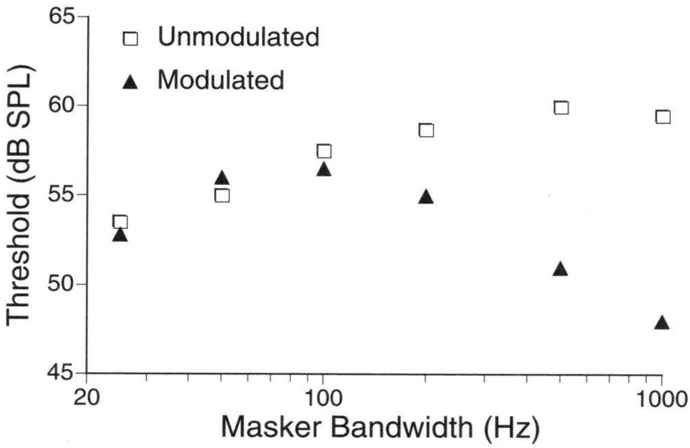

FIGURE 11.7. Thresholds for detecting a 1000 Hz tone as a function of the bandwidth of a noise masker. One masker is unmodulated and the other is comodulated noise. Data from Moore (1999). Copyright 1999 by Nature Publishing Group.

in humans, is an "across-channel" processing that facilitates auditory scene analysis (Hall et al. 1984). Normally, when a target sound is presented with a concurrent masker, masking becomes more pronounced with increasing masker bandwidth up to the critical band (Fletcher 1940), and beyond the critical band the masking effect remains unchanged (Fig. 11.7). When the masker is comodulated with the signal (e.g., having a tight temporal correlation), an increase in masker bandwidth initially also produces an increase in masking up to the critical band, but with a further increase in masker bandwidth, information in the flanking bands produces a masking release, thereby improving signal detection. For humans, the magnitude of masking release ranges from a few dB to as large as 15 dB, depending on the signal type and the properties of the background noise (Buus 1985; McFadden 1986; Schooneveldt and Moore 1987; Hall and Grose 1988; Moore and Schooneveldt 1990; Moore 1990, 1999; Fantini et al. 1993). CMR has since been demonstrated in starlings (Klump and Langemann 1995; Langemann and Klump 2001), chinchillas (Niemic et al. 2000; Niemic 2001), and gerbils (Klump et al. 2001).

The CMR phenomenon has not been demonstrated for anurans, in spite of its obvious advantage for listening in a frog chorus. In a typical frog chorus, background noise is often correlated in time due to the repetitive nature of frogs' vocalizations and surrounding biotic (e.g., insect sounds) and abiotic sounds, and of the frog's tendency to emit calls antiphonally. Spectrotemporal correlations of background noise and target signal can in theory give rise to CMR.

Goense and Feng (2003) studied single neurons in the torus of *Rana pipiens pipiens* to determine whether they exhibit CMR-like properties. They examined how the attributes of background noise influenced the detection threshold for a synthetic mating call (i.e., a series of tone pulses at unit's CF at a pulse rate of

20 Hz) for single units in the torus; both signal and noise were broadcast from a common loudspeaker in free-field. The noise was centered at a unit's CF, with bandwidth ranging from 100 to 4000 Hz at constant spectral level, 15 dB above the unit's threshold for the signal in quiet. They found that unmodulated noise generally increases the signal detection threshold. Sinusoidally modulated noise with a 7 to 9 Hz modulation rate at equal peak intensity (and/or equal overall energy) also elevates this threshold, but to a lesser extent. For both types of noise, an increase in noise bandwidth typically increases the detection threshold. A number of torus neurons exhibit CMR-like properties. In particular, when modulated noise is the masker, an increase in noise bandwidth initially elevates the signal detection threshold, but a further broadening of noise bandwidth leads to a reduction in detection threshold, much like the psychophysical data for modulated masker shown in Figure 11.7.

Bibikov and colleagues (Bibikov 2002; Bibikov and Grubnik 1990, 1996; Bibikov and Nizamov 1996) have suggested the benefit of sustained and slowly fluctuating background sound for detection of amplitude-modulated (AM) signals. They studied the responses of single neurons in the torus of ranid frogs to AM signals (10%, 20 Hz sinusoidal AM) in isolation as well as in the presence of low-frequency (0 to 50 Hz) bandpass noise. In many units, recordings from an adapted state show that addition of modulated noise enhances the unit's discharge rate as well as time-locking to the 20 Hz amplitude modulation. They proposed that this enhancement is attributed to stochastic resonance. Their experiments also reveal that, in the adapted state, time-locking to the signal envelope is strong, compared to when the signal is concurrently gated with transient noise. This indicates that the male's antiphonal calling may indirectly confer a benefit to signal detection by maintaining a continuous presence of conspecific calls within a chorus (that consequently produces neural adaptation).

In general, sharing of a breeding habitat with other sympatric species and insects produces either a sustained background noise, or comodulation of background noise over a broad frequency band at the lowest common denominating modulation frequency. The CMR-like response properties described above can potentially improve signal detection by taking advantage of the broadband nature of the comodulating background noise. Improvement in signal detection through CMR may therefore confer a selective advantage on individuals in multispecies assemblages of anurans and other calling animals.

6.3 Localization of Sources

A frog's ability to localize sound, as measured by phonotaxis, also seems to be compromised in the presence of background sounds (Schwartz and Gerhardt 1995; Farris et al. 2002). Whereas such literature in human psychophysics is rich, the level of our understanding in frogs is rudimentary by comparison. For example, there has not been any study of localization acuity in the presence of background sound (one or more distractors). Similarly, whereas the precedence effect is well characterized in humans (Litovsky and Shinn-Cunningham 2001), this has not been directly addressed in frogs, in spite of its relevance to hearing in choruses. In the precedence effect (Haas effect, or law of the first wavefront), when two identical sounds with a slight time difference are broadcast from two loudspeakers in free-field, one hears and locates one sound only, namely, the dominant sound that precedes the lagging sound. Whereas only one sound is heard when the lag time is small, both sounds are heard at their respective spatial locations when the lag time is long. In the presence of multiple distractors, one's localization performance is degraded when the number of distractors is increased from 0 to 2 (Langendijk et al. 2001). Similar sort of studies are much needed in frogs if we are to understand the constraints for hearing in a frog chorus. The only relevant data on this topic have come from echo suppression experiments of Galazyuk et al. (2005) that is described in Section 6.5.

Similar to the behavioral characterization of sound localization in real-world settings, there is a dearth of literature on the neural basis of sound localization in the presence of background sound. In anurans, physiological studies of sound localization have been focused exclusively on coding of sound direction in the absence of background noise (see Fay and Feng 1987; Eggermont 1988; Rose and Gooler, Chapter 9). In the cat auditory cortex, coding of sound direction is maintained in the presence of background noise as long as the signal is clearly audible (Furukawa and Middlebrooks 2001); this result is compatible with the psychophysical data in humans (Good and Gilkey 1996; Good et al. 1997; Lorenzi et al. 1999). It is possible that central neurons in the frog auditory system may behave similarly, but direct experimental validation is necessary.

6.4 *Identification of Signals*

Because many males participate in a frog chorus, there is a significant overlap between a male's call with those of his neighbors. A female presumably performs signal sorting, by grouping sounds emanating from one source into one stream and segregating it from streams associated with neighboring sources (i.e., auditory grouping and stream segregation). Farris et al. (2002) attempted to address the issue of auditory grouping in Túngara frogs (*Physalaemus pustulosus*). They exposed females to each of the two components of the species' advertisement call, that is, chuck and whine alone, or to combinations of chucks and whines at varying spatial separations. When presented alone, whines are attractive for females. In contrast, a chuck does not elicit phonotaxis when presented alone, but when added to a whine it increases the call's attractiveness (Ryan and Rand 1990). When combinations of chucks and whines were presented with wide angular separations, females approach the loudspeaker broadcasting the chuck (Farris et al. 2002). The authors interpret this as evidence for auditory grouping. However, it is also plausible that females perceive only a single image of the auditory scene, and this image is dominated by the chuck (see Section 5.2), therefore the females select the loudspeaker that produces the chuck. The second interpretation probably cannot be ruled out in light of the negative evidence for auditory grouping described earlier (see also Section 5.2 of Wells and Schwartz, Chapter 3).

Whereas the neural basis of detection of signals in noise has begun to be unraveled (see Section 4.1), that of signal identification is completely unexplored. This is a much harder problem to tackle because frogs are difficult to condition and have limited motor capacities that are suitable for conditioning experiments. Furthermore, physiological recording from freely moving frogs is technically challenging. Here the modeling approach may be a more promising avenue for gaining insight into this problem. Liu et al. (2000, 2001) developed a binaural model system that simulated the subtraction operation of the frog's auditory periphery and of the central binaural inhibitory interactions. They showed that such a binaural system can effectively localize the directions of multiple concurrent sounds, and extract one of the sounds from the ambiance. The caveat is that this model requires the binaural network to contain a topographic map of interaural time differences akin to those present in the avian laminaris nucleus and the mammalian medial superior olivary nucleus. Such a map has not been demonstrated in the frog's binaural system to date. Therefore, a more realistic modeling approach is needed to gain insight into the means by which the frog's auditory system extracts (and identifies) a signal in noise.

6.5 Environmental Effects on Hearing

As described previously, the environment that surrounds a frog chorus also has an influence over signal detection, discrimination, and localization. In particular, acoustic reverberation can distort the fine structure and the envelope of sounds thereby compromising sound communication. The physiological basis of such influences is not well understood. The two recent studies below show that whereas echoes compromise (or smear) temporal encoding of some auditory midbrain neurons (which allows animals to "perceive" their presence), other midbrain neurons effectively suppress echoes such that signals can be extracted with little interference.

A preliminary study by Ratnam et al. (2004) tested the hypothesis that reverberations from foliage, tree trunks, and surface barriers increase the attack and decay times of the trill components of frog calls, and thus impair the ability of auditory midbrain neurons to detect and discriminate calls. They indicated that, in the torus of *Rana pipiens pipiens*, for neurons that show strong time-locking to "anechoic" synthesized trills, reverberation reduces their time-locking ability. The reduction in time-locking capacity is due largely to the presence of spikes in the silent periods between trills, as it is more pronounced with an increase in reverberation time; the onset synchronization is unchanged and remains robust. Also, reverberation does not change the unit's spike rate, and in some cases it even increases the unit's spike rate. These results suggest that echoes and reverberation mainly influence call discrimination, and not call detection.

Interestingly, other torus neurons in *Rana pipiens pipiens* show an unusual response characteristic, that is, paradoxical latency shift that is characterized by an increase in the first spike latency with increasing sound level (Galazyuk and Feng 2001; Galazyuk et al. 2005). Because paradoxical latency shift is attributed to high-threshold early inhibition, these neurons display "echo-suppression" properties. Specifically, when presented to a pair of discrete sounds (a leading sound pulse that is followed by an echo, i.e., a weaker sound replica), response to the echo is totally suppressed at short echo delays. These neurons are thus well suited for suppressing interfering echoes. At this time, it is unclear whether the temporal smearing and echo-suppression properties appear de novo in the auditory midbrain, or reflect properties already present at lower centers in the auditory brainstem. Future studies are needed for a comprehensive understanding on environmental effects of hearing both behaviorally and physiologically.

Also, as described previously, although the distance effect on the acoustics is well understood, our knowledge of its effects on acoustically guided behavior is limited (see Section 3), and the underlying physiology is completely unknown. There is a critical need to gain an understanding of how distance influences the analysis of complex auditory scenes.

6.6 Gating Mechanisms and Descending Control

Mate choice in a chorus involves a decision-making process. Females not only must analyze the complex auditory scene to identify who is calling and from where, they also must select the males with whom to mate. Once mate choice is decided, they must initiate phonotaxis, or rely on auditory memory to approach the chosen male. In a cocktail party a human listener can attend to one or two talkers at any instant, and auditory attention (a form of active hearing important for focused listening) is regulated by a descending circuit. The frog torus receives both ascending and descending projections (see Wilczynski and Endepols, Chapter 8). Furthermore, Endepols and Walkowiak (1999, 2001) have shown that when the descending inputs are stimulated, toral neurons display long-lasting inhibition indicating that their auditory responses are under descending regulation. The descending system is presumably involved in modulation of acoustically guided behaviors such as attention, motivation, and turning on and off of phonotaxis and vocalization, but the precise roles and mechanisms of the descending control remain to be determined.

7. Conclusions

Acoustic communication in a frog's natural environment is challenging because sound is susceptible to distortion during transmission and to masking by chorus noise. At present, knowledge of how anurans solve these communication problems is limited. This chapter describes the many interesting behavioral and physiological questions that remain unresolved. In particular, behavioral evaluations of detection, recognition, or localization have been made extensively, but mostly in quiet backgrounds. Further studies are much needed to gain insight into the auditory performances in the presence of multiple competing sounds (resembling the frog's natural listening environments). The extent of involvement of other sensory cues in localization and recognition of mates also needs to be reexamined.

Future experiments should examine frog's behaviors either in total darkness or in an optically uniform arena. Another issue needing further study is whether frogs perform auditory grouping and source segregation, that is, whether they hear different callers within a chorus as separate perceptual objects or whether the entire auditory scene is perceived as a single object as in insects (Wendler 1989; Helversen and Helversen 1995; Helversen et al. 2001). This issue must be resolved before the search for neuronal correlates of stream segregation and auditory grouping can begin.

References

Beckers OM, Schul J (2004) Phonotaxis in *Hyla versicolor* (Anura, Hylidae): The effect of absolute call amplitude. J Comp Physiol A 190:869–876.

Bibikov NG (2002) Addition of noise enhances neural synchrony to amplitude-modulated sounds in the frog's midbrain. Hear Res 173:21–28.

Bibikov NG, Grubnik ON (1990) Detection of a periodic component of amplitude modulation against a background of noise by neurones of the torus semicircularis of the lake frog. Sens Sys 4:28–34.

Bibikov NG, Grubnik ON (1996) Enhancement of neural discharge synchronization with stimulus envelope in the course of long-term adaptation. Sens Sys 10:5–18.

Bibikov NG, Nizamov SV (1996) Temporal coding of low-frequency amplitude modulation in the torus semicircularis of the grass frog. Hear Res 101:23–44.

Bosch J, Riva DLI (2004) Are frog calls modulated by the environment? An analysis with anuran species from Bolivia. Can J Zool 82:880–888.

Bradbury JW, Vehrencamp SL (1998) Principles of Animal Communication. Sunderland MA: Sinauer.

Bregman A (1990) Auditory Scene Analysis: The Perceptual Organization of Sound. Cambridge, MA: MIT Press.

Brush JS, Narins PM (1989) Chorus dynamics of a neotropical amphibian assemblage: Comparison of computer simulation and natural behavior. Anim Behav 37:33–44.

Brzoska J (1980) Quantitative studies on the elicitation of the electrodermal response by calls and synthetic acoustical stimuli in *Rana lessonae* Camerano, *Rana ridibunda* Pallas and the hybrid *Rana "esculenta"* L. (Anura, Amphibia). Behav Processes 5:113–141.

Brzoska J, Walkowiak W, Schneider H (1977) Acoustic communication in the grass frog (*Rana t. temporaria* L.): Calls, auditory thresholds and behavioral responses. J Comp Physiol A 118:173–186.

Buchanan BW (1998) Lo-illumination prey detection by squirrel treefrogs. J Herpetol 32:270–274.

Buus S (1985) Release from masking caused by envelope fluctuations. J Acoust Soc Am 78:1958–1965.

Capranica RR, Moffat AJM (1983) Neurobehavioral correlates of sound communication in anurans. In: Ewert JP, Capranica RR, Ingle DJ (eds) Advances in Vertebrate Neuroethology. New York: Plenum, pp. 701–730.

Chek AA, Bogart JP, Lougheed SC (2003) Mating signal partitioning in multi-species assemblages: A null model test using frogs. Ecol Lett 6:235–247.

Deily JA, Schul J (2004) Recognition of calls with exceptionally fast pulse rates: Female phonotaxis in the genus *Neoconocephalus* (Orthoptera: Tettigoniidae). J Exp Biol 207:3523–3529.

Drewry GE, Rand AS (1983) Characteristics of an acoustic community: Puerto Rican frogs of the genus Eleutherodactylus. Copeia 1983:941–953.

Dunia R, Narins PM (1989) Tone-derived vs. tone-in-noise-derived filter functions of frog auditory nerve fibers: A comparison. Hear Res 37:241–254.

Eggermont JJ (1988) Mechanisms of sound localization in anurans. In: Fritsch B, Ryan MJ, Wilczynski W, Hetherington TE, Walkowiak W (eds), The Evolution of the Amphibian Auditory System. New York: Wiley, pp. 307–336.

Ehret G, Capranica RR (1980) Masking patterns and filter characteristics of auditory nerve fibers in the green treefrog (*Hyla cinerea*). J Comp Physiol A 141:1–12.

Ehret G, Gerhardt HC (1980) Auditory masking and effects of noise on responses of the green treefrog (*Hyla cinerea*) to synthetic mating calls. J Comp Physiol A 141:13–18.

Elepfandt A, Eistettler A, Fleig E, Günther M, Hainich S, Hepperle S, Traub B (2000) Hearing threshold and frequency discrimination in the purely aquatic frog *Xenopus laevis* (Pipidae): Measurement by means of conditioning. J Exp Biol 203:3621–3629.

Endepols H, Walkowiak W (1999) Influence of descending forebrain projections on processing of acoustic signals and audiomotor integration in the anuran midbrain. Eur J Morphol 37:182–184.

Endepols H, Walkowiak W (2001) Integration of ascending and descending inputs in the auditory midbrain of anurans. J Comp Physiol A 186:1119–1133.

Fantini DA, Moore BCJ, Schooneveldt (1993) Comodulation masking release as a function of type of signal, gated or continuous masking, monaural or dichotic presentation of flanking bands and center frequency. J Acoust Soc Am 93:2106–2114.

Farris HE, Rand AS, Ryan MJ (2002) The effects of spatially separated call components on phonotaxis in tungara frogs: Evidence for auditory grouping. Brain Behav Evol 60(3):181–188.

Fay RR, Feng AS (1987) Mechanisms for directional hearing among non-mammalian vertebrates. In: Yost WA, Gourevitch G (eds) Directional Hearing. New York: Springer-Verlag, pp. 179–213.

Fay RR, Popper AN (1999) Comparative Hearing: Fish and Amphibians. New York: Springer-Verlag.

Feng AS, Narins PM, Xu CH (2002) Vocal acrobatics in a Chinese frog, *Amolops tormotus*. Naturwissenschaften 89:352–356.

Feng AS, Schellart NAM (1999) Central auditory processing in fish and amphibians. In: Popper AN, Fay RR (eds) Comparative Hearing: Fish and Amphibians. New York: Springer-Verlag, pp. 218–268.

Feng AS, Shofner WP (1981) Peripheral basis of sound localization in anurans. Acoustic properties of the frog's ear. Hearing Res 5:201–216.

Fletcher H (1940) Auditory patterns. Rev Mod Phys 12:47–65.

Forrest TG (1994) From sender to receiver: Propagation and environmental effects on acoustic signals. Am Zool 34:644–654.

Freedman EG, Ferragamo M, Simmons AM (1988) Masking patterns in the bullfrog (*Rana catesbeiana*). II: Physiological effects. J Acoust Soc Am 84:2081–2091.

Furukawa S, Middlebrooks JC (2001) Sensitivity of auditory cortical neurons to locations of signals and competing noise sources. J Neurophysiol 86:226–240.

Galazyuk AV, Lin WY, Llano D, Feng AS (2005) Leading inhibition to neural oscillation is important for time domain processing in the auditory midbrain. J Neurophysiol 94: 314–326.

Galazyuk AV, Feng AS (2001) Oscillation may play a role in time domain central auditory processing. J Neurosci 21 (RC147): pp. 1–5.

Garcia EJ, Narins PM (2001) Shared acoustic resources in an Old World frog community. Herpetologica 57:104–116.

Gerhardt HC (2001) Acoustic communication in two groups of closely related treefrogs. Adv Study Behav 30:99–167.

Gerhardt HC, Huber F (2002) Acoustic Communication in Insects and Anurans. Chicago: University of Chicago Press.

Gerhardt HC, Klump GM (1988a) Phonotactic responses and selectivity of barking treefrogs (Hyla gratiosa) to chorus sounds. J Comp Physiol A 163:795–802.

Gerhardt HC, Klump GM (1988b) Masking of acoustic signals by the chorus background noise in the green tree frog: A limitation on mate choice. Anim Behav 36:1247–1249.

Gerhardt HC, Schwartz JJ (2001) Auditory tuning and frequency preferences in anurans. In: Ryan MJ (ed) Anuran Communication. Washington: Smithsonian Institution Press, pp. 73–85.

Gerhardt HC, Tanner SD, Corrigan CM, Walton HC (2000) Female preference functions based on call duration in the gray tree frog *(Hyla versicolor)*. Behav Ecol 11:663–669.

Goense JBM, Feng AS (2003) Effects of noise bandwidth and modulation on signal detection for single neurons in the frog auditory midbrain. ARO Abstr 26:176.

Good MD, Gilkey RH (1996) Sound localization in noise: The effect of signal-to-noise ratio. J Acoust Soc Am 99:1108–1117.

Good MD, Gilkey RH, Ball JM (1997) The relation between detection in noise and localization in noise in the free field. In: Gilkey RH, Anderson TR (eds) Binaural and Spatial Hearing in Real and Virtual Environments. Mahwah, NJ: Erlbaum, pp. 349–376.

Grafe, TU (1997) Cost and benefits of mate choice in the lek-breeding frog, *Hyperolius marmoratus*. Anim Behav 53:1103–1117.

Greenfield MD (1994) Synchronous and alternating choruses in insects and anurans: Common mechanisms and diverse functions. Amer Zool 34:605–615.

Greenfield MD, Rand AS (2000) Frogs have rules: Selective attention algorithms regulate chorusing in *Physalaemus pustulosus* (Leptodactylidae). Ethology 106:331–347.

Hall JW, Grose JH (1988) Comodulation masking release: Evidence for multiple cues. J Acoust Soc Am 84:1669–1675.

Hall JW, Haggard MP, Fernandes MA (1984) Detection in noise by spectro-temporal pattern analysis. J Acoust Soc Am 76:50–56.

Helversen Dv, Helversen Ov (1995) Acoustic pattern recognition in orthopteran insects: Parallel or serial processing. J Comp Physiol A 177:767–774.

Helversen Dv, Schul J, Kleindienst HU (2001) Male recognition mechanism for female responses implies a dilemma for their localization in a phaneropterine bushcricket. J Comp Physiol A 186:1153–1158.

Helversen Dv, Wender G (2000) Coupling of visual to auditory cues during phonotactic approach in the phaneropterine bushcricket *Poecilimon affinis*. J Comp Physiol A 186: 729–736.

Hödl W, Amézquita A (2001) Visual signalling in anuran amphibians. In: Ryan MJ (ed) Anuran Communication. Washington: Smithsonian Institution Press, pp. 121–141.

Jørgensen MB, Gerhardt HC (1991) Directional hearing in the gray treefrog *Hyla versicolor*: Eardrum vibrations and phonotaxis. J Comp Physiol A 169:177–183.

Kime NM, Turner WR, Ryan MJ (2000) The transmission of advertisment calls in central American frogs. Behav Ecol 11:71–83.

Klump GM, Dooling RJ, Fay RR, Stebbins WC (1995) Methods in Comparative Psychoacoustics. Basel: Birkhäuser Verlag.

Klump GM, Langemann U (1995) Comodulation masking release in a songbird. Hear Res 87:157–164.

Klump GM, Langemann U, Friebe A, Hamann I (2001) An animal model for studying across-channel processes: CMR and MDI in the European starling. In: Breebart DJ, Houtsma AJM, Kohlrausch A, Prijs VF, Schoonhoven R (eds) Physiological and Psychophysical Bases of Auditory Function. Maastricht: Shaker, pp. 266–272.

Lang F (2000) Acoustic communication distances of a gomphocerine grasshopper. Bioacoustics 10:233–258.

Langemann U, Klump GM (2001) Signal detection in amplitude-modulated maskers: I. Behavioral auditory threshold in a songbird. Eur J Neurosci 13:1025–1032.

Langendijk EHA, Kistler DJ, Wightman FL (2001) Sound localization in the presence of one to two distractors. J Acoust Soc Am 109:2123–2134.

Lardner B, bin Lakim M (2002) Tree-hole frogs exploit resonance effects. Nature 420:475.

Larsen LO, Pedersen JN (1982) The snapping response of the toad *Bufo bufo*, towards prey dummies at very low light intensities. Amphibia-Reptilia 2:321–327.

Lin WY, Feng AS (2001) Free-field unmasking response characteristics of frog auditory nerve fibers: Comparison with the responses of midbrain auditory neurons. J Comp Physiol A 187:699–712.

Lin WY, Feng AS (2003) GABA is involved in spatial unmasking in the frog auditory midbrain. J Neurosci 23:8143–8151.

Litovski RY, Shinn-Cunningham BG (2001) Investigation of the relationship among three common measures of precedence: Fusion, localization dominance, and discrimination suppression. J Acoust Soc Am 109:346–358.

Littlejohn MJ (1977) Long range acoustic communication in anurans: An integrated and evolutionary approach. In: Taylor DH, Guttman SI (eds) The Reproductive Biology of Amphibians. New York: Plenum, pp. 263–334.

Liu C, Wheeler BC, O'Brien WD Jr, Bilger RC, Lansing CR, Feng AS (2000) Localization of multiple sound sources with two microphones. J Acoust Soc Am 108:1888–1905.

Liu C, Wheeler BC, O'Brien WD Jr, Bilger RC, Lansing CR, Jones DL, Feng AS (2001) A two-microphone dual delay-line approach for extraction of a speech sound in the presence of multiple interferers. J Acoust Soc Am 110:3218–3231.

Lorenzi C, Gatehouse S, Lever C (1999) Sound localization in noise in normal-hearing listeners. J Acoust Soc Am 105:1810–1820.

McFadden D (1986) Comodulation masking release: Effects of varying the level, duration, and time delay of the cue band. J Acoust Soc Am 80:1658–1667.

Megela AL, Capranica RR (1982) Differential patterns of physiological masking in the anuran auditory nerve. J Acoust Soc Am 71:641–645.

Michelsen A (1978) Sound reception in different environments. In: Ali AB (ed) Perspectives in Sensory Ecology. New York: Plenum, pp. 345–373.

Michelsen A, Rohrseitz K (1997) Sound localisation in a habitat: An analytical approach to quantifying the degradation of directional cues. Bioacoustics 7:291–313.

Moore BCJ (1990) Co-modulation masking release: Spectro-temporal pattern analysis in hearing. Br J Audiol 24:131–137.

Moore BCJ (1999) Modulation minimizes masking. Nature 397:108–109.

Moore BCJ, Schooneveldt GP (1990) Comodulation masking release (CMR) as a function of bandwidth and time delay between on-frequency and flanking band maskers. J Acoust Soc Am 88:725–731.

Morris MR (1991) Female choice of large males in the treefrog *Hyla ebraccata*. J Zoology Lond 223:371–378.

Moss CF, Simmons AM (1986) Frequency selectivity of hearing in the green treefrog, *Hyla cinerea*. J Comp Physiol A 159:257–266.

Murphy CG (2003) The cause of correlations between nightly numbers of male and female barking treefrogs (*Hyla gratiosa*) attending choruses. Behav Ecol 14:274–281.

Murphy CG, Gerhardt HC (2002) Mate sampling by female barking treefrogs (*Hyla gratiosa*). Behav Ecol 13:472–480.

Narins PM (1982) Effects of masking noise on evoked calling in the Puerto Rican coqui (Anura: Leptodactylidae). J Comp Physiol A 147:439–446.

Narins PM (1987) Coding of signals in noise by amphibian auditory nerve fibers. Hear Res 26:145–154.

Narins PM, Feng AS, Lin W, Schnitzler HU, Denzinger A, Suthers RA, Xu CH (2004) Old world frog and bird vocalizations contain prominent ultrasonic harmonics. J Acoust Soc Am 115:910–913.

Narins PM, Grabul DS, Soma KK, Gaucher P, Hödl W (2005) Cross-modal integration in a dart-poison frog. Proc Natl Acad Sci USA 102:2425–2429.

Narins PM, Hödl W, Grabul DS (2003) Bimodal signal requisite for agonistic behavior in the dart-poison frog, *Epipedobates femoralis*. Proc Natl Acad Sci USA 100:577–580.

Narins PM, Wagner I (1989) Noise susceptibility and immunity of phase locking in amphibian auditory-nerve fibers. J Acoust Soc Am 85:1225–1264.

Narins PM, Zelick R (1988) The effects of noise on auditory processing and behavior in amphibians. In: Fritzsch B, Ryan MJ, Wilczynski W, Hetherington TE, Walkowiak W (eds) The Evolution of the Amphibian Auditory System. New York: Wiley, pp. 511–536.

Niemic AJ (2001) The effects of increasing masker temporal regularity on co-modulation masking thresholds in chinchillas. ARO Abstr 24:85.

Niemic AJ, Florin Z, Winter A (2000) The use of spectral and temporal cues by chinchillas in co-modulation masking experiments. ARO Abstr 23:27.

Padgham M (2004) Reverberation and frequency attenuation in forests—Implications for acoustic communication in animals. J Acoust Soc Am 115:402–410.

Passmore NI, Capranica RR, Telford SR, Bishop PJ (1984) Phonotaxis in the painted reed frog (*Hyperolius marmoratus*). The localization of elevated sound sources. J Comp Physiol A 154:189–197.

Penna M (2004) Amplification and spectral shifts of vocalizations inside burrows of the frog *Eupsophus calcaratus* (Leptodactylidae). J Acoust Soc Am 116:1254–1260.

Penna M, Solís R (1998) Frog call intensities and sound propagation in the South American temperate forest region. Behav Ecol Sociobiol 42:371–381.

Pollack GS (1988) Selective attention in an insect auditory neuron. J Neurosci 8: 2635–2639.

Ratnam R, Feng AS (1998) Detection of auditory signals by frog inferior colliculus neurons in the presence of spatially separated noise. J Neurophysiol 80:2848–2859.

Ratnam R, Iyer N, Goense J, Feng AS (2004) Effect of reverberation on neural response to amplitude modulated signals. ARO Abstr 27:113 (#336).

Rheinlaender J, Gerhardt HC, Yager DD (1979) Accuracy of phonotaxis by the green treefrog (*Hyla cinerea*). J Comp Physiol A 133:247–255.

Richards DG, Wiley RH (1980) Reverberations and amplitude fluctuations in the propagation of sound in a forest: implications for animal communication. Am Nat 115: 381–399.

Rinberg D, Davidowitz H (2003) Wind spectra and the response of the cercal system in the cockroach. J Comp Physiol A 189:867–876.

Römer H, Krusch M (2000) A gain control mechanism for processing of chorus sounds in the afferent auditory pathway of the bushcricket *Tettigonia viridissima* (Orthoptera, Tettigoniidae). J Comp Physiol A 186:181–191.

Römer H, Lewald J (1992) High-frequency sound transmission in natural habitats: Implications for the evolution of insect acoustic communication. Behav Ecol Sociobiol 29: 437–444.

Ryan MJ, Cocroft RB, Wilczynski W (1990) The role of environmental selection in intraspecific divergence of mate recognition signals in the cricket frog, *Acris crepitans*. Evolution 44:1869–1872.

Ryan MJ, Rand AS (1990) The sensory basis of sexual selection for complex calls in the túngara frog, *Physalaemus pustulosus* (sexual selection for sensory exploitation). Evolution 44:305–314.

Ryan MJ, Wilczynski W (1991) Evolution of intraspecific variation in the advertisement call of a cricket frog (*Acris crepitans*, Hylidae) Biol J Linn Soc 44:249–271.

Scharf B (1970) Critical bands. In: Tobias JV (ed) Foundations of Modern Auditory Theory. New York: Academic, pp. 159–202.

Schooneveldt GP, Moore BCJ (1987) Comodulation masking release as a function of signal frequency, flanking band frequency, masker bandwidth, and flanking band level. J Acoust Soc Am 82:1944–1956.

Schul J, Bush SL (2002) Non-parallel coevolution of sender and receiver in the acoustic communication system of treefrogs, Proc Roy Soc B 269:1847–1852.

Schwartz JJ, Buchanan BW, Gerhardt HC (2001) Female mate choice in the gray treefrog (*Hyla versicolor)* in three experimental environments. Behav Ecol Sociobiol 49: 443–455.

Schwartz JJ, Gerhardt HC (1989) Spatially mediated release from auditory masking in an anuran amphibian. J Comp Physiol A 166:37–41.

Schwartz JJ, Gerhardt HC (1995) Directionality of the auditory system and call pattern recognition during acoustic interference in the gray tree frog *Hyla versicolor*. Audit Neurosci 1:195–206.

Schwartz JJ, Gerhardt HC (1998) The neuroethology of frequency preferences in the spring peeper. Anim Behav 56:55–69.

Schwartz JJ, Huth K, Lasker J (2004) Impact of the chorus environment on temporal processing of advertisement calls by gray treefrogs. Abstract 147th Meeting Acoust Soc Amer 115:2374.

Schwartz JJ, Wells KD (1983) An experimental study of acoustic interference between two species of neotropical treefrogs. Anim Behav 31:181–190.

Strother WF (1962) Hearing in frogs. J Aud Res 2:279–286.

Wang J, Narins PM (1996) Directional masking of phase locking in the amphibian auditory nerve. J Acoust Soc Am 99:1611–1620.

Wendler G (1989) Acoustic orientation of crickets (Gryllus campestris) in the presence of two sound sources. Naturwissenschaften 76:128–129.

Wilczynski W, Ryan MJ (1999) Geographic variation in animal communication systems, In: Foster SA, Endler JA (eds) Geographic Variation of Behavior, pp. 234–261. New York: Oxford University Press.

Wiley RH, Richards DG (1978) Physical constraints on acoustic communication in the atmosphere: Implications for the evolution of animal vocalizations. Behav Ecol Sociobiol 3:69–94.

Wollerman L (1999) Acoustic interference limits call detection in a neotropical frog, *Hyla ebraccata*. Anim Behav 57:841–851.

Wollerman L, Wiley RH (2002) Background noise from a natural chorus alters female discrimination of male calls in a Neotropical frog. Anim Behav 63:15–22.
Zelick R, Narins PN (1985) Temporary threshold shift, adaptation, and recovery characteristics of frog auditory nerve fibers. Hear Res 17:161–176.
Zhang HM, Xu J, Feng AS (1999) Effects of GABA-mediated inhibition on direction-dependent frequency tuning in the frog inferior colliculus. J Comp Physiol A 184:85–98.

Index

Springer Handbook of Auditory Research *(continued from page ii)*

Volume 22: Evolution of the Vertebrate Auditory System
Edited by Geoffrey A. Manley, Arthur N. Popper, and Richard R. Fay

Volume 23: Plasticity of the Auditory System
Edited by Thomas N. Parks, Edwin W. Rubel, Arthur N. Popper, and Richard R. Fay

Volume 24: Pitch: Neural Coding and Perception
Edited by Christopher J. Plack, Andrew J. Oxenham, Richard R. Fay, and Arthur N. Popper

Volume 25: Sound Source Localization
Edited by Arthur N. Popper and Richard R. Fay

Volume 26: Development of the Inner Ear
Edited by Matthew W. Kelley, Doris K. Wu, Arthur N. Popper, and Richard R. Fay

Volume 27: Vertebrate Hair Cells
Edited by Ruth Anne Eatock, Richard R. Fay, and Arthur N. Popper

Volume 28: Hearing and Sound Communication in Amphibians
Edited by Peter M. Narins, Albert S. Feng, Richard R. Fay, and Arthur N. Popper

For more information about the series, please visit www.springer-ny.com/shar.

Printed in the United States of America.